Pollution Ecology of
Estuarine Invertebrates

WATER POLLUTION

A Series of Monographs

EDITORS

K. S. SPIEGLER
Department of Chemistry
and Chemical Engineering
Michigan Technological
University, Houghton
and
Department of Mechanical
Engineering
University of California,
Berkeley
Berkeley, California

J. I. BREGMAN
WAPORA, Inc.
6900 Wisconsin Avenue, N.W.
Washington, D. C.

Pollution Ecology of Estuarine Invertebrates

EDITED BY

C. W. Hart, Jr.
National Museum of Natural History
Smithsonian Institution
Washington, D.C.

Samuel L. H. Fuller
The Academy of Natural Sciences of Philadelphia
Philadelphia, Pennsylvania

1979

ACADEMIC PRESS
A Subsidiary of Harcourt Brace Jovanovich, Publishers

New York   London   Toronto   Sydney   San Francisco

ACADEMIC PRESS, INC.
111 Fifth Avenue, New York, New York 10003

United Kingdom Edition published by
ACADEMIC PRESS, INC. (LONDON) LTD.
24/28 Oval Road, London NW1 7DX

Library of Congress Cataloging in Publication Data
Main entry under title:

Pollution ecology of estuarine invertebrates.

Includes index.
1. Estuarine ecology. 2. Marine invertebrates.
3. Aquatic animals, effect of water pollution on.
I. Hart, C. W. II. Fuller, Samuel L. H.
QH541.5.E8P64   592'.05'2636   79–18157
ISBN 0–12–328440–6

PRINTED IN THE UNITED STATES OF AMERICA

79 80 81 82   9 8 7 6 5 4 3 2 1

To Thomas C. Cheng,
who first suggested that these
volumes be attempted

Contents

Chapter 8   Larval Decapods (Arthropoda: Crustacea: Decapoda)
Charles E. Epifanio

Chapter 9   Cyathura (Arthropoda: Crustacea: Isopoda: Anthuridae)
W. D. Burbanck and Madeline P. Burbanck

Chapter 10   Isopods Other Than Cyathura (Arthropoda: Crustacea: Isopoda)
Charles E. Powell, Jr.

Chapter 11   Amphipods (Arthropoda: Crustacea: Amphipoda)
Donald J. Reish and J. Laurens Barnard

List of Contributors

Numbers in parentheses indicate the pages on which the authors' contributions begin.

J. LAURENS BARNARD (345), Department of Invertebrate Zoology, National Museum of Natural History, Smithsonian Institution, Washington, D.C.

MADELINE P. BURBANCK (293), Department of Biology, Emory University, Atlanta, Georgia 30322

W. D. BURBANCK (293), Department of Biology, Emory University, Atlanta, Georgia 30322

JOHN A. COUCH (235), Gulf Breeze Environmental Research Laboratory, U.S. Environmental Protection Agency, Gulf Breeze, Florida 32561

JOHN K. DAWSON (145), Harbors Environmental Projects, Institute for Marine and Coastal Studies, Allan Hancock Foundation, University of Southern California, Los Angeles, California 90032

THOMAS W. DUKE (171), Gulf Breeze Environmental Research Laboratory, U.S. Environmental Protection Agency, Gulf Breeze, Florida 32561

CHARLES E. EPIFANIO (259), College of Marine Studies, University of Delaware, Lewes, Delaware 19958

J. M. FERRIS (1), Department of Entomology, Purdue University, West Lafayette, Indiana 47906

V. R. FERRIS (1), Department of Entomology, Purdue University, West Lafayette, Indiana 47906

C. W. HART, JR. (127), National Museum of Natural History, Smithsonian Institution, Washington, D.C. 20560

DABNEY G. HART (127), Metrek Division of the MITRE Corporation, McLean, Virginia 22102

WINSTON MENZEL (371), Department of Oceanography, The Florida State University, Tallahassee, Florida 32306

CHARLES E. POWELL, JR. (325), The Academy of Natural Sciences of Philadelphia, Philadelphia, Pennsylvania 19105

DONALD J. REISH (77, 345), Department of Biology, California State University, Long Beach, California 90840

DOROTHY F. SOULE (35), Harbors Environmental Projects, Institute for Marine and Coastal Studies, Allan Hancock Foundation, University of Southern California, Los Angeles, California 90032

JOHN D. SOULE (35), Harbors Environmental Projects, Institute for Marine and Coastal Studies, Allan Hancock Foundation, and School of Dentistry, University of Southern California, Los Angeles, California 90032

AUSTIN B. WILLIAMS (171), National Marine Fisheries Service Systematics Laboratory, National Museum of Natural History, Smithsonian Institution, Washington, D.C. 20560

Preface

The companion to this volume, "Pollution Ecology of Freshwater Invertebrates,"* was undertaken several years ago in an attempt to compile what was known of the "normal" ecology of certain freshwater invertebrate groups, to bring into focus the most recent systematic interpretations, to discuss the current knowledge of the pollution ecology of those groups, and to point up the pertinent ancillary background material. In that volume, as in this, all of the major invertebrate groups are not represented. For instance, oysters are not covered here simply because adequate coverage would comprise a volume in itself and, because of their intrinsic commercial value, many well-known volumes have already been devoted to them. With other groups there is a scarcity of people who wish to commit themselves, and there is still taxonomic chaos at the alpha level.

However, we look on this as a good beginning, because in studying water pollution from a biological standpoint, knowledge of as many groups as possible must be available so that information on species compositions, population sizes, and the physical–chemical environments to which they are exposed can be balanced and synthesized.

As with the freshwater volume, this is not intended to be a final word. These chapters do not point to any specific index that will define the degree of pollution. They do, however, attempt to pull together existing data, introduce new ideas and information, and synthesize, where possible, the latest systematic interpretations.

C. W. HART, JR.

---

*C. W. Hart, Jr., and Samuel L. H. Fuller, 1974, Academic Press, Inc., New York.

Pollution Ecology of Estuarine Invertebrates

CHAPTER 1

Thread Worms (Nematoda)

V. R. FERRIS and J. M. FERRIS

I. Introduction

We attempt to summarize here what is known, not only of the pollution ecology of estuarine nematodes, but also of the pollution ecology of freshwater nematodes. Because of the nature of estuaries, it would have been difficult to treat estuarine nematodes alone, and, in fact, we include also considerable information which might properly belong in a treatise on pollution ecology of marine nematodes! Not only is there overlap of species, but also the nematode ecology of the complex estuarine habitat can best be understood in relation to what is known of nematode ecology in the less violently fluctuating freshwater

and marine environments. In addition, it is difficult to circumscribe the boundaries of an estuarine system, as it gradually changes from the inner estuary,
subject to freshwater runoff, to the stable conditions of the open sea (Perkins,
1974). Indeed, studies of the nematodes of such systems often document a
transition from freshwater to marine species, as one progresses through the
system (Schütz, 1966; Nicholas, 1975). Before the effects of any type of pollution in an aquatic environment can be understood in relation to a faunal group,
the effects on the group of various environmental components in unpolluted
habitats must be sorted out and understood. For the most part this is the point at
which those interested in nematode pollution ecology find themselves. Much
information on ecology is accumulating, and some data on the effects of pollution are developing.

II. Systematics

It is beyond the scope of this chapter to treat in any detail the morphology and
systematics of aquatic nematodes. Many of the authors whose work with ecological systems will be discussed have also contributed heavily to the systematic
literature (e.g., Lorenzen, 1971a,b, 1972). For citations and lists of references to
taxonomic papers, see Hope and Murphy (1972) and Gerlach and Riemann
(1973, 1974). For illustrations of marine species, see Wieser (1953a, 1954,
1956, 1959a); for illustrations, references, and a key to freshwater species, see
Ferris *et al.* (1973) and the papers by Pillai and Taylor (1968a,b,c). A brief
discussion of some of the morphological features of aquatic nematodes, particularly as they relate to ecology, is necessary. For purposes of orientation we
include also a list of orders of nematodes which contain most of the freshwater
and estuarine species, together with examples of genera contained within those
orders (Table I). Many genera are found only in marine or brackish habitats, and
such species in Table I are so indicated. Wieser (1953a, 1959a) divided marine
nematodes into four morphological groups, based on the structure of their buccal
cavities, and considered these groups to represent four different feeding types.
These divisions have been accepted in the main by many workers in aquatic
nematology and have played an important part in the ecological thinking of
marine and estuarine nematologists. Wieser's feeding types are as follows:

Group 1-A: Selective Deposit Feeders.   No true buccal cavity, though sometimes traces of it. Food obtained mainly by sucking power of esophagus, and
consistency of food material soft, e.g., detritus and bacteria.

Group 1-B: Nonselective Deposit Feeders.   Cup-shaped, conical, or cylin-
drical buccal cavity, without armature. Food obtained as in previous group, but
with help from movements of lips and anterior part of buccal cavity. Food the
same as previous group, plus larger objects, e.g., diatoms.

TABLE I

ORDERS THAT CONTAIN MOST FRESHWATER AND ESTUARINE NEMATODE SPECIES, WITH EXAMPLES
OF GENERA CONTAINED

Adenophorea:	Desmodorida (*cont.*)	Dorylaimida (*cont.*)
Araeolaimida	*Microlaimus* (M)	*Eudorylaimus*
Anaplectus	*Monoposthia* (M)	*Labronema*
Anonchus	*Prodesmodora*	*Laimydorus*
Aphanolaimus	*Spirinia* (M)	*Mesodorylaimus*
Araeolaimus (M)[a]		*Miconchus*
Axonolaimus (M)	Chromadorida	*Mononchulus*
Bastiania	*Achromadora*	*Mononchus*
Camacolaimus (M)	*Chromadora* (M)	*Mylonchulus*
Chronogaster	*Chromadorina*	*Nygolaimus*
Cylindrolaimus	*Chromadorita*	*Oxydirus*
Euteratocephalus	*Comesoma* (M)	*Paractinolaimus*
Haliplectus (M)	*Ethmolaimus*	*Prionchulus*
Leptolaimus	*Euchromadora* (M)	*Thornia*
Nemella (M)	*Gomphionema* (M)	
Odontophora (M)	*Hypodontolaimus*	Secernentea:
Paraphanolaimus	*Monochromadora*	Tylenchida
Paraplectonema	*Neochromadora* (M)	*Aphelenchoides*
Plectus	*Paracyatholaimus*	*Atylenchus*
Rhabdolaimus	*Sabatieria* (M)	*Dolichodorus*
Teratocephalus		*Hemicycliophora*
	Enoplida	*Hirschmanniella*
Desmoscolecida	*Anoplostoma* (M)	*Tylenchus*
Desmoscolex	*Anticoma* (M)	
Tricoma (M)	*Bathylaimus* (M)	Rhabditida
	Cryptonchus	*Acrobeloides* (R)[b]
Monhysterida	*Deontostoma* (M)	*Acrostichus* (D)[c]
Desmolaimus	*Enoplus* (M)	*Butlerius* (D)
Linhomoeus (M)	*Ironus*	*Cephalobus* (D)
Monhystera	*Metoncholaimus* (M)	*Cylindrocorpus* (D)
Monhystrella	*Oncholaimus*	*Demaniella* (D)
Odontolaimus	*Prismatolaimus*	*Diploscapter* (R)
Siphonolaimus (M)	*Tobrilus*	*Eucephalobus* (R)
Sphaerolaimus (M)	*Tripyla*	*Fictor* (D)
Terschellingia (M)	*Tripyloides* (M)	*Goffartia* (D)
Theristus	*Trischistoma* (M)	*Mononchoides* (D)
Xyala (M)	*Viscosia* (M)	*Panagrolaimus* (R)
		Paroigolaimella (D)
Desmodorida	Dorylaimida	*Pelodera* (R)
Ceramonema (M)	*Anatonchus*	*Rhabditis* (R)
Desmodora (M)	*Aulolaimoides*	*Rhabdontolaimus* (D)
Draconema (M)	*Dorylaimus*	*Turbatrix* (R)

[a]Marine or brackish habitats only. All other genera are found at least occasionally in fresh water.
[b]Belongs to Rhabditoidea.
[c]Belongs to Diplogasteroidea.

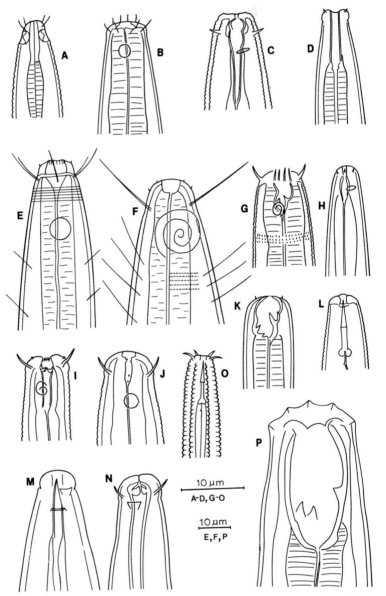

Fig. 1. Examples of buccal types among aquatic nematode genera. A, *Aphanolaimus;* B, *Monhystera;* C, *Anaplectus;* D, *Rhabditis;* E, *Theristus;* F, *Sabatieria;* G, *Achromadora;* H, *Goffartia;* I, *Ethmolaimus;* J, *Prodesmodora;* K, *Butlerius;* L, *Hirschmanniella;* M, *Mesodorylaimus;* N, *Ironus;* O, *Atylenchus;* P, *Miconchus.* A, B, Group 1-A; C–F, H, Group 1-B; G, I–J, Group 2-A; K–P, Group 2-B. See text for further explanation.

Group 2-A: Epigrowth Feeders. Small armature in buccal cavity. Food (e.g., algae) scraped off larger surfaces, or else food object pierced and cell contents sucked through hole.

Group 2-B: Predators and Omnivores. Large and powerful armature in buccal cavity. Prey swallowed whole or pierced by spear or teeth.

The above classifications have proved useful in grouping species in ecological studies. In Fig. 1, which includes illustrations of buccal types for a number of aquatic genera, groupings for the above classification are indicated.

As observations of actual feeding practices increase, refinements in our understanding of Wieser's groupings are possible. For example, *Monhystera filicaudata,* with a small unarmed buccal cavity and previously thought to be restricted to detritus and bacteria, feeds in the laboratory on detritus, bacteria, and dinoflagellates (Tietjen, 1967). *Rhabditis marina,* thought to be a nonselective deposit feeder, was found by tracer-feeding experiments to digest only two species of bacteria, and could be maintained in culture only in association with one species. Intact algae were found in the gut, suggesting a restriction in the diet based on inability to digest all ingested potential food organisms (Tietjen *et al.,* 1970). Similarly, tracer-feeding experiments with an epigrowth feeder (*Chromadora macrolaimoides*) indicated considerable selectivity. Of a total of 20 species of algae and 14 species of bacteria, only two species of diatoms were capable of sustaining indefinite growth (Tietjen and Lee, 1973; Tietjen *et al.,* 1970). Further examples of selective feeding by marine nematodes are given in Tietjen and Lee (1977a). Thus species which appear to be competitors for the same food source because of similarity in buccal armature may actually be dividing the food resource. Several investigators (Tietjen, 1969; Hopper and Meyers, 1967; Warwick and Buchanan, 1970) have commented on the importance of considering Wieser's predator group 2-B as an omnivore group, with some members possibly only possessing latent ability to function as predators. Many species in this group are considered to feed on diatoms, bacteria, flagellates, and other organisms in a nonselective manner. It is not always possible to place a species categorically in one or another of Wieser's groups. The combining of groups 1-A and 1-B into a single group characterized as deposit feeders, without oral armature, might ease the problem of proper assignment.

III. Importance and Function of Nematodes in Aquatic Ecosystems

For any taxonomic group, two questions regarding pollution ecology of the group might be asked: (1) What is the importance of the group to the functioning of the particular aquatic ecosystem as a whole? (2) Can knowledge about members of the group be used in some way as indicators of pollution? The first question, with respect to nematodes, will be dealt with first; and the second question will be considered following the discussion of what is known about pollution ecology of aquatic nematodes.

A. DENSITY

Nematodes seem to be the dominant group within the shallow water meiofauna of littoral, estuarine, and coastal sediments—and also comprise a large part of the meiofauna at ocean depths below 500 m (Tietjen, 1969, 1971; Perkins, 1974; Nicholas, 1975). Although many authors report finding large numbers of nematodes in such sediments, and also in freshwater sediments, there is usually great variation in replicate samples so that data for mean densities can be given only within wide limits (Filipjev, 1918, 1921; Hirschmann, 1952; Wieser, 1952, 1959a, 1960; Capstick, 1959; King, 1962; Teal and Wieser, 1966; McIntyre, 1969; Gerlach, 1952a, 1971; Biró, 1968, 1972, 1973; Tietjen, 1969, 1971, 1977; Hopper and Meyers, 1967; Schiemer *et al.*, 1969; McCloskey, 1970; Warwick and Buchanan, 1970; Meyers *et al.*, 1970; McIntyre and Murison, 1973; Ferris *et al.*, 1972; Juario, 1975; Nicholas, 1975; McLachlan *et al.*, 1977). Comparisons are difficult because of the various ways in which the data are expressed. Nicholas (1975) summarized data from a number of studies and expressed them as maximum numbers of nematodes found per square meter of substrate. He included figures for littoral, sublittoral, and estuarine habitats that range from 0.82×10^6–4.8×10^6 nematodes per square meter of substrate. Numbers of species range from 41 to over 100.

Comparable data from a variety of sources are scarce for freshwater habitats. For streams and rivers this may stem partly from difficulty in distinguishing between nematodes which are actually living in the aquatic habitat and those which have simply washed in from terrestrial areas. Ferris and Ferris (1972) recovered maximum totals of 0.2×10^6 nematodes per square meter, comprising 30 species, in the benthos of fast-moving areas of a small Indiana (U.S.A.) stream: whereas in a pooled area of a slow-moving stream they recovered 2×10^6–3×10^6 individuals per square meter, comprising about 100 species. When the species composition of the latter group was analyzed, however, many of the species were determined to be terrestrial species that had washed into the stream (Ferris *et al.*, 1972). A similar problem in interpreting data probably exists for some estuarine habitats.

Biró (1973) found a maximum of 0.8×10^5–1×10^5 individuals per square meter, comprising 31 species, in the open-water sediments of Lake Balaton (Hungary), the largest inland body of water in central Europe. Frequently at his sampling sites 1×10^4–2×10^4 individuals per square meter were found. Schiemer *et al.* (1969) reported 1×10^5–10×10^5 individuals per square meter from the softer muds of Neusiedler Lake, which is lcoated on the border between Austria and Hungary, and is similar to Balaton Lake in many ways (Biró, 1973).

B. TROPHIC POSITION

As discussed above in the consideration of systematics and buccal armature, some aquatic nematodes are known to feed on bacteria and/or algae, whereas others may be considered predators, parasites, or omnivores. Nematodes are believed to occupy key roles in aquatic ecosystems as consumers of primary decomposers, and at the second trophic level, as grazers on primary producers. At the third and higher trophic levels, as predators and parasites, their role seems less important (Nicholas, 1975; Juario, 1975).

C. REPRODUCTION AND LIFE CYCLES

In order to estimate the importance of aquatic nematodes within an ecosystem, a knowledge of the numbers of generations produced annually is desirable. Many investigators have contributed data on free-living marine and estuarine species (Hopper and Meyers, 1966a; Lee *et al.*, 1970; Gerlach, 1971, 1972; Gerlach and Schrage, 1971, 1972; Skoolman and Gerlach, 1971; Tietjen and Lee, 1972, 1973, 1977b; Hopper and Cefalu, 1973; Hopper *et al.*, 1973).

Basically, two methods exist for determining these data as follows: (1) Size distributions within samples taken throughout the year can be analyzed for size classes, and the information deduced from these data; or (2) the life cycles may be determined from laboratory cultures and the data extrapolated to field conditions. Problems exist with both methods. With shorter-lived species especially, the generations overlap and are not distinguishable from field sample data. Even in longer-lived species, the life span of individuals may greatly exceed generation times. In the laboratory it is difficult to simulate natural conditions; however, attempts to vary environmental factors in experiments have resulted in new information concerning reproduction.

To summarize briefly some of the findings, Hopper and Meyers (1966a) reported life cycles of about 1 month for laboratory-reared colonies of chromadorid, monhysterid, and oncholaimid nematodes. Gerlach (1971, 1972) summarized data from many sources and observed that laboratory experiments are usually conducted at higher temperatures than one might expect to find in marine environments. Below 7°C generation time for *Monhystera disjuncta* is greatly extended, e.g., to 80 days at 1°C. Under natural conditions Gerlach (1972) estimated that this species might produce 12 generations per year. Other species are known to require much longer periods for their life cycles at 7°C within the laboratory, some requiring more than one year (Gerlach and Schrage, 1972). Generally speaking, one expects very small nematodes to have a short life cycle and large species to take much longer, but Gerlach and Schrage (1972) found no strict relation between generation time and body weight. Usually the life cycle is shortened with increasing temperature, but lengthened at temperatures at the upper limits (Hopper *et al.*, 1973). Variations in salinity also influence the nematode life cycle and probably also spatial and temporal variations of food organisms (Tietjen *et al.*, 1970; Tietjen and Lee, 1972, 1973, 1977b).

Few data exist for reproductive potential and life cycles of nematode species found in freshwater habitats, although many freshwater nematode species have been described. [See Gerlach and Riemann (1973, 1974) and Ferris *et al.* (1973) for references to taxonomic literature.] In the early 1960s concern over finding nematodes in the municipal water supplies of the United States (Chang *et al.*, 1960; Chang, 1961; Chang and Kabler, 1962; Calaway, 1963) led to a flurry of studies financed by federal agencies on the biology and fate of nematodes in sewage effluent. Although the finding of nematodes in potable water supplies is still of some interest (Schiemer, 1975; Tombes *et al.*, 1977), many of the existing data on reproductive potential of freshwater aquatic species were developed as a result of the earlier excitement (Chaudhuri, 1964; Chaudhuri *et al.*, 1964; Goodrich *et al.*, 1968; Pillai and Taylor, 1968a,b,c). Most of the species recovered from filter beds of water treatment facilities, and hence of interest to the agencies that supported such research, were rhabditids which can be cultured on a variety of bacteria, amebae, or small prey nematodes. All have life cycles of about 2 days at high temperatures (around 30°C) and several days to 1 week or longer at lower temperatures (around 15°C). Many of the rhabditids found in freshwater are taxonomically close to species which have been studied extensively by nematologists and others interested in developing techniques for axenic cultivation (Cryan *et al.*, 1963; Nicholas, 1962, 1975) and for which much data on reproduction and life cycle exist. In general, total generation time is less than 1 week at about 25°C and somewhat longer at lower temperatures.

Lists of genera and species recovered from natural freshwater environments include many of those found in sewage treatment plants; but they also include a wide variety of other species in a number of nematode orders (Ferris *et al.*, 1972, 1973; Fisher, 1968; Hirschmann, 1952; Zullini, 1976a). Based on buccal armature, and by analogy to similar species found in marine habitats, it can be assumed that some of these at least feed on algae. Others feed on bacteria, a few on roots of higher aquatic plants, and a number probably should be classified as omnivores.

The rhabditid species which feed on bacteria and which have short life cycles of a few days are among those which seem to build up quickly in enriched freshwater habitats—as are diplogasteroids that feed on microfauna (including nematodes) and that are known to have short life cycles (Ferris *et al.*, 1972; Zullini, 1976a). Species of orders other than Rhabditida also are an important part of the nematode community of freshwater habitats. These belong to all of the feeding groups and by analogy to better-known soil forms and marine species, and probably have life cycles that range from about 1 month to as long as 1 year (Ferris *et al.*, 1972; Zullini, 1976a). Because this composition of aquatic nematode communities is probably a reflection of both past and present benthic conditions, a more precise knowledge of the biology of these species would be a great help in interpreting community structure data as a tool for evaluating disturbance to freshwater habitats.

D. BIOENERGETICS

Most estimates of the importance of nematodes to an ecosystem are derived from assessments of the contributions of the nematodes to the total respiration of the sediments in which they are found. Unfortunately, rigorous methods for quantitative assessment of the importance of decomposer animals have not yet been developed (Nicholas, 1975). In order to arrive at estimates of respiration, laboratory data on respiration for a given species are used, together with field survey data converted to biomass estimates. To derive yearly estimates, many assumptions about environmental variables may be needed. Variables that greatly affect respiration data include variations of temperature and season; variations in activity among nematode species and within populations; and variations in food supplies which may cause an animal to have to devote more or less energy to obtaining food as compared with that devoted to growth and reproduction (Teal, 1962; Gerlach, 1971, 1972; Gerlach and Schrage, 1971, 1972; Skoolman and Gerlach, 1971; Schiemer and Duncan, 1974; Marchant and Nicholas, 1974; Nicholas, 1975).

A total respiration of 13.9 cm^3 of O_2/m^2/hour in November at 20°C was found for a Massachusetts estuarine habitat. A figure of 5.1 cm^3 of O_2/m^2/hour was obtained for the same area at the same temperature in June, when the nematode biomass was less than half that found in November. These figures for total respiration are comparable to estimates for soil species (Nicholas, 1975; Wieser and Kanwisher, 1961).

Using estimates from laboratory cultures, Gerlach (1972) derived a model to estimate the life cycle turnover rate for two species of estuarine nematodes. Assuming a hypothetical mortality, he estimated the life cycle turnover rate to be 3.0 for *Chromadorita tenuis* and 2.2 for *Monhystera disjuncta*. Minimum generation time for the latter species was estimated at 13 days in the laboratory, but it was assumed that under natural conditions an average of 12 generations per year might be produced, resulting in a yearly turnover rate of 26, or about ten times more than the average macrofauna turnover rates. Because small animals of the meiofauna metabolize more per unit weight than do larger animals, Gerlach

(1971) calculated that a meiobenthic standing stock which comprised 3% of the total biomass would contribute about 15% of the food chain in terms of animals and spawn which are preyed upon or consumed by scavengers and decomposers.

Data exist for ingestion and assimilation rates of *Plectus palustris,* a bacteria-feeding freshwater benthic species. At bacterial concentrations and temperatures comparable to those which might be found in used-water treatment facilities, the nematode ingested a quantity of bacteria equal to 650% of the animal body weight. It was estimated that only 12% of the ingested material was assimilated (Duncan *et al.,* 1974; Schiemer, 1975).

IV. Pollution Ecology

Before considering what is known about pollution ecology of aquatic nematodes, it is important to look first at what is known about nematode ecology of unpolluted aquatic environments. Data range from mere lists of species found in various habitats to the results of complicated quantitative techniques for analysis of communities. The use of quantitative techniques has helped in the difficult task of evaluating the importance of dominant features of the aquatic environment, which must be studied in the field. A major problem in studying biota of aquatic habitats is extreme variability among samples, resulting in a need for very large numbers of samples if the usual statistical tests are to be used to demonstrate validity of results. Many workers have found that if several samples are collected a fair picture of the abundance of the commoner species may be obtained (Hynes, 1970). Much information has come from the use of distribution-free statistics, diversity measures, indices of affinity, trellis diagrams, ordination, and various clustering techniques (Wieser, 1960; Hopper and Meyers, 1967; Warwick and Buchanan, 1970; Tietjen, 1971, 1977; Ward, 1973; Ferris *et al.,* 1972; Heip and Decraemer, 1974; Warwick and Gage, 1975).

Data for normal and polluted areas of freshwater habitats will be discussed first, followed by information for shallow marine areas. The more complex estuarine environment will be discussed last.

A. FRESHWATER NEMATODES

1. Ecology of Normal Habitats

Little critical work has been done on the ecology of nematodes of freshwater habitats. Much of what is known has been learned from "control" samples in studies of river waters above and below sewage treatment plant outfall areas. A few studies have occurred on relatively clean streams or lakes, or parts of rivers (Filipjev, 1929–1930; Fisher, 1968; Biró, 1968, 1972, 1973; Chaudhuri *et al.,*

1964; Baliga *et al.*, 1969; Schiemer *et al.*, 1969; Ferris *et al.*, 1972; Nicholas, 1975; Zullini, 1976a).

In the Lake Balaton (Hungary) study Biró (1973) found about 90% of the individuals to be comprised of five species in the genera *Paraplectonema, Paraphanolaimus, Theristus,* and *Monhystera* (all in feeding group 1-B) and *Ironus* (group 2-B), with seasonal differences in the proportions present. A water temperature below or above 12°C seemed to be a critical factor in determining predominance of species. Composition of the nematode fauna varied in different regions of the lake. In general, densities were greater in spring and autumn than in summer months. Biró (1973) confirmed the findings of Schiemer *et al.* (1969), which were based on observations in Neusiedler Lake (Austria), that *Paraplectonema pedunculatum* does not occur in very muddy substrates or very sandy ones. Biró (1973) found that more *P. pedunculatum* occurred where small segment particles ($<2~\mu$m) comprise about 30% of the sediment particles than in sediments where the proportion of the small particles was 40 or 10%. Biró (1973) was unable to demonstrate in Lake Balaton the relationship of two other species, *Tobrilus gracilis* and *Monhystera paludicola,* to mud depth that was shown by Schiemer *et al.* (1969), probably because the compact bottom sediments of Lake Balaton differ from the soft muds of Neusiedler Lake. Schiemer and Duncan (1974) noted that *Tobrilus gracilis* in Neusiedler Lake lives predominantly in oxygen-deficient mud layers, and showed by laboratory experiments that metabolism of the species is partially anaerobic, even when oxygen is available.

In a study of nematode occurrence in a small Illinois stream above and below the outfall of a waste treatment plant, Baliga *et al.* (1969) found more benthic nematodes in autumn months than in summer. They attributed this increase to a reduction in flow or velocity of the river. In their study, variations in chemical oxygen demand (COD), total nitrogen, and pH of the benthos were found to have no direct relationship to nematode concentration. These authors concluded that a sluggish velocity in their stream promoted settling of particles and an environment conducive to reproduction. Furthermore, a reduction in velocity kept nematodes from being flushed away. Unfortunately, these authors did not characterize the nematode populations taxonomically.

In a recent study of the Seveso River in Italy, Zullini (1976a) found an absence of seasonal fluctuations, with little month-to-month variation in numbers or species of nematodes. Zullini found Adenophorea as a group (predators and ominvores in Wieser's classification) more abundant in cleaner upstream areas of the Seveso River. The genus *Tobrilus* (believed to be predaceous) of the Enoplida occurred most frequently in unpolluted waters, although other authors have associated *Tobrilus* species with river pollution (Schiemer, 1975). Zullini (1976a) calculated diversity indices (presumably Shannon–Weaver indices as discussed in Zullini, 1976b) for all of his river sites and found the highest index at the station he designated as being clean and unpolluted.

In a 2-year study of two relatively clean small streams in Tippecanoe County, Indiana, community structure of nematodes was used as a means of interpreting ecological conditions within the stream habitats (Ferris *et al.,* 1972; Callahan, 1976; Callahan *et al.,* 1977). In addition to the nematode data collected each month from April to October at 16 sites during each of the 2 years, data for physicochemical parameters were collected at each site at biweekly intervals during the 2-year period. The data included values for nitrate, nitrite, dissolved oxygen, CO_2, pH, total hardness, phosphates, and water temperature. In addition, organic matter was estimated, and relative quantities of particles of different sizes were determined for the substrate at each site and examined by texture analysis (Krumbein, 1939). Of the 156 nematode species collected and identified, 74 were considered to be benthic inhabitants, and the rest to be terrestrial inhabitants of shore areas. Prominence values based on density and frequency were calculated for each species, and coefficients of similarity–dissimilarity figured for each site. A dendrogram, which clustered the sites on the basis of their similarities, gave results similar to those obtained from an ordination, which grouped sites on the basis of their dissimilarities (Ferris *et al.,* 1972). In the three-dimensional community ordination, similar groupings of sites were obtained whether data from all species were used, or data from the 74 benthic species only (Figs. 2 and 3). Good reproducibility of results between years was obtained, indicating that site influences were strong enough to produce aggregations which were more similar during different years than were the within-year similarities among sites. This finding was in contrast to published data from other studies of freshwater biota which used cluster analyses and other groups of organisms (Roback *et al.,* 1969; Cairns and Kaesler, 1969). Inspection of the

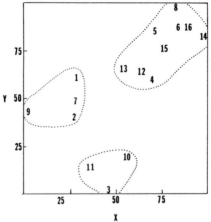

Fig. 2.   Ordination based on 2 years of data for 74 freshwater benthic nematode species at 16 stream sites, Tippecanoe County, Indiana. *X-Y* plane.

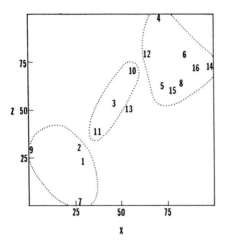

Fig. 3. Ordination based on 2 years of data for 74 freshwater benthic nematode species at 16 stream sites, Tippecanoe County, Indiana. *X-Z* plane.

nematode data revealed that groupings were closely related to buildup of a relatively few nematode species in several taxonomic orders and trophic groups. Thus groupings similar to those previously obtained with all species were also obtained when the data were limited to those from the 17 dominant benthic species and even from only nine dominant species.

Buildup of four species greatly influenced groupings: two species of *Monhystera* (Adenophorea: Monhysterida), which belongs to Wieser's feeding group 1-B, nonselective microbivores; a species of *Mesodorylaimus* (Adenophorea: Dorylaimida), which is probably an algal feeder of feeding group 2-B, but may be predaceous or omnivorus; and a species of *Tylenchus* (Secernentea: Tylenchida), also feeding group 2-B, which probably feeds on mosses and/or algae. (Several other nematode species, including predaceous species and microbivores, also appeared to influence grouping.) Percentage of the total count comprised of the five predominant species at each site apportioned to each of Wieser's four feeding groups is shown in Fig. 4. The sites are grouped in the figure to coincide with the groupings of Figs. 2 and 3. The total number of nematodes $\times$ $10^3/m^2$ to a depth of 3 cm at each site is given at the lower right of each circle diagram.

Numbers of species of benthic nematodes at each site were related to median particle size (Mdϕ) of the substrate ($\rho = 0.66$, $P < 0.01$), with a larger number of species correlated with smaller particle size. Buildup of numbers of individuals was related to Mdϕ, QDϕ (heterogeneity of particle size), and the presence of nitrite-nitrogen. Average number of individuals per species, tested with Mdϕ gave a ρ value of 0.60 ($P < 0.02$) with higher average counts

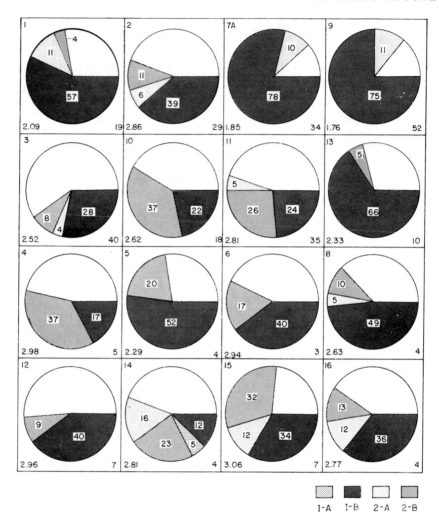

Fig. 4. Summary of data for 74 freshwater benthic nematode species at 16 stream sites, Tippecanoe County, Indiana. Percentage of total count at each site comprised of five predominant species, apportioned to Wieser's four feeding groups. Sites arranged to coincide with groupings of Figs. 2 and 3. Site number at upper left of each circle, H' at lower left, and total count $\times 10^3/m^2$ to a depth of 3 cm at lower right.

associated with small particles; and a value of 0.49 ($P < 0.05$) when tested with QDϕ, with higher counts associated with greater heterogeneity of particle size. Average count per species, tested with mean nitrite values, gave a ρ of 0.60 ($P < 0.02$), with higher counts correlated with larger means for nitrite-nitrogen.

Because the Shannon–Weaver diversity index (Poole, 1974; Pielou, 1977) is widely used in aquatic ecology, H' values were calculated for each site, and these are given at the lower left of each circle diagram in Fig. 4. These values could not be correlated with any other site values, nor could they be related in any way to the groupings of Figs. 2 and 3. Although the site with the highest total count for the 2-year period (site 9) had the lowest H' value, other sites that had high H' values (sites 4, 12, 15) had total counts near the median. The kind of inverse relationship between abundance and diversity shown by Porter (1977) for freshwater algae, could not be demonstrated.

2. Ecology of Polluted Habitats

Nematodes from filter beds have been of interest to nematologists and others since the pioneer studies of N. A. Cobb (Cobb, 1918; Peters, 1930a,b; Calaway, 1963, 1969; Kokordák, 1969, 1970; Goodrich *et al.*, 1968; Pillai and Taylor, 1968a,b,c; Murad, 1970). The concern which arose during the 1960s over the importance of the discovery of nematodes in municipal water supplies in the United States led to investigations also of the nematodes in river areas above and below sewage treatment plants in an effort to locate the source of the contaminations.

Nematodes of wastewater plants belong predominantly to the families Rhabditidae and Diplogasteridae of the Rhabditida (Secernentea), and in lesser numbers to the Dorylaimida (Adenophorea). All of the Rhabditidae can be considered to belong to Wieser's group 1-B, nonselective microbivores; many of the Diplogasteridae also belong to group 1-B and others to group 2-A, epigrowth feeders which scrape algae or feed on small microfauna. The Dorylaimida found in this habitat are predators and belong to Wieser's group 2-B. Calaway (1963) concluded that nematodes are found in sufficient numbers in wastewater treatment plants to permit the conclusion that their work is valuable and suggested that a function of the nematodes in wastewater treatment is to keep the filter beds porous and accessible to oxygen by feeding on biological films. According to Schiemer (1975) the major influence is on decomposition processes as a result of their metabolism; and on bacteria as a result of their grazing.

Zullini (1976b) found nematodes more numerous in semisolid substrates of the early stages of sewage treatment than in later liquid phases. He discerned no seasonal rhythms, but found the populations to vary in an irregular manner, as did Peters (1930a). The Rhabditidae prevailed in areas of raw waste, whereas Diplogasteridae prevailed in later stages of treatment. Zullini (1976b) suggested that diplogasterid predators need more oxygen and "better environmental conditions." He also noted that within the Diplogasteridae those species with wider buccal cavity require more oxygenated and less polluted habitats. Zullini (1976b) thought that this might constitute another example of the generalization of Wieser and Kanwisher (1961) that high oxygen consumption is associated with a large buccal cavity possibly because nonselective feeders use more energy in obtaining sufficient useable food. The drying beds, composed of waste by-products of the treatments, contained predominantly Rhabditidae. Diversity indices, calculated by the Shannon–Weaver formula, could not be interpreted precisely, but did seem to increase with the progress of the cleaning treatment.

In the study of nematode concentrations in the Illinois stream, Baliga *et al.* (1969) found more nematodes below than above the outfall of a waste treatment plant, and concluded that their data showed that nematodes were being added to the water at the effluent point. An alternative view might be that nutrient enrichment produced a buildup at that point, of nematodes already present. The absence of taxonomic data in this study precludes further discussion of the matter. Zullini (1976a) found in his Seveso River study (Italy) that numbers of nematodes greatly increased at sites receiving not only sewage discharges, but also sites receiving heavy industrial pollution of several kinds. At such sites, numbers of species as well as diversity indices, tended to decrease. In addition to nematode data, Zullini (1976a) collected data for O_2, COD, inorganic salts pollution, and data from a fish survival test from five sites along the river. In addition, he measured sedimentary material. As was found in sewage treatment plants, members of the Rhabditidae and Diplogasteridae predominated at the polluted stations, particularly those species that are microbivores. Zullini (1976a) concluded that rhabditoids predominated in areas of very high pollution, whether sewage or factory discharges, whereas diplogasteroids were most numerous in conditions of medium to high pollution. Zullini designated indicator species as follows: *Paroigolaimella bernensis* in moderately polluted waters; *Acrostichus nudicapitatus* in more polluted areas; and *Rhabditis oxycerca* in the most polluted areas. In an earlier study of nematode associations in freshwater habitats, Hirschmann (1952) suggested that different rhabditid species associations might be indicative of various degrees of pollution.

Although the two streams investigated by Ferris *et al.* (1972) and Callahan (1976) were both relatively unpolluted, both flow through farmland and forested areas, and one also flows through a small housing development. Sampling sites were chosen with these differences in mind. The ordering of sites in the community ordination (Figs. 2 and 3) can be related to the ecological conditions at the sites. Those sites where one or more species reached high numbers tended to cluster into two groups situated at the left of the *X-Y* plane. These sites were those which received fertilizer, feedlot, or residential runoff and seemed clearly to be areas of slight nutrient enrichment or mild perturbation. The rest of the sites, which generally had lower counts of individuals, grouped together at the upper right of the *X-Y* plane (Fig. 2) and tended to be the more undisturbed sites. Ferris *et al.* (1972) concluded that analysis of benthic nematode community structure can probably become a useful tool for evaluating disturbance to freshwater stream habitats, and suggested that composition of nematode communities is a reflection of both past and present conditions. Their demonstration that sufficient information for analysis can be obtained from data on a few dominant species could simplify procedures and increase the applicability and usefulness of the technique.

B. COASTAL MARINE NEMATODES

1. Factors That Influence Nematode Ecology of Unpolluted Habitats

Information has accumulated from a number of laboratories on the ecology of marine nematodes. The studies often include nematodes of deep sea as well as coastal areas. In some cases a single study will deal with the above habitats plus estuarine areas. It is difficult to separate the information by habitat, and there will be some overlap in the discussion below with the discussion of factors known to influence nematode ecology in brackish habitats.

a. Salinity. In studies that include nearshore habitats, salinity seems to be the most important factor in determining nematode composition. As might be expected, individual species vary in their sensitivity to saline, with some marine species known to tolerate salinities as low as 15%. Tietjen and Lee (1972, 1977b) found that for two species of marine nematodes (*Monhystera denticulata* and *Chromadorina germanica*) generation times were more significantly lengthened by exposure to lower salinities than to higher ones. Ott and Schiemer (1973) noted that nematodes from anaerobic layers of sediments are more sensitive to salinity changes than those from upper layers. Zonation as a result of salinity is likely to occur in waters which range between 3 and 30% salinity (Gerlach, 1953a; Wieser, 1959b; Nicholas, 1975). In a study of nearshore benthic fauna in Loch Etive (Scotland), Warwick and Gage (1975) used an inverse reciprocal averaging ordination which showed that salinity is the major factor producing zonation. Inland saline waters are said to have species typical of fresh water, brackish water, or seawater, depending on the degree of salinity (Meyl, 1955; Nicholas, 1975).

b. Grain Size. Within an area of uniform salinity, grain size of sediments is a dominant factor in determining the composition of nematode communities, as well as communities of other meiofauna (Wieser, 1959c, 1960; Tietjen, 1971, 1977; Nicholas, 1975; Ward, 1973, 1975). Tietjen (1971) summarized the literature which indicates the influence of sediment type on distribution of meiofauna. In addition to a possible direct influence on meiofauna, a correlation often exists between grain size and organic content of substrate, and hence the influence of grain size might be an indirect result of the nature of available food supply (Tietjen, 1971, 1977; Marcotte and Coull, 1974; Ward, 1975). Wieser (1960) found two distinct meiofaunal communities in the benthos of Buzzards Bay, Massachusetts—one characteristic of sandy habitats and the other characteristic of silty areas. Heip and Decramer (1974) found diversity lowest (Shannon–

Weaver index) at stations with low median grain size and a high percentage of silt in their study of nematode communities at five stations in the southern North Sea, as did Tietjen (1977) in Long Island Sound. Ward (1973) found generic and dominance diversity lowest in the mudiest habitats studied in Liverpool Bay (Britain).

Warwick and Buchanan (1970) used several quantitative techniques in a study of nematode communities at three sampling stations off the Northumberland coast (Britain). These included an inshore station with a depth of 35 m; a middle station, depth 54 m; and an offshore station, depth 80 m. Sediment composition varied considerably among the stations. Williams' index of diversity was highest at the sandiest station and lowest at the most silty station. However, overlap of eurytopic species between mud and sand made it difficult to categorize distinct nematode communities in terms of species composition as Wieser (1960) did for Buzzards Bay. By eliminating eurytopic species from consideration, Warwick and Buchanan (1970) were able to characterize a definite "mud fauna," probably typical of a wide geographic area. Tietjen (1977), utilizing a cluster analysis, found two basic faunistic units in the sediments of Long Island Sound—a mud unit characterized by high species dominance, low species diversity, and low species endemism; and a sand unit characterized by low species dominance, high species diversity, and high species endemism.

Nematodes seem to be restricted mainly to the upper few centimeters of sediment regardless of sediment type and regardless of depth of water, although vertical gradients do seem to vary with season in shallow waters (Wieser and Kanwisher, 1961; Teal and Wieser, 1966; Muus, 1967; Tietjen, 1969, 1971; Boucher, 1972). Differential sensitivities to reduced oxygen supplies, abundance of interstitial water, and qualitative differences in food material at different depths could all be factors in determining vertical gradients (Tietjen, 1969, 1971; Boucher, 1972; Ott, 1972b). Ott (1972a) and Ott and Schiemer (1973) believe a specific and diverse anaerobic nematode fauna may exist.

c. Other Factors. Data on other factors that influence nematode communities in coastal marine environments are less clearcut. It is often difficult to separate the effects of various factors from each other (e.g., depth and temperature) and, in addition, the effects of a given variable might be quite different in shallow water than in deep areas. Other factors that might be important include depth of water, carbon, temperature, and season.

Warwick and Buchanan (1970) concluded from their studies of nematode communities from 35 to 80 m that species composition was less dependent on depth than on sediment type. Tietjen (1971) found differences in nematode communities off North Carolina between the shallower sediments (50–500 m) and the deeper sediments (600–2500 m). The fauna at 600–750 m represented a transition between the two regions, but was more closely related to the deepwater fauna. Reasons for the change in fauna with depth could include changes in sediment composition, food, temperature, or a combination of factors. He found a significant correlation between organic carbon present and percentage of total nematode fauna which were deposit feeders. Organic content of the sediments increased significantly below 500 m depth and reached its maximum between 800 and 1500 m. Tietjen (1977) found a major difference between the nematode populations of North Carolina and those of shallower sediments (4–38 m) of Long Island Sound to be the high incidence of species dominance (low equitability) in the muds of Long Island Sound. Tietjen suggests that in shallower muds, where densities may be high as a result of abundant food materials, competition for some common resource may result in exclusion of many species. In deep-sea muds, under more benign conditions, the lower densities may result in interaction and less competition, with more niche fractionation. This would result in less dominance by a few species and a higher species equitability.

Few data exist on the effect of water temperatures on the composition of nematode communities. Tietjen (1971) noted that of 104 previously described nematode species which he encountered in his study of deep-sea meiobenthos (all described from shallow water), 51 were found at depths greater than 500 m, where temperatures are colder by 10°C and vary less than in areas above 500 m. Considerable data exist on the *in vitro* temperature tolerances and responses of individual nematode species (Gerlach and Schrage, 1971, 1972; Hopper *et al.*, 1973; Tietjen and Lee, 1972, 1977b). For many species, optimum temperature, resulting in the most rapid growth and development, is around 25°C. The actual time to complete the life cycle varies from a few days to several weeks or more, depending on the species. At temperatures colder than the optimum, the number of days to complete the life cycle may be greatly increased. Gerlach and Schrage (1972) found that some species reared in laboratory culture at 7°C (to simulate North Sea conditions) may take more than 2 years to complete their life cycles.

The effect of seasons on fluctuations within marine nematode communities varies with depth and with substrate type. Warwick and Buchanan (1971) at their 80-m depth station found species composition to remain stable throughout the year, and the age structure constant from month to month. Heip and Decraemer (1974) found that where diversity was high (at their offshore stations, 7–15 m depth), the diversity increased throughout the year; at a shallow station where diversity was medium, it stayed constant; and where it was low, diversity was lower in June than in other months. Hopper and Meyers (1967) found seasonal shifts in benthic nematode communities within a subtropical seagrass bed (less than 1 m depth) in Biscayne Bay, Florida.

2. Community Structure

Although many coastal marine nematode species are cosmopolitan, lists of species which occur in comparable habitats in different areas often have little similarity to each other. The predominant species, in particular, often differ.

There is evidence of ecological displacement and sometimes species from the same genus replace each other (Nicholas, 1975). Tietjen (1977) found the two most abundant families in the muds of Long Island Sound, the Comesomatidae (order: Chromadorida) and Linhomoeidae (order: Monhysterida) are commonly the predominant families found in muds in other parts of the world, with the most common genera in Long Island Sound typically predominant in other muds. Likewise, the families that become increasingly abundant with increase in grain size in sediments of Long Island Sound also characterized the shift to muddy sands and sands of other geographic areas. All of the factors discussed above probably interact to determine the spatial and temporal distribution of nematodes, with many factors relating directly to availability of appropriate food organisms. Tietjen *et al.* (1970) suggest predation, antibiotics, and other biotic factors, as well as presence or absence of limiting food organisms, as features of the environment which might limit growth and reproduction of *Rhabditis marina* in nature. Salt/sand ratios appear to be a useful measure for interpreting the composition of the nematode fauna (Nicholas, 1975; Tietjen, 1977; Wieser, 1960).

Of the 212 species found in Tietjen's (1971) deep-sea study, 79 were found at depths no greater than 500 m, and 59 were restricted to depths greater than 500 m. Species that were predominant in the sands (*Xyala striata, Desmodora varioannulata, Ceramonema reticulatum, Microlaimus problematicus, Monoposthia duodecimilata*) were absent below 500 m. Other species, which predominated in the clayey silts, were absent above 600 m. *Sabatieria americana* was the only species to attain predominance in the sandy silts of the transition zone. In the shallower sediments, 50% of the nematode fauna were deposit feeders and 40% epigrowth feeders (probably using benthic microalgae to a depth of 100 m). In deeper sediments, deposit feeders, probably feeding on bacteria and organic detritus, accounted for 80%, whereas epigrowth feeders accounted for only 8% of the nematode fauna.

In his study of nematodes of shallow sediments of Long Island Sound, Tietjen (1977) found 229 species of which one, *Sabatieria pulchra*, predominated in many of the mud samples. No species clearly dominated the sand stations. Ten of the species were eurytopic, four species attained their maximum abundances in muds, and 20 appeared to prefer sandy sediments. A total of 94 species were found only in sandy substrates whereas only five species were restricted to muds.

In Hopper and Meyer's (1967) shallow seagrass community, four species of nematodes were predominant and regularly comprised between 87 and 95% of the total number of nematodes collected during the winter and spring. The four predominant species were *Metoncholaimus scissus, Theristus fistulatus, Spirinia parasitifera,* and *Gomphionema typica*. Population densities reached peaks in November–December and again in February–March. Tietjen and Lee (1973) found several species of the Chromadoridae to dominate the nematode assemblages in the aufwuchs of macrophytes in the North Sea Harbor, New York. Warwick and Buchanan (1970) found that the relative abundance of the four different feeding types was quite consistent at their three stations off Northumberland. Nonselective deposit feeders were the predominant group at each station, with epigrowth feeders second in importance.

3. Ecology of Polluted Habitats

Ecological studies undertaken for the purpose of demonstrating the effects of pollution on marine nematodes are few, although several investigators have noted possible effects of industrial and other kinds of pollution on results obtained in their ecological or systematic investigations (Lorenzen, 1971a,b, 1972, 1974; Marcotte and Coull, 1974; Vitiello and Vivier, 1974; McLachlin et al., 1977). As with freshwater habitats, analysis of field data from polluted areas becomes very complicated because the interacting effects of many naturally occurring features of the aquatic environment on community structure and population changes are not well understood, as discussed above. Hopper and Meyers (1967) and Meyers et al. (1970) observed that the nematode fauna of their shallow subtropical seagrass community seemed to be extremely variable and affected by a complexity of environmental factors which remained a mystery to the investigators, and which, they suggested, might lead to various "micro-differences" in habitats. These problems must be kept in mind in the interpretation of results of any experiments designed to study aquatic pollution ecology.

Lorenzen (1974) in a study near Helgoland, Germany, found no change in sublittoral nematode densities, nor in number of species present, following one year of titanium waste disposal, and concluded that for the short term, at least, the waste disposal had no effect on benthic nematode populations. Likewise, Vitiello and Vivier (1974) found an abundance of nematodes in sediments of a deep zone off the coast of France, used for dumping industrial wastes from aluminum production, and concluded that the presence of the nematodes and other meiofauna proved the nontoxicity of the waste materials. Heip and Decraemer (1974) in their study of the southern North Sea did not attempt quantitative estimations of total density, but calculated Shannon–Weaver diversity indices from data which indicated the proportion of individuals belonging to each species in each sample. The lowest diversities were obtained at two stations directly influenced by the outflow of the IJzer River, which they describe as heavily polluted, going to the northeast with the current along the Belgian coast. Another nearshore station with shallow depth, but not in the path of the river outflow, had a higher diversity index. The highest indices were at offshore stations. The authors suggested that the river output of organic material, particularly in June when diversity indices were lowest, contributed to the low indices of the two nearshore stations in the path of the outflow. They caution, however, that it is difficult to separate the effect of the substrate grain size and silt content on nematode diversity from the effects of other factors in any sort of pollution study (see Section IV,B,1,b). They suggest that, if possible, areas of similar substrate should be compared in such studies.

Although finding comparable study areas which differ only in the pollution factors being studied is difficult, Tietjen (1977) appears to have succeeded in doing so in his study of benthic nematodes of Long Island Sound. He was able to separate his 18 stations into four sedimentary environments and to examine the quantitative and qualitative aspects of nematode population structure in relation to the environmental impact of heavy metals, organic carbon, and other variables. Tietjen (1977) concluded that no relationship could be shown between species diversity indices (Shannon–Weaver) and the heavy metal content of the sediments, despite the high concentrations in some sediments. An affinity dendrogram of faunal similarities (Fig. 5) divided the 18 stations rather clearly into two distinct groups, which Tietjen (1977) called a mud unit and a sand unit (see Section IV,B,1,b). Anomalies occur in the groupings in that two stations which must be classified as sands (stations 14 and 25) are in the mud unit, and one muddy sand (station 86) is in the sand unit. Inspection of Tietjen's (1977) Table 2, however, which lists heavy metal concentrations, indicates that all of the stations in the

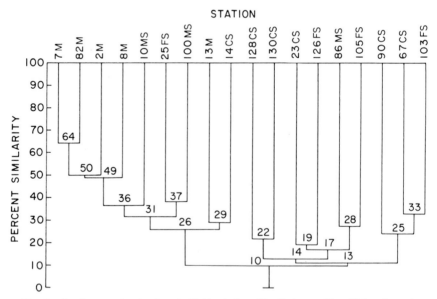

Fig. 5. Dendrogram showing faunal affinities indicated by Czekanowski coefficient for stations in Long Island Sound. Scale is percent similarity with values of junctions given. M, muds; MS, muddy sands; FS, fine sands; CS, medium-coarse sands (From Tietjen, 1977, reproduced by permission.)

mud unit on the dendrogram are stations with high concentration of heavy metals ($\bar{X}$ = 743 ppm, range = 427-1184). By contrast, the stations grouped in the sand unit have significantly ($P < 0.001$) less heavy metals concentration ($\bar{X}$ = 135 ppm, range = 20-253). Although the nematode fauna of Long Island Sound is similar in a general way to nematode faunas of other less impacted shallow areas, the clustering method used seems to have revealed subtle differences in community composition which can be related to the pollution by heavy metals. The mud unit stations also have significantly more ammonia ($P < 0.01$) and a significantly higher percent organic carbon ($P < 0.05$).

C. ESTUARINE NEMATODES

1. Factors That Influence Nematode Ecology of Unpolluted Habitats

Because estuaries are a meeting place of fresh water and seawater, the estuarine environment fluctuates more than either ocean or freshwater habitats. The result is a complex interplay of local factors including strength of tides and exposure at low tide, wave action, cyclical changes in salinity, changes in sediment type, and variations in organic matter (Perkins, 1974; Nicholas, 1975). From a growing number of studies of nematodes from estuaries, salt marshes, and other brackish habitats, it is apparent that generalizations are difficult because estuaries are so different from each other.

a. Salinity. In a given estuary the degree of salinity may vary from 0.5 to over 40% and depends on a variety of factors including land runoff, range of tides, topography, soil type, and even wind patterns. As mentioned above, in all aquatic environments where salinity is variable, this seems to be the principal factor governing nematode distribution (Gerlach, 1953a; Wieser, 1959b; Everard, 1960; Riemann, 1966; Warwick, 1971; Warwick and Gage, 1975; Nicholas, 1975; Ivanega, 1976). In the Exe Estuary (Britain) Warwick (1971) found dorylaimids, rhabditids, and a species of *Tripyla* in areas of low salinity, whereas other species including *Anoplostoma viviparum, Hypodontolaimus geophilus, Sabatieria vulgaris,* and *Theristus oxycerus,* decrease in numbers in such areas. Other species, however, remained fairly uniform along the transect of his stations, and he suggested that these species are very euryhaline. Warwick (1971) suggested that as yet uninvestigated physiological adaptations must be responsible for the differences. The reciprocal averaging ordination employed by Warwick and Gage (1975) revealed a distinct shallow water brackish component of the Loch Etive (West Scotland) nematode fauna which could be compared to nematode assemblages found in estuaries. These authors concluded that the distribution limits of this brackish component are controlled by the extreme rather than the average value of the salinity.

b. Grain Size. As with freshwater and coastal marine habitats, sediment grain size is an important factor governing distribution and buildup of nematodes in estuaries. An associated factor is the degree to which estuarine sediments retain water. Several authors have suggested that a major portion of the relationship between estuarine nematode distribution and sediment grain size can be attributed ultimately to difference in buccal morphology and hence differences in feeding preferences (Wieser and Kanwisher, 1961; Tietjen, 1969; Warwick, 1971; Perkins, 1974; Nicholas, 1975).

There is a consensus that mud supports deposit feeders and that the omnivores (or predators) increase in wet sands. Epigrowth feeders predominate in well-drained sands. Many estuarine habitats are blends of the above substrates and their nematode faunas vary accordingly. In addition, it is generally agreed that the species of mud faunas tend to be very small (with some exceptions) and with short setae, whereas species of sand faunas vary in size, depending on the amount of interstitial space, and have longer setae (Gerlach, 1953a; Wieser, 1959a; Warwick, 1971; Perkins, 1974).

c. Oxygen. Clean estuaries may have depleted oxygen near the bottom because of decaying organic material. In *Zostera* beds and salt marsh pools extreme diurnal shifts may occur from super-saturation by day to a deficiency at night, as a result of plant photosynthesis and respiration. Oxygen levels in estuaries may depend on concentrations of various minerals, pH, vertical stratification of salinity and topographic features, and on the degree of mixing with oxygenated water from rivers or sea.

Despite the local variations, it seems likely that most nematodes would usually be found in the upper layers of sediment where oxygen is sufficient for aerobic respiration. This seems to be the case, with most investigators finding the bulk of the benthic nematodes to be in the upper 3–5 cm of sediments as is true for other aquatic habitats (Nicholas, 1975). However, other reports indicate that many estuarine nematodes do seem to have a tolerance for anaerobic conditions (Wieser and Kanwisher, 1961; Teal and Wieser, 1966; Tietjen, 1969; Perkins, 1974). It seems true, however, that nematodes are found at greater depths in sandy habitats than in other substrates (Nicholas, 1975). McIntyre and Murison (1973) indicated that nematodes in the upper intertidal zone migrate into the sand as far as 20–30 cm as waves advance, and return later to the surface. They also found nematodes throughout a greater depth of sand in winter than in summer.

d. Seasons. Several studies have indicated a seasonality in the nematode communities of estuaries and related habitats of temperate climates. The changes correlate with yearly temperature cycles and are related to food availability. In New England estuaries Tietjen (1969) found spring and summer peaks in total population and observed significant seasonal changes in species composition within the estuaries. The epigrowth feeders reached maximum densities in spring and summer, and deposit feeders and omnivores in autumn and winter. Wieser and Kanwisher (1961) found seasonal variation in the nematode populations of a small Massachusetts salt marsh. By contrast Warwick (1971) found no marked seasonal variation in either species composition or densities in the Exe Estuary (Britain), although many short-term changes occurred during the year in which samples were taken.

2. Community Structure

It is difficult to generalize about community structure of nematodes in estuaries because densities and composition can both be expected to change markedly within the reaches of a single estuary. Even though a "brackish fauna" can be identified, these species will be accompanied by less eurytopic species at any given station sampled. On the shore of the Gulf of Finland, 60% of the nematode community was found to be euryhaline marine species, 30% brackish water and ground-water species, and 10% terrestrial species. Farther back from the shore in estuarine headwaters, 10% were euryhaline marine species, 50% brackish water and ground-water species, and 40% terrestrial species (Gerlach, 1953b; Perkins, 1974).

As discussed above, sediment composition has a profound effect on community structure. In the Blyth Estuary (Britain), Capstick (1959) found eight species of nematodes in muddy sand, with a total abundance of 1320/10 cm³. In fine sand he found 13 species with an abundance of 300/10 cm³. In a Georgia salt marsh, nematodes reached the highest density of 1630/cm² in the *Spartina* zone (where deposition of fine sediment occurred most rapidly). Where currents prevented sedimentation of the finer materials, and in the more sandy sediments, nematode populations were as low as 46/cm² (Teal and Wieser, 1966; Perkins, 1974). Warwick (1971) recognized six major habitats in the Exe Estuary (Britain) which differed in grain composition and drainage, and found each to be characterized by a typical association of 20–40 common species. In general, finer estuarine sediments do not have the diversity of species characteristic of somewhat coarser sediments (Perkins, 1974).

3. Ecology of Polluted Habitats

There have been few studies of pollution effects on nematodes in estuaries. The fluctuations in salt content in estuaries present colonizers with such severe problems that a decrease in species number with increased distance from the sea is almost a certainty. The normal situation in an estuary can be compared to the abnormal situation following pollution in other water areas—a decline in the number of species accompanied by an increase in the abundance of individual species (Perkins, 1974). Greatly increased pollution (or, in the case of an estuary, a much lowered salinity) may be followed by a decline in abundance of even the resistant species.

When estuaries do become polluted, oxygen concentration may fall and the substrate may become anaerobic. Even normal transport of nutrient materials to estuaries can produce eutrophication. An increasing problem in estuaries is the transport of agricultural pesticides accumulated by runoff (Perkins, 1974).

The problem in studying pollution in estuaries is to separate the effects of the pollution from normal features of the estuary. Most observations to date tend to be of an anecdotal nature. Nelson *et al.* (1972) collected a new species of *Enoplus* from the top centimeter of brownish mud in the intertidal mud flats of Burrard Inlet (Canadian Pacific coast) and felt that the close proximity of various industrial sites indicated that the area was probably very polluted. They suggested that perhaps the new species might be considered an indicator species. However, they noted that the normal variations of the area in temperature and salinity were extreme as a result of their intertidal position and the influx nearby of two freshwater streams.

In the study of Heip and Decraemer (1974) of nematode communities in the southern North Sea, two sites were directly influenced by the outflow of the IJzer River, a heavily polluted stream. Inasmuch as these two sites had lower diversity indices than the other sites, and the diversity index was lower in June than in other months of the year, these authors suggest a strong possibility that the nematode community was reacting to an increased output of organic material from the river in June. Again, however, it is difficult to separate the effect of silt per se from the effects of other factors which might be carried in the river effluent. These authors identified species of the genera *Sabatieria* and *Theristus* as generalists which can adapt to a wide array of environmental conditions, and which predominated at the two low-diversity stations. Other (different) species predominated at high-diversity stations.

V. Conclusions

The case for the importance of the role of nematodes in normal aquatic ecosystems seems well established, based primarily on data for marine species. Thus investigations into the effects of pollution on this important component of normal unpolluted aquatic ecosystems can be justified and supported on the same basis as are studies of other important components of such ecosystems.

The use of nematodes as indicators of pollution is more complex. As has been discussed above, although many nematodes are cosmopolitan, local conditions are sufficiently variable to cause the dominant species to differ, although the same ecological categories of species may be present in comparable habitats. The concept that the degree of spatial and temporal homogeneity of a habitat can be determined by an analysis of the species present and their relative abundances has been established for various groups of aquatic organisms (Patrick, 1961; Sanders, 1969; Marcotte and Coull, 1974; Wieser, 1975; Porter, 1977). Indications are that this principle seems to hold for nematodes (Wieser, 1960; Hopper and Meyers, 1967; Tietjen, 1969, 1977; Ferris *et al.*, 1972). Areas that can be regarded as heterogeneous, with a large number of stable, predictable niches, are usually rich in species with low densities. Homogeneous areas are indicated by larger numbers of one species, or a small group of closely related species (Pianka, 1974).

Analysis of community structure to determine useful information about the aquatic nematode habitat seems to hold promise. Warwick and Buchanan (1970) showed a relationship between diversity (Williams' index) and marine sediment grain size; and Heip and Decraemer (1974) showed a similar relationship between diversity (Shannon–Weaver index) and median grain size and percentage of silt. Although Tietjen (1977) showed that species diversity, richness and equitability were all significantly related to silt–clay content of sediments of Long Island Sound, no relationship could be shown between diversity indices (Shannon–Weaver) and sediment metal or organic carbon content. The Shannon–Weaver index (H') did not prove useful in the study of freshwater nematodes of Ferris *et al.* (1972). It should be recalled that H' is an estimate of the diversity of the total population of individuals in the species pool, and measures the uncertainty of predicting the species of an individual drawn at random. The H' value depends on the number of species and the distribution of individuals among the species. Thus, sites with the same number of species with the same distribution would have the same diversity index (H'), even though identity of the species differed and the total numbers differed. Diversity, as measured by H', may be less appropriate for species-rich nematode communities than for other aquatic assemblages (Poole, 1974).

The cluster analyses employed by Ferris *et al.* (1972) and by Tietjen (1977) require calculations of coefficients of similarity based on densities of individual species held in common by the sites, and produced clusters that can be interpreted in terms of features of the environment. Warwick and Gage (1975) had success with reciprocal averaging ordination. Classification techniques which they used, based on different faunal indices of affinity between stations, yielded different dendrograms, with an index based only on presence-absence data yielding results most similar to reciprocal averaging (Warwick and Gage, 1975). Those techniques, whether classifications or ordinations, which require some measure of intergroup similarity (or dissimilarity), may be more useful for relating nematode communities to aquatic pollution than are techniques based on information analysis, such as the Shannon–Weaver index (Pielou, 1977).

It seems evident from all freshwater, coastal marine, and estuarine nematode investigations that in shallow silty habitats one or a few species reach high percentage abundances, whereas in sandy substrates more species have lower maxima, probably indicating a greater diversity of niches in the sandy substrates

(Nicholas, 1975; Tietjen, 1977). The enrichment of an aquatic area by diverse substances which can be classed as mild pollutants seems also to result in an increase in dominance of a few species. Several studies indicate that a similar effect occurs following heavier pollution (Heip and Decraemer, 1974; Tietjen, 1977; Zullini, 1976a). Probably very heavy pollution would result in a decline in numbers of individuals as well as numbers of species. The importance of obtaining baseline data on natural variability within a target habitat cannot be overemphasized.

Acknowledgments

The authors are grateful to Dr. John H. Tietjen and Bruce E. Hopper, who reviewed parts of the manuscript and made many helpful suggestions. The authors' original research discussed in this chapter was supported by Office of Water Resources Research Project No. A-015-IND (Agreement 14-31-0001-3514) and by National Science Foundation Grant GZ-416. We also acknowledge perspectives in our thinking gained during periods of support from National Science Foundation Grants BMS72-02296 and DEB77-12656. Journal Paper No. 7093, Purdue University Agricultural Experiment Station.

References

Baliga, K. Y., Austin, J. H., and Engelbrecht, R. S. (1969). Occurrence of nematodes in benthic deposits. *Water Res.* **3**, 979–993.

Biró, K. (1968). The nematodes of Lake Balaton. II. The nematodes of the open water mud in the Keszthely Bay. *Ann. Biol. Tihany* **35**, 100–116.

Biró, K. (1972). Nematodes of Lake Balaton. III. The fauna in late-summer. *Ann. Biol. Tihany* **39**, 89–100.

Biró, K. (1973). Nematodes of Lake Balaton. IV. Seasonal qualitative and quantitative changes. *Ann. Biol. Tihany* **40**, 135–158.

Boucher, G. (1972). Distribution quantitative et qualitative des nématodes d'une station de vase terrigène côtière de Banyuls-sur-Mer. *Cah. Biol. Mar.* **13**, 457–474.

Cairns, J., Jr., and Kaesler, R. L. (1969). Cluster analysis of Potomac River survey stations based on protozoan presence–absence data. *Hydrobiologia* **34**, 414–432.

Calaway, W. T. (1963). Nematodes in wastewater treatment. *J. Water Pollut. Control Fed.* **35**, 1006–1015.

Calaway, W. T. (1969). *Mononchoides* associated with waste water treatment. *J. Nematol.* **1**, 5–6.

Callahan, C. A. (1976). An evaluation of nematode community structure as a method for quantifying and interpreting ecological changes in water resource environments. Ph.D. Thesis, pp. 1–97. Purdue University, Lafayette, Indiana.

Callahan, C. A., Ferris, J. M., and Ferris, V. R. (1977). Evaluation of nematode community structure as a method for quantifying ecological change in stream environments. *J. Nematol.* **9**, 265–266.

Capstick, C. K. (1959). The distribution of free-living nematodes in relation to salinity in the middle and upper reaches of the River Blyth estuary. *J. Anim. Ecol.* **28**, 189–210.

Chang, S. L. (1961). Viruses, amebas, and nematodes and public water supplies. *J. Am. Water Works Assoc.* **53**, 288–296.

Chang, S. L., and Kabler, P. W. (1962). Free-living nematodes in aerobic treatment plant effluent. *J. Water Pollut. Control Fed.* **34**, 1256–1261.

Chang, S. L., Woodward, R. L., and Kabler, P. W. (1960). Survey of free-living nematodes and amebas in municipal supplies. *J. Am. Water Works Assoc.* **52**, 613-617.

Chaudhuri, N. (1964). Occurrence and controlled environmental studies of nematodes in natural and waste waters. Ph.D. Thesis, pp. 1-160. University of Illinois, Chicago.

Chaudhuri, N., Siddiqi, R., and Englebrecht, R. S. (1964). Source and persistence of nematodes in surface waters. *J. Am. Water Works Assoc.* **56**, 73-88.

Cobb, N. A. (1918). Nematodes of the slow sand filter-beds of American cities. *Contrib. Sci. Nematol.* **7**, 189-212.

Cryan, W. S., Hansen, E., Sayre, F. W., Martin, M., and Yarwood, E. A. (1963). Axenic cultivation of the dioecious nematode *Panagrellus redivivus*. *Nematologica* **9**, 313-319.

Duncan, A., Schiemer, F., and Klekowski, R. Z. (1974). A preliminary study of feeding rates on bacterial food by adult females of a benthic nematode, *Plectus palustris* de Man 1880. *Pol. Arch. Hydrobiol.* **21**, 249-258.

Everard, C. O. R. (1960). The salinity tolerance of *Panagrolaimus rigidus* (Schneider, 1866), Thorne, 1937, and *Panagrolaimus salinus* Everard, 1958. (Nematoda: Panagrolaiminae). *Ann. Mag. Nat. Hist.* [*13*] **3**, 53-59.

Ferris, V. R., and Ferris, J. M. (1972). Nematode community dynamics in natural and disturbed environments. *Proc. Int. Symp. Nematol., 11th, 1972* pp. 21-22.

Ferris, V. R., Ferris, J. M., and Callahan, C. A. (1972). Nematode community structure: A tool for evaluating water resource environments. *Purdue Univ. Water Resour. Res. Cent., Tech. Rep.* **30**, 1-40.

Ferris, V. R., Ferris, J. M., and Tjepkema, J. P. (1973). Genera of freshwater nematodes (Nematoda) of eastern North America. Biota of freshwater ecosystems. *Environ. Prot. Agency, Ident. Manual* No. 10, pp. 1-38.

Filipjev, I. N. (1918). "Free Living Marine Nematodes of the Sevastopol Area," Issue 1. Russ. Acad. Sci., Petrograd (Transl. 1968, I. P. S. T., Jerusalem).

Filipjev, I. N. (1921). "Free Living Marine Nematodes of the Sevastopol Area," Issue 2. Russ. Acad. Sci., Petrograd (Transl. 1970, I. P. S. T., Jerusalem).

Filipjev, I. N. (1929-1930). Les nématodes libres de la baie de la Neva et de l'extremité orientale du golfe de Finlande. *Arch. Hydrobiol.* **21**, 1-64.

Fisher, K. D. (1968). Population patterns of nematodes in Lake Champlain. *Nematologica* **14**, 7.

Gerlach, S. A. (1953a). Die Biozönotische Gliederung der Nematodenfauna an den Deutschen Küsten. *Z. Morphol. Oekol. Tiere* **41**, 411-512.

Gerlach, S. A. (1953b). Die Nematodenfauna der Uferzonen und des Küstengrundwasers am finnischen Meerbysen. *Acta Zool. Fenn.* **73**, 1-32.

Gerlach, S. A. (1971). On the importance of marine meiofauna for benthos communities. *Oecologia* **6**, 176-190.

Gerlach, S. A. (1972). Die Produktionsleistung in der Helgoländer Bucht. *Verh. Dtsch. Zool. Ges.* **65**, 1-13.

Gerlach, S. A., and Riemann, F. (1973). The Bremerhaven checklist of aquatic nematodes. A catalogue of Nematoda Adenophorea excluding the Dorylaimida. Part 1. *Veroeff. Inst. Meeresforsch. Bremerhaven, Suppl.* **4**, No. 1, 1-404.

Gerlach, S. A., and Riemann, F. (1974). The Bremerhaven checklist of aquatic nematodes. A catalogue of Nematoda Adenophorea, excluding the Dorylaimida. Part 2. *Veroeff. Inst. Meeresforsch. Bremerhaven, Suppl.* **4**, No. 2, 405-734.

Gerlach, S. A., and Schrage, M. (1971). Life cycles in marine meiobenthos. Experiments at various temperatures with *Monhystera disjuncta* and *Theristus pertenuis* (Nematoda). *Mar. Biol.* **9**, 274-280.

Gerlach, S. A., and Schrage, M. (1972). Life cycles at low temperatures in some free-living marine nematodes. *Veroeff. Inst. Meeresforsch. Bremerhaven* **14**, 5-11.

Goodrich, M., Hechler, H. C., and Taylor, D. P. (1968). *Mononchoides changi* n. sp. and *M. bollingeri* n. sp. (Nematoda: Diplogasterinae), from a waste treatment plant. *Nematologica* **14**, 26–36.

Heip, C., and Decraemer, W. (1974). The diversity of nematode communities in the southern North Sea. *J. Mar. Biol. Assoc. U. K.* **54**, 251–255.

Hirschmann, H. (1952). Die Nematoden der Wassergrenze mittelfrankischer Gewasser. *Zool. Jahrb., Abt. Syst. Oekol. Geogr. Tiere* **81**, 313–407.

Hope, W. D., and Murphy, D. G. (1972). A taxonomic hierarchy and checklist of the genera and higher taxa of marine nematodes. *Smithson. Contrib. Zool.* **137**, 1–101.

Hopper, B. E., and Cefalu, R. C. (1973). Free-living marine nematodes from Biscayne Bay, Florida. VII. Enoplidae: *Enoplus* species in Biscayne Bay with observations on the culture and bionomics of *E. paralittoralis* Wieser, 1953. *Proc. Helminthol. Soc. Wash.* **40**, 275–280.

Hopper, B. E., and Meyers, S. P. (1966a). Aspects of the life cycle of marine nematodes. *Helgol. Wiss. Meeresunters.* **13**, 444–449.

Hopper, B. E., and Meyers, S. P. (1966b). Observations on the bionomics of the marine nematode, *Metoncholaimus* sp. *Nature (London)* **209**, 899–900.

Hopper, B. E., and Meyers, S. P. (1967). Population studies on benthic nematodes within a subtropical seagrass community. *Mar. Biol.* **1**, 85–96.

Hopper, B. E., Fell, J. W., and Cefalu, R. C. (1973). Effect of temperature on life cycles of nematodes associated with mangrove (*Rhizophora mangle*) detrital system. *Mar. Biol.* **23**, 293–296.

Hynes, H. B. N. (1970). "The Ecology of Running Waters." Univ. of Toronto Press, Toronto.

Ivanega, I. G. (1976). Composition and distribution of the nematode fauna in the Dnestr estuary. *Zool. Zh.* **55**, 1250–1252.

Juario, J. V. (1975). Nematode species composition and seasonal fluctuation of a sublittoral meiofauna community in the German Bight. *Veroeff. Inst. Meeresforsch. Bremerhaven* **15**, 283–337.

King, C. E. (1962). Some aspects of the ecology of psammolittoral nematodes in the North Eastern Gulf of Mexico. *Ecology* **43**, 515–523.

Kokordák, J. (1969). Über die Möglichkeit einer Auznützung der Nematoden als Indikatoren der Tätigkeit der mechanisch-biologischen Kläranlage. *Folia Vet.* **13**, 123–132.

Kokordák, J. (1970). Über die Möglichkeit einer Auznützung von Nematoden und anderen Organismen bei der Auswertung der Effektivität von mechanisch-biologischen Kläranlage. *Folia Vet.* **14**, 83–102.

Krumbein, W. C. (1939). Graphic presentation and statistical analysis of sedimentary data. *In* "Recent Marine Sediments" (P. D. Trask, ed.), pp. 558–591. Am. Assoc. Pet. Geol., Tulsa, Oklahoma.

Lee, J. J., Tietjen, J. H., Stone, R. J., Muller, W. A., Rullman, J., and McEnery, M. (1970). The cultivation and physiological ecology of members of salt marsh epiphytic communities. *Helgol. Wiss. Meeresunters.* **20**, 136–156.

Lorenzen, S. (1971a). Die Nematodenfauna im Verklappungsgebiet für Industrieabwässer nordwestlich von Helgoland. I. Araeolaimida und Monhysterida. *Zool. Anz.* **187**, 223–248.

Lorenzen, S. (1971b). Die Nematodenfauna im Verklappungsgebiet für Industrieabwässer nordwestlich von Helgoland. II. Desmodorida und Chromadorida. *Zool. Anz.* **187**, 283–302.

Lorenzen, S. (1972). Die Nematodenfauna im Verklappungsgebiet für Industrieabwässer nordwestlich von Helgoland. III. Cyantholaimidae, mit einer Revision von *Pomponema* Cobb, 1917. *Veroeff. Inst. Meeresforschung. Bremerhaven* **13**, 285–306.

Lorenzen, S. (1974). Die Nematodenfauna der sublitoralen Region der Deutschen Bucht, insbesondere im Titan-Abwassergebiet bei Helgoland. *Veroeff. Inst. Meeresforschung. Bremerhaven* **14**, 305–327.

McCloskey, L. R. (1970). The dynamics of the community associated with a marine scleractinean coral. *Int. Rev. Ges. Hydrobiol.* **55**, 13–81.

McIntyre, A. D. (1969). Ecology of marine meiobenthos. *Biol. Rev. Cambridge Philos. Soc.* **44**, 245–290.

McIntyre, A. D., and Murison, D. J. (1973). The meiofauna of a flatfish nursery ground. *J. Mar. Biol. Assoc. U. K.* **53**, 93–118.

McLachlan, A., Winter, P. E. D., and Botha, L. (1977). Vertical and horizontal distribution of sub-littoral meiofauna in Algoa Bay, South Africa. *Mar. Biol.* **40**, 355–364.

Marchant, R., and Nicholas, W. L. (1974). An energy budget for the free-living nematode *Pelodera* (Rhabditidae). *Oecologia* **16**, 237–252.

Marcotte, B. M., and Coull, B. C. (1974). Pollution, diversity and meiobenthic communities in the North Adriatic (Bay of Piran, Yugoslavia). *Vie Milieu, Ser. B* **24**, 281–300.

Meyers, S. P., Hopper, B. E., and Cefalu, R. (1970). Ecological investigations of the marine nematode *Metoncholaimus scissus. Mar. Biol.* **6**, 43–47.

Meyl, A. H. (1955). Freilebende Nematoden aus binnenländischen Salzbiotopen zwischen Braunschweig und Magdeburg. *Arch. Hydrobiol.* **50**, 568–614.

Murad, J. L. (1970). Population study of nematodes from drying beds. *Proc. Helminthol. Soc. Wash.* **37**, 10–13.

Muus, B. J. (1967). The fauna of Danish estuaries and lagoons: distribution and ecology of dominating species in the shallow reaches of the mesohaline zone. *Medd. Dan. Fisk.-Havunders.* **5**, 1–316.

Nelson, H., Hopper, B., and Webster, J. M. (1972). *Enoplus anisopiculus,* a new species of marine nematode from the Canadian Pacific coast. *Can. J. Zool.* **12**, 1681–1684.

Nicholas, W. L. (1962). A study of a species of *Acrobeloides* (Cephalobidae) in laboratory culture. *Nematologica* **8**, 99–109.

Nicholas, W. L. (1975). "The Biology of Free-Living Nematodes." Oxford Univ. Press (Clarendon), London and New York.

Ott, J. (1972a). Determination of fauna boundaries of nematodes in an intertidal sand flat. *Int. Rev. Ges. Hydrobiol.* **57**, 645–663.

Ott, J. (1972b). Studies on the diversity of the nematode fauna in intertidal sediments. *In* "European Symposium in Marine Biology" (B. Battaglia, ed.), pp. 275–285. Piccin Editore, Padova.

Ott, J., and Schiemer, F. (1973). Respiration and anaerobiosis of free living nematodes from marine and limnic sediments. *Neth. J. Sea Res.* **7**, 233–243.

Patrick, R. (1961). A study of the numbers and kinds of species found in rivers in Eastern United States. *Proc. Acad. Nat. Sci. Philadelphia* **113**, 215–258.

Perkins, E. J. (1974). "The Biology of Estuaries and Coastal Waters." Academic Press, New York.

Peters, B. G. (1930a). A biological investigation of sewage. *J. Helminthol.* **8**, 133–164.

Peters, B. G. (1930b). Some nematodes met with in a biological investigation of sewage. *J. Helminthol.* **8**, 165–184.

Pianka, E. R. (1974). "Evolutionary Ecology." Harper, New York.

Pielou, E. C. (1977). "Mathematical Ecology." Wiley, New York.

Pillai, J. K., and Taylor, D. P. (1968a). *Butlerius micans* n. sp. (Nematoda: Diplogasterinae) from Illinois, with observations on its feeding habits and a key to the species of *Butlerius* Goodey, 1929. *Nematologica* **14**, 89–93.

Pillai, J. K., and Taylor, D. P. (1968b). Biology of *Paroigolaimella bernensis* and *Fictor anchicoprophaga* (Diplogasterinae) in laboratory culture. *Nematologica* **14**, 159–170.

Pillai, J. K., and Taylor, D. P. (1968c). Emendation of the genus *Demaniella* Steiner, 1914 (Nematoda: Diplogasterinae), with observations on the biology of *D. basili* n. sp. *Nematologica* **14**, 285–294.

Poole, R. W. (1974). "An Introduction to Quantitative Ecology." McGraw-Hill, New York.

Porter, K. G. (1977). The plant-animal interface in freshwater ecosystems. *Am. Sci.* **65**, 159-170.
Riemann, F. (1966). Die interstitielle Fauna in Elbe-Aestuar Verbeitung und Systematik. *Arch. Hydrobiol., Suppl.* **31**, 1-279.
Roback, S. W., Cairns, J., Jr., and Kaesler, R. L. (1969). Cluster analysis of occurrence and distribution of insect species in a portion of the Potomac River. *Hydrobiologia* **34**, 484-502.
Sanders, H. L. (1969). Benthic marine diversity and the stability-time hypothesis. *Brookhaven Symp. Biol.* **22**, 71-81.
Schiemer, F. (1975). Nematoda. *In* "Ecological Aspects of Used-Water Treatment" (C. R. Curds and H. A. Hawkes, eds.), Vol. 1, pp. 269-288. Academic Press, New York.
Schiemer, F., and Duncan, A. (1974). The oxygen consumption of a freshwater benthic nematode, *Tobrilus gracilis* (Bastian). *Oecologia* **15**, 121-126.
Schiemer, F., Löffler, H., and Dollfus, H. (1969). The benthic communities of Neusiedlersee (Austria). *Verh., Int. Ver. Theor. Angew. Limnol.* **17**, 201-208.
Schütz, L. (1966). Ökologische Untersuchungen über die Benthosfauna in Nordostseekanal. II. *Int. Rev. Ges. Hydrobiol.* **51**, 633-685.
Skoolmun, P., and Gerlach, S. A. (1971). Jahreszeitliche Fluktuationen der Nematodenfauna im Bezeitenbereich des in Wesser-Ästuars (Deutsche Bucht). *Veroeff. Inst. Meeresforsch. Bremerhaven* **13**, 119-138.
Teal, J. M. (1962). Energy flow in the salt marsh ecosystem of Georgia. *Ecology* **43**, 614-624.
Teal, J. M., and Wieser, W. (1966). The distribution and ecology of nematodes in a Georgia salt marsh. *Limnol. Oceanogr.* **11**, 217-222.
Tietjen, J. H. (1967). Observations on the ecology of the marine nematode *Monhystera filicaudata* Allgen, 1929. *Trans. Am. Microsc. Soc.* **86**, 204-206.
Tietjen, J. H. (1969). The ecology of shallow water meiofauna in two New England estuaries. *Oecologia* **2**, 251-291.
Tietjen, J. H. (1971). Ecology and distribution of deep-sea meiobenthos off North Carolina. *Deep-Sea Res.* **18**, 941-957.
Tietjen, J. H. (1977). Population distribution and structure of the free-living nematodes of Long Island Sound. *Mar. Biol.* **43**, 123-136.
Tietjen, J. H., and Lee, J. J. (1972). Life cycles of marine nematodes. Influence of temperature and salinity on the development of *Monhystera denticulata* Timm. *Oecologia* **10**, 167-176.
Tietjen, J. H., and Lee, J. J. (1973). Life-history and feeding habits of the marine nematode *Chromadora macrolaimoides* Steiner. *Oecologia* **12**, 303-314.
Tietjen, J. H., and Lee, J. J. (1977a). Feeding behavior of marine nematodes. *In* "Ecology of Marine Benthos" (B. C. Coull, ed.), pp. 21-35. Univ. of South Carolina Press, Columbia.
Tietjen, J. H., and Lee, J. J. (1977b). Life histories of marine nematodes. Influence of temperature and salinity on the reproductive potential of *Chromadorina germanica. Mikrof. Meeresb.* **61**, 263-270.
Tietjen, J. H., Lee, J. J., Rullman, J., Greengart, A., and Trompeter, J. (1970). Gnotobiotic culture and physiological ecology of the marine nematode *Rhabditis marina* Bastian. *Limnol. Oceanogr.* **15**, 535-543.
Tombes, A. S., Abernathy, A. R., and Welch, D. M. (1977). Occurrence of free-living nematodes in finished water. *Am. Zool.* **17**, 950.
Vitiello, P., and Vivier, M. H. (1974). Donńees quantitatives sur la meiofaune d'une zone profonde de déversements industriels. *Un. Océanogr. Fr.* **6**, 13-16.
Ward, A. R. (1973). Studies on the sublittoral free-living nematodes of Liverpool Bay. I. The structure and distribution of the nematode populations. *Mar. Biol.* **22**, 53-66.
Ward, A. R. (1975). Studies on the sublittoral free-living nematodes of Liverpool bay. II. Influence of sediment composition on the distribution of marine nematodes. *Mar. Biol.* **30**, 217-225.
Warwick, R. M. (1971). Nematode associations in the Exe estuary. *J. Mar. Biol. Assoc. U. K.* **51**, 439-454.

Warwick, R. M., and Buchanan, J. B. (1970). The meiofauna off the coast of Northumberland. I. The structure of the nematode population. *J. Mar. Biol. Assoc. U. K.* **50**, 129-146.

Warwick, R. M., and Buchanan, J. B. (1971). The meiofauna off the coast of Northunberland. II. Seasonal stability of the nematode population. *J. Mar. Biol. Assoc. U. K.* **51**, 355-362.

Warwick, R. M., and Gage, J. D. (1975). Nearshore zonation of benthic fauna, especially Nematoda, in Loch Etive. *J. Mar. Biol. Assoc. U. K.* **55**, 295-311.

Wieser, W. (1952). Investigations on the microfauna inhabiting seaweeds on rocky coasts. IV. *J. Mar. Biol. Assoc. U. K.* **31**, 145-174.

Wieser, W. (1953a). Free-living marine nematodes. I. Enoploidea. *Acta Univ. Lund.* [*N.F.2*] **49(6)**, 1-155.

Wieser, W. (1953b). Die Beziehung zwischen Mundhohlengestalt, Ernahrungsweise und Vorkonimen beifreilebenden marinen Nematoden. *Ark. Zool.* [*2*] **4**, 439-484.

Wieser, W. (1954). Free-living marine nematodes. II. Chromadoroidea. *Acta Univ. Lund* [*N.S.*] **50(16)**, 1-148.

Wieser, W. (1956). Free-living marine nematodes. III. Axonolaimoidea and Monhysteroidea. *Acta Univ. Lund.* [*N.S.*] **52(13)**, 1-115.

Wieser, W. (1959a). Free-living marine nematodes. IV. General part. Reports of Lund University Chile Expedition, 1948-1949. *Acta Univ. Lund.* [*N.F.2*] **55(5)**, 1-111.

Wieser, W. (1959b). "Free-Living Nematodes and Other Small Invertebrates of Puget Sound Beaches." Univ. of Washington Press, Seattle.

Wieser, W. (1959c). The effect of grain size on the distribution of small invertebrates inhabiting the beaches of Puget Sound. *Limnol. Oceanogr.* **4**, 181-194.

Wieser, W. (1960). Benthic studies in Buzzards Bay. II. The meiofauna. *Limnol. Oceanogr.* **5**, 121-137.

Wieser, W. (1975). The meiofauna as a tool in the study of habitat heterogeneity. Ecophysiological aspects. *Cah. Biol. Mar.* **16**, 647-670.

Wieser, W., and Kanwisher, J. (1961). Ecological and physiological studies on marine nematodes from a small salt marsh near Woods Hole, Massachusetts. *Limnol. Oceanogr.* **6**, 262-270.

Zullini, A. (1976a). Nematodes as indicators of river pollution. *Nematol. Medit.* **4**, 13-22.

Zullini, A. (1976b). Nematodes of some activated sewage treatment plants. *Ateneo Parmense, Acta Nat.* **12**, 271-283.

CHAPTER 2

Bryozoa (Ectoprocta)

DOROTHY F. SOULE and JOHN D. SOULE

I. Introduction

In recent years environmental studies around the world have focused on ports, estuaries, and inshore marine waters because of the increasing evidence that such waters have been the sources of irreplaceable biological communities. Society has in the past casually accepted the use of estuaries for transportation of sometimes hazardous materials and as giant sewerage systems for domestic, industrial, and terrigenous effluents; society has used shallow wetlands as a source of cheap land by dredging and filling for marine and nonmarine related industry; it has used the inshore waters for heat exchangers for power plants and gasification plants. The degradation of estuaries in the past was regarded as an inevitable consequence, subservient to the needs of society.

However, estuaries and other inshore areas provide organic input that supports the nursery grounds for phytoplankton, holoplankton, and meroplankton. These are essential in turn to many of the food chains on which commercially valuable fish and shellfish depend. The meroplankton consists of eggs, larvae, and sometimes other juvenile stages of animals that will become sessile or attached as adults. This includes the larvae of some species of adult Bryozoa, which are often prominent members of the communities on substrata in estuaries, ports, and harbors.

Since the coming of environmental legislation in the United States in 1970, many Environmental Impact Assessments, Reports, and Studies (EIA, EIR, EIS) have been prepared for projects on estuaries and shorelines. In seeking baseline data on invertebrate species, including bryozoans, it has become apparent that few of the historical biological investigations have included physical and chemical data that could be related to species distribution. Conversely, in many more recent studies major groups of species are not identified because of short time lines, funding limitations, and limitations on the availability of the expertise to identify species, particularly colonial animals such as bryozoans, hydroids, and tunicates. The dearth of national support for systematics–ecology research in the last 35 years, and hence a lack of value placed on that sector by higher education, has led to a great shortage of expertise and of progress in the basic science aspects of ecology such as taxonomy (Lee, 1978) and the necessarily related physiology and genetics. Carriker (1976) emphasized the critical need for estuarine research in systematics, so that constructive plans can be made for reversing the degradation of the biota.

In many respects the study of the Bryozoa is still in the "alpha" stage. Nonspecialist biologists may recognize no more than a few species, while among specialists the higher categories are not always well understood, and many families, genera, and species are inadequately characterized. The ecological roles of Bryozoa, such as food for other invertebrates and fishes (Ryland, 1970) or reef stabilization (Soule and Soule, 1974; Cuffey, 1974) are not sufficiently known. The genetic relationships among supposedly cosmopolitan species have not been investigated, except in a few species (Schopf and Gooch, 1971; Schopf, 1974), and the research on physiological responses to physical and chemical parameters also has been limited. Thus, while Bryozoa are common components of estuarine and inshore communities, little application of information on their presence or absence can be made as yet. This constitutes neglect of a great potential biological resource for environmental evaluation.

Since few nonspecialists are familiar with the bryozoans as an important component of invertebrate communities, a brief review of the major groups follows.

II. Bryozoan Systematics

A. WHAT'S IN A NAME?

The colonial invertebrate organisms known presently as Bryozoa ("mossanimals") were historically subjected to a wide variety of interpretations and classifications—as plants (Bryophyta) and later as animal–plants, the Zoophyta of the eighteenth and nineteenth century scientists. Hyman (1959) recounted some of the historical perceptions such as the illustration in 1558 by Rondelet of a bryozoan as a "sea plant"; the belief of Imperato, circa 1599, that corals were plants that contained "pores" (bryozoans) on the surface; and the observations in 1723 of Peyssonel that corals were animal secretions inhabited by "insects" (the bryozoan tentacle crowns he saw extruded).

Bryozoans were placed with coelenterate polyps such as corals and hydroids until the existence of a one-way digestive tract in bryozoans was confirmed by Blainville in 1820 and Grant in 1827; subsequently the bryozoans were referred to a variety of other categories. According to Hyman (1959), in 1843 Milne-Edwards and others placed bryozoans in phylum Molluscoidea with the tunicates, while in 1880 Lang placed bryozoans in phylum Vermes as a class called Prosopygii, which included sipunculids, priapulids, phoronids, and brachiopods.

In 1831 Ehrenberg distinguished the group as Bryozoa, whereas the animals had been loosely categorized as Polyzoa by Thompson in 1830. This led to an extensive and entertaining battle in the literature, mostly along chauvinistic lines, which lasted over 100 years (Waters, 1880; Hincks, 1880; Jones, 1880; Stebbing, 1911; Harmer, 1911, 1947; Brown, 1958; Soule, 1958). Nitsche (1869) and others recognized that there were two rather different groups among the living bryozoan species; he named those with the anal opening within the tentacle circle the Entoprocta, and those with the anal opening outside the tentacle circle the Ectoprocta. Hyman (1959) followed the designations indicated by Nitsche, and elevated the Entoprocta and Ectoprocta to phylum status, precipitating another round of literary opinions which continues today (Schopf, 1967, 1968; Mayr, 1968, 1969; Soule and Soule, 1968; Cuffey, 1969; Zimmer, 1973; Nielsen, 1977a; Beatty and Blackwelder, 1974; Ghiselin, 1977; Bushnell, 1974; Woollacott and Zimmer, 1978).

Hyman (1959) placed the phylum Ectoprocta in a grouping designated as enterocoelous coelomates. She then clustered the Phoronida, Ectoprocta, and Brachiopoda as the lophophorate phyla (having a lophophore, or crown of ciliated tentacles surrounding the mouth). The far less numerous Entoprocta were placed at a much lower phylogenetic level near the nematodes, as pseudocoelomates.

Although Mayr (1969) stipulated that an old phylum name should not be replaced when one major group is removed from it (e.g., the Entoprocta from Bryozoa) there are precedents on Hyman's side of the argument; for example, the disappearance of the old phyla, Pisces, Vermes, and Molluscoidea. Since international rules do not apply to higher categories, this is a largely philosophical argument that has perhaps been carried on beyond its useful context. More important is the possible confusion engendered by the fact that a number of European investigators do not agree that the two groups are widely separated and continue to use the term phylum Bryozoa for both groups. Others restrict phylum Bryozoa to the Ectoprocta alone; in either case it is currently necessary to define the scope of the term when it is used initially. As the present authors stated (Soule and Soule, 1968), usage by the specialists involved in research will largely dictate the term applied. Since the establishment of the International Bryozoology Association in 1968, the use of phylum Ectoprocta by specialists has largely disappeared. There are far more ectoproct species than there are entoprocts, and some researchers are active in both groups (Nielsen, 1977a,b). The jury is still out on whether a phylum level separation more accurately represents the status of the two groups.

B. Current Classifications

The living Bryozoa *sensu stricto* (=Ectoprocta) have generally been divided into two classes or groups: the freshwater Phylactolaemata and the marine Gymnolaemata (Allman, 1856). Bushnell (1974) dealt with the relation of the Phylactolaemata and pollution.

There are an estimated 3500–4000 living estuarine and marine species included in the class Gymnolaemata by most authors, and many more species in the fossil record. The order Cheilostomata contains by far the most living species. Cheilostome colonies are composed of individual zooids that are mostly calcified (some with elaborate surficial patterns and accessory parts), and have subterminal apertures for extruding the tentacles. Species of the cheilostome suborder Anasca have membranous frontal walls to control extrusion of the tentacles, while species of the cheilostome suborder Ascophora have a special sac (the asc) beneath the calcified frontal for the same purpose. When present, brooding structures (ooecia, =ovicells) in Cheilostomata serve single parental individuals, although contributions to the ovicells may be made by contiguous individuals (Woollacott and Zimmer, 1972; Soule, 1973).

The order Ctenostomata contains soft-bodied individuals with terminal apertures; if brooding occurs it is generally within the tentecle sheath or body cavity (Ström, 1978). The two orders have similarities that lead some authors to link them as Cteno-Cheilostomata, or Stenostomata (Marcus, 1938; Silen, 1942; Woollacott and Zimmer, 1978).

The Cyclostomata differ from the above groups in that they are heavily calcified, have terminal apertures, and often have colonial brood chambers (gonozooecia). The Cyclostomata are apparently the only living group surviving from the extensive bryozoan fauna that flourished in the Paleozoic as the Cystoporata, Trepostomata, and Cryptostomata. Some authors therefore unite the cyclostomes with the fossil groups at the class level as Stenolaemata. This is perhaps a more accurate assessment of relationships, but the reference literature on living bryozoan faunal assemblages is most likely to be found under the names Cheilostomata, Ctenostomata, and Cyclostomata.

C. FUNCTIONAL MORPHOLOGY

1. Colonies

Because the morphologies of the major groups are so different (Hyman, 1959) it might be anticipated that their occurrences in invertebrate faunal communities would be affected by morphology. This would apply to estuarine occurrences in particular.

For example, different species may form branching colonies that may be erect or flaccid, depending in part on their degree of calcification. Other species may be encrusting or at least recumbent upon substrates. However, a single species may either be erect or recumbent, for example, depending on the substrate available, the turbulence of waters, and perhaps other factors as well (see Boardman *et al.*, 1973).

In earlier times efforts were made to classify bryozoans on the basis of colony form, but the traits were soon found to cross taxonomic lines extensively. Colony form is useful, however, in locating bryozoans in invertebrate communities in the field, prior to microscopic examination necessary for species identification (Soule and Soule, 1975).

    a. Recumbent or incrusting forms

      i. Soft-bodied species

        (a) Stolonate: forms with creeping tubular branches or stolonate connections extended over a firm substratum. Certain Ctenostomata have true stolons, which are individual zoids lacking polyps and modified to form the extensions (*Bowerbankia*). Some Cheilostomata are stoloniform but the extensions are formed by the basal portions of normal zoids (*Aetea*). Penetrating forms construct stolonate colonies by burrowing through mollusk shells (Cheilos-tomata: *Penetrantia*; Ctenostomata: *Terebripora, Immergentia*).

  (b) Fleshy (carnose): species that form gelatinous colonial masses, sometimes of considerable size (Ctenostomata: *Alcyonidium*).

  ii. Incrusting, more or less calcified, species

    (a) Flexiform: species that form a lightly calcified flat crust over a soft, flexible substratum such as algae blades (*Membranipora*).

    (b) Rigid: heavily calcified species which usually form unilaminar or multilaminar crusts over solid, inflexible surfaces such as shell and stone. Most of the Cheilostomata are of this type. A few species form heavy, knobby incrustations over originally flexible surfaces such as twigs, turning them into solid structures.

    (c) Tubular: A number of Cyclostomata species form colonies of recumbent (adnate) tubules; some tubules have erect terminal apertures (*Stomatopora, Diaperoecia*).

  b. Erect, leafy, branching, or bushlike forms

    (a) Branching—species with various modes of branching, some jointed; attached to the substratum loosely by rootlets. Anascan Cheilos-tomata such as *Bugula* and *Scrupocellaria* are typical. Most are not resistant to turbulent waters. Easily confused with certain hydroids.

    (b) Foliaceous, also known as flustraform—species that form leaflike colonies which resemble small cabbage plants or some lichens (Anascan: *Thalamoporella*).

    (c) Fenestrate—erect species, with a "chicken-wire" appearance, which form large colonies; usually typical of deep-water forms. Reteporidae such as *Phidolopora* are typical.

    (d) Tubular—certain Cyclostomata form colonies in which tubules extend upward from the basal disc of the colony (*Lichenopora*).

2. Individuals

The morphology of the individual polypides (tentacle crown, mouth, and complex digestive tract) attached inside either the calcareous or soft-bodied zooecia is quite similar in spite of the striking external differences (Hyman, 1959; Ryland, 1970). Subtle differences such as length of tentacles may help to explain how two similar species apparently share the same ecological niche or microenvironment. Woollacott and Zimmer (1977) contains a number of good papers reviewing the literature and recent research on the functional morphology of bryozoans. Hyman (1959) strongly emphasized the coelomate nature of ecto-proct bryozoans, but Nielsen (1977b) disagrees and further research is certainly in order, making use of ultrastructural studies and biochemical techniques, on the question. The boxlike or tubular exterior body (zooecium) of individuals is usually less than 1 mm in length and is lined with body wall tissues and associated musculature. Lutaud (1977) summarized research on the individual and colonial nervous systems of bryozoans; there is no excretory system.

3. Larvae

Larval morphology differs greatly in the bryozoan groups, even though the larval form is known for only 45 genera or species (Woollacott and Zimmer, 1977). The cyphonautes larva resembles a small clam, complete with shells and a digestive tract, and it was found in the plankton for years before it was recognized as a bryozoan larva. Only a few species of the genera *Conpeum, Electra,* and *Membranipora* (Cheilostomata, Anasca) and *Alcyonidium, Farella,* and *Hypophorella* (ctenostomes) have this truly planktotrophic form. A few others resemble the cyphonautes but lack the feeding structures (*Flustrellidra* and *Pherusella,* Ctenostomata).

Other bryozoan larvae are shell-less and coronate, resembling trochophores somewhat, although the ciliary bands and gut are placed differently. These forms apparently are not planktonic or cannot remain planktonic for extended periods of time. Some of the coronate bryozoans have an oblate larva, completely covered with cilia, which settles within a few hours at most (*e.g., Bugula*). Soule and Soule (1978) reviewed the limited information on bryozoan larval longevity, which ranged from 1 hour to 2 months. According to Marcus (1940), cyphonautes larvae may have a planktotrophic life of up to 2 months. The nonfeeding (lecithotropic) larvae have a limited planktonic or mobile period from 1 hour up to several days (Lynch, 1949; Maturo, 1959; Mawatari, 1951; Ryland, 1960, 1962).

The larval period of a number of invertebrates is prolonged by a number of factors; Scheltema (1968) summarized some of the more recent data. Barrois (1877) attributed prolonged swimming of bryozoan larvae to atmospheric conditions, as did Thorson and Bougis many years later (Soule and Soule, 1977). Many invertebrates have temperature optima for rapid metamorphosis, and deviations, either higher or lower, prolong the period. The stimuli that induce reproduction may include changes in temperature rather than the particular numerical temperature value itself. Effects of temperature will be discussed in Section V,D.

Lynch (1959) summarized much of the experimental literature on factors that prolonged larval settling. These included long periods of reduced light; lowered pH; sea water higher in magnesium, potassium chloride, or sodium chloride than the normal milieu; and the presence of various organic compounds. Lynch also noted a number of substances that hastened metamorphosis of bryozoan and other invertebrate larvae. Some of the above substances or conditions could easily be found in estuaries, associated with either natural events or man-made pollution.

Although the experimental procedures undertaken in some of the early work appear to be naive by today's standards, the references still are almost the only source of information on the effects of potential pollutants on particular bryozoan and other invertebrate species.

III. Habitats

A. SUBSTRATE SELECTION

A number of observers have noted that some invertebrate larvae seem able to test the substrate prior to final attachment and subsequent metamorphosis (Wilson, 1951, cf. Lynch, 1958a,b). Lack of suitable substrate may prolong the larval period (Eiben, 1976). There have been some indications that the larvae prefer to settle on a substrate covered by a bacterial film (ZoBell and Allan, 1935; Daniel, 1960), while others have found that clean substrates were preferred by test organisms (Crisp and Ryland, 1960), and Scheer (1945) found that plates covered with a diatom film were preferred. Soule and Soule (1974) showed colony attachment on a diatom-coated surface under scanning electron microscope.

Bryozoans compete with corraline algae, sponges, and colonial tunicates for initial colonization of substrates including rocks, metal, and coral reefs. In estuarine environments bryozoans find substrates more limited, since they generally do not tolerate sediment and unstable substrates well. In harbors, encrusting, branching, and erect forms find many opportunities among wood, concrete, and steel pilings, riprap rock and breakwaters, floating and fixed docks, and ship hulls to settle and flourish provided that the water quality is adequate (Soule and Soule, 1974, 1977). Some harbors in the tropics have flourishing coral patch reefs, as do Medang, New Guinea, and even Honolulu.

B. ESTUARINE HABITATS

The physical configurations of estuaries generally include salinity gradients based on continual or almost continual freshwater flow from rivers and streams, interacting with a salinity wedge or tidal exchange (Dyer, 1972). However, fluctuation in salinity is not the only variable to which sessile and attached estuarine organisms must adjust. Temperature usually varies more widely in shallow areas where ambient air temperatures undergo extensive variation, both diurnally and seasonally. Flushing and mixing characteristics are extremely important, for they largely determine the carrying capacity of the waters to accept organic input, whether it is from natural sources or from industrial and domestic waste effluents. The biological oxygen demand (BOD) and chemical oxygen demand (COD) of inputs to an estuary would quickly exhaust the dissolved oxygen in the water column if physical flushing and mixing did not occur. While phytoplankton produce much of the dissolved oxygen in the sea, phytoplankton can also deplete the water of oxygen if a rapidly reproducing population becomes confined to a given area by "wind cells" that restrict mixing. The cells respire and when a phytoplankton "bloom" dies and is biodegraded by microhetero- trophs oxygen drops rapidly.

It is unfortunate that few field investigations of estuarine organisms have included extensive physical measurements or chemical analyses of pollutant burdens. Most studies of estuarine bryozoans have not been carried out in the context of broad-spectrum investigations of other biota and concurrent physical conditions. Conversely, laboratory studies can only simulate effects of one or two parameters at a time and may miss much of the synergism that takes place in nature.

C. SALINITY AND DISTRIBUTION

Winston (1977) reviewed the literature on estuarine ectoprocts and tabulated data on the species found in diminished salinity in the tropical, warm-temperate, and boreal-antiboreal provinces. She located 22 studies from warm-temperate estuaries and 21 studies from boreal-antiboreal areas, but only 13 from tropical estuaries. Based on her review, she concluded that only 7% of the cyclostome genera occurred in brackish water (from only two of the five suborders), but 55% of the ctenostome genera, 12% of the cheilostome anascan genera, and 9% of the ascophoran genera have been found in brackish water. However, the species that do occur sometimes are found in very large numbers; *e.g., Zoobotryon verticil- latum, Bugula* spp. (Conversely, our personal observations are that the ascopho- rans and cyclostomes are most prominent in tropical coral reef habitats, and very few ctenostomes and anascans are found there.)

A few nominally cosmopolitan species have been identified as fouling, euryhaline organisms in the eastern, middle, and western Pacific and in Atlantic and Mediterranean waters (Soule and Soule, 1978). These include the cteno- stomes *Bowerbankia gracilis* and *Zoobotryon verticillatum*; the anascans *Bugula avicularia, Bugula neritina, Scruparia ambigua, Scrupocellaria bertholleti, Scrupocellaria scruposa;* and the ascophorans *Savignyella lafonti, Cryptosula pallasiana,* the *Schizoporella* complex often included as *S. unicornis* or *S. errata*), and the *Watersipora* species, which have been identified in various ways (Soule and Soule, 1975).

D. SALINITY TOLERANCE

Since bryozoans have no excretory system, osmoregulation would have to occur at the cellular or tissue level, probably in the tentacles in particular. The soft-bodied, tubular ctenostomes would be able to tolerate swelling of tissues (edema) to some extent when exposed to reduced salinity, whereas the tubular cyclostomes would not have room to swell without creating problems due to tissues pressed against the exoskeleton. In the cheilostomes, the anascans have a membranous frontal wall and they, too, could probably withstand some edema. The ascophorans would probably be in less favorable condition since frontal walls are calcified and the ascus, which is the hydrostatic sac that controls the extrusion of the tentacles, lies beneath the frontal. Little or no research has been done on the cellular mechanisms for osmoregulation in bryozoans.

Ryland (1970) summarized the associations of particular species with reduced salinities in fjords of Norway, in the Baltic Sea, Kiel Bay, the Mediterranean Sea, Black Sea, Caspian Sea, and the British Isles.

In the Suez Canal, he noted that five species had been recorded by Hastings (1927) in hypersalinities up to 49‰. Hypersalinities may be features of large shallow estuaries in warm and tropical climates with high evaporation rates.

Fiala-Medioni (1973) listed the bryozoans in the estuary at Banyuls sur Mer and reported on the reproductive periods for 115 species. This will be an important future reference since that area is becoming impacted by pollution in the Mediterranean. Evaluation of recovery by various authorities after control often neglects to substantiate later whether the community that was previously present is in fact similar to the succeeding community.

Winston (1977) used the ''Venice System'' (from the 1958 Symposium on Classification of Brackish Waters, 1959) to list species recorded in the literature from the various brackish water environments where salinities were either given or could be estimated from other hydrographic sources. The classification is as follows:

Venice system (‰)	Remane system
Hyperhaline = 40	
Euhaline = 30–40	Euryhaline 1
Mixohaline = 30	
Polyhaline = 30–18	
Pliohaline = 18–8	Euryhaline 2 ⎤
Miohaline = 8–3	Euryhaline 3 ⎦ Mesohaline
Oligohaline = 3–0.5	= Euryhaline 4
Limnetic = 0.5	

There is, of course, considerable peril in assuming that species lists are either complete or correct when making generalizations such as those required to quantify the comparative numbers of species and higher categories. In some cases, bryozoan specialists are not involved in identifications for general invertebrate surveys; sometimes the literature available to generalists is inadequate, and sometimes comparative type collections are inaccessible to generalist and

californica, Bugula neritina, Hippopodina feegeensis, Savignella lafonti, Schizoporella unicornis, Zoobotryon verticillatum, and two species apparently endemic to Hawaii, *Scrupocellaria sinuosa* and *Watersipora edmondsoni.* At Kaneohe Bay floating, tangled masses of the soft *Amathia* and *Zoobotryon* were stiffened by the tiny colonies of the more rapid *Savignyella.*

Evans *et al.* (1972) presented a quick survey of the watersheds for Pearl Harbor and Kaneohe Bay, Oahu, and the biological conditions and salinities in May–June 1971 at diver-surveyed stations in Pearl Harbor. *Bugula neritina* was reported in the East Lock and the Middle Lock. Long (1970) recorded several species in the fouling community outside Pearl Harbor.

In tropical estuaries in Mexico studies were made at Mazatlán Harbor, Mexico, in the summer of 1973. The area was of interest because it was the locality for some of Busk's (1855) type species from Carpenter's collections in 1852. Bryozoans were collected in the Gulf of California and at Mazatlán in 1957 (Soule, 1959, 1961, 1963) and at Scammon Lagoon, Baja California in 1959 (Soule and Soule, 1964). However, the 1973 studies were directed toward comparing the relatively underdeveloped Mazatlán harbor with the urbanized Los Angeles–Long Beach Harbor (Soule *et al.,* 1977). Temperature, salinity, dissolved oxygen, and pH were measured at 1-m intervals through the water column; sediment grain size was measured and the distribution of invertebrates, including bryozoans, was recorded in Mazatlán. Since that time considerable development has occurred; inner harbor species have disappeared, mangroves have been destroyed, and oxygen levels have been depleted (J. Hendricks, personal communication).

C. European Studies

In European waters the bryozoan studies at Roscoff, France, are important once again since the area was inundated with oil from the *Amoco Cadiz* tanker disaster. Prenant and Bobin (1956, 1966) gave depth and distribution records for the species discussed and illustrated; some of these are common to other estuarine waters, as well as having occurred at Roscoff. It will be important to follow the species succession in recovering from the incident.

Another paper not previously cited in surveys is that of Carrada (1973), who reported on the bryozoans collected on algal substrates on the northwest coast of Spain in the San Sebastian area (the Vigo estuary and adjacent littoral lagoons). Carrada compared the 13 species he collected (three ctenostomes, nine cheilostomes, and one cyclostome) with those reported by Barroso in 1912 from Santander. *Alcyonidium hirsutum* and *Bowerbankia gracilis* were the lagoonal species found by Carrada but not by Barroso.

Geraci and Relini (1971) investigated bryozoans in relation to pollution in the Port of Genoa, Italy. They found remarkable differences in the composition of populations and settlement according to hydrologic and topographic conditions. The species more resistant to unfavorable conditions were the ctenostomes *Zoobotryon verticillatum* and *Bowerbankia* sp., and the cheilostome anascans *Bugula neritina* and *Bugula stolonifera*. *Bugula stolonifera* was collected at the area most heavily affected by domestic sewage nearly all year round.

V. Disturbances, Natural and Man-Made

A. NATURAL EVENTS

Instability is a characteristic of most estuaries and natural "catastrophic" events can upset the usually fluctuating biological patterns of estuaries beyond seasonal changes in temperature and runoff. For example, Andrews (1973) discussed the effects of tropical storm Agnes on epifaunal invertebrates in Virginia estuaries. He noted that the storm caused heavy freshwater runoff into Chesapeake Bay in June 1972, following a wet year in 1971. Salinities were depressed and remained so until the spring of 1973. The most common bryozoans on settling plates immediately after the storm were the apparently euryhaline (opportunistic?) species *Membranipora tenuis* and *Electra crustulenta*. The incidence of *Anguinella palmata* appeared to increase, while *Victorella pavida*, *Alcyonidium verrilli*, and *Bowerbankia gracilis* appeared to decrease in numbers or occurrences in the months following the study.

B. MAN-MADE EFFECTS

Leathem *et al.* (1973) sampled 103 stations on Delaware Bay before and after hydraulic dredging and spoil disposal. Using a Petersen grab sampler before the dredging, they found four species of Bryozoa in December 1971: *Alcyonidium polyoum*, *Conopeum tenuissimum*, *Electra hastingsae*, and *Membranipora tenuis*. After spoil disposal was completed in March 1972, the same species plus *Alcyonidium verrilli* were found. However, in June 1972 the *Alcyonidium* sp. had disappeared and *Aeverrillia setigera* had arrived. Since *Alcyonidium* are incrusting and have a soft gelatinous surface, deposition of fine sediment on frontals could seriously affect colonies, whereas *Aeverrillia* forms minute, stolonate colonies. The difficulty with this simple assumption is an important one; there was no prior seasonal or annual record for comparison. *Alcyonidium polyoum* is generally regarded as a cool water species (or winter?) whereas *Aeverrillea* occurs in warmer waters (or summer?) as far north as Massachusetts (Osburn, 1953), so that the change in species cannot be effectively evaluated.

While the greatest recent impetus for biological inventories has been applied environmental studies, it is unfortunate that studies are almost always too short to gain insight into seasonality for even a single year. In addition the variation between years can be as great or greater than some seasonal fluctuations.

C. POLLUTANTS

Symposia on the effects of pollutants and volumes of contributed papers have proliferated in recent years, but the availability of such reports is often limited to participants and not as widely circulated in libraries as desirable. One very useful source on the "state of the art" is the literature review issue of the *Journal of the Water Pollution Control Federation* every June. Examination of the 441 citations listed and summarized in the section on marine and estuarine pollution, 401 citations under thermal effects, and 466 citations for heavy metals shows hundreds of references dealing with pollution. However, no references were cited that deal directly with the effects of pollutants on bryozoans.

Bryozoans are not commonly used for laboratory experimental animals but records on tests of other invertebrates can be useful. Bryozoans are among the organisms most resistant to antifouling measures such as copper-containing paints for boat hulls. Lee and Trot (1973) compared antifouling paints and hull wood of various resistances to fouling in Hong Kong Harbor, a subtropical water regime. *Bugula neritina* and an unidentified incrusting bryozoan were present on all test and control surfaces. *Bugula* dominated community succession at 5–7 weeks.

Soule and Soule (1977) reviewed the research on effects of various parameters on fouling, and pointed out that the role of microflora such as bacteria and diatoms in attachment had not been sufficiently investigated. Apparently, chelating agents are able to suppress ionic activity, and it is the ionic form of heavy metals that produces toxic effects. Chlorophyll is a chelating agent and thus microalgae could have a significant role in preparing a surface for bryozoan or other colonization. Toxicity of nickel, mercury, lead, copper, vanadium, cadmium, and manganese are known to respond to chelation (Gornitz, 1972). However, Al-Ogily (1977) examined bryozoans and marine algae for antibiotics associated with antifouling roles. Pinter (1969) previously noted that certain marine algae secrete substances that prevent bryozoan settling.

The biological impact of a large-scale desalinization plant at Key West, Florida was studied by Chesher (1975), who concluded that all flora and fauna in the harbor receiving waters were adversely affected by the discharge. Effects of discharges from antifouling maintenance procedures were worse than routine operation effluents, and this was largely due to corroded copper particles in the discharge. Chesher reported that Miller, in 1946, reported that *Bugula neritina* could live but not grow in ionic copper concentrations of 0.2–0.3 ppm and that larval stages died at concentrations in excess of 0.3 ppm. Some species of marine invertebrates concentrate copper more rapidly than those in dark areas. Bryozoans were absent from settling panels until June 1971, when the badly corroded copper–nickel trays were removed, after which bryozoans were found on some panels. Foraminifera, serpulids, sabellids, and barnacles were prolific, however.

Straughan (1975) examined bryozoans for "hyperplasia" of ovicells in species chronically exposed to natural oil seepage at Coal Oil Point, California. Contrary to the earlier reports of Powell *et al.* (1970), no evidence of "hyperplasia" was seen. Ross (Ross and McCain, 1976) agreed with our opinion that Powell observed a normal phenomenon of the species of *Schizoporella* that could not be considered to be hyperplasia.

Soule *et al.* (1978) studied the impact of a Bunker C spill following explosion of the tanker *Sansinena* in Los Angeles Harbor in December 1976. Although the study involved physical, chemical, and multiple biological parameters, bryozoans were collected on settling racks in meroplankton sampling and also collected from nearby breakwater rocks.

Although *Membranipora villosa* occurs in the harbor year around on settling racks, it was not found in the impact area at all for at least 7 months after the incident. Eight species of ectoprocts and one entroproct were collected in early January, 2 weeks after the explosion, as colonies scraped from the breakwater. Only three of the species, *Celleporaria brunnea, Membranipora tuberculata,* and *Thalamoporella californica,* were present in April; by June these species had been replaced by dense growths of *Scrupocellaria diegensis.* One year after the explosion five species were found, but *Bugula neritina* had not returned to the rocks.

Bugula neritina did return in May of 1977, its usual period to recur, at the boat launching ramp at nearby Cabrillo Beach, even though floating oil sheens were common. *Watersipora arcuata, Celleporaria brunnea, Cryptosula pallasiana,* and *Bowerbankia gracilis* usually appear in the harbor in May or June; they were not seen at the ramp until October of 1977. While not strictly an estuarine location, residential and storm runoff occurs in the beach area.

D. THERMAL IMPACTS

Temperature is one of the most important natural, controlling parameters in determining distribution patterns, reproductive cycles, and growth patterns. Large-scale zones determine biogeographic patterns such as those described by Ekman (1953) and Hesse *et al.* (1937). Small-scale thermal patterns influence distribution in shallow water areas so that fluctuations in diurnal and seasonal temperatures will exert a strong effect on populations and communities.

Ryland (1970) noted that bathymetric boundaries of bryozoans also correlated with temperature and salinity fluctuations. Ström (1978) summarized some of the information on temperature and breeding seasons, while Jebram (1977) remarked on the effects of temperature in altering colony formation and growth rates in laboratory cultures. Schopf (1977) reported that regional differences in gene frequencies of *Schizoporella errata* (formerly *S. unicornis* at Woods Hole, Massachusetts), paralleled temperature differences in the Cape Cod area.

Long-term variation in annual temperatures has been documented for eastern Pacific coastal waters by examining the annual data reports from Scripps Institute of Oceanography. These long-term shifts can cause extensive changes in benthic, planktonic, and fouling communities, which are not due to any man-made impact. Therefore it is difficult to predict exact impacts of man-made changes such as elevated temperatures due to power plant heat exchangers or lowered temperatures due to liquified natural gas (LNG) gasification plants.

Soule (1974) discussed the fluctuations along the Pacific coast and in Los Angeles Harbor. In one year the minimum low temperature might be higher than the maximum for the same month in another year. Records kept by the authors for 8 years on bryozoans in Los Angeles Harbor (unpublished data) showed distinct differences in fauna. *Membranipora villosa* was found on settling racks, almost alone, during the first 6 months of 1975 and 1976; both years had warmer winters than 1971 and 1974, when diversity was greater.

Studies on thermal tolerance, acclimation, and acclimatization were initiated largely on the east coast of the United States in relation to commercial oyster culture (Loosanoff, 1969). Prosser (1969) pioneered in postulating enzymatic mechanisms for temperature adaptation, while Vernberg and Vernberg (1969) initiated studies on the effects of temperature on invertebrate respiration. Symposia on the physiological responses of marine biota to pollutants have been edited by Vernberg and Vernberg (1974) and Vernberg *et al.* (1977), which give indications of the effects on invertebrates. Such research has not yet been done with bryozoans, and offers good indications of the kind of research needed for the group.

VI. Analytical Approaches

APPLIED STUDIES

The majority of estuarine bryozoan surveys in the literature have been made from field collections that were mostly lacking in physical and chemical data and therefore the organisms could not be correlated with such parameters. Evaluation of impacts of pollutants in urban estuaries really requires multidisciplinary investigations.

Applied studies for public agencies and private industries have been almost the only source of funds in the United States in recent years. There have been a number of surveys for public permitting agencies, but such documents often are not widely available, and valuable information may go unrecognized. Many

documents do not indicate the scientists who were responsible for species identifications, and therefore new or unusual distribution records are sometimes
lacking in credibility.

The agencies and industries have been amenable to the application of rapidly
developing computer techniques for data analysis as well as for data storage, and
funds have consequently been available to explore new methods. While these
activities have been taking place in the United States, and to some extent in other
countries, it has not been possible to identify those studies that have dealt with
bryozoans. Therefore, the authors take the liberty of presenting results of some of
their own research in Los Angeles–Long Beach Harbors (San Pedro Bay),
California.

1. Case Study: Los Angeles–Long Beach Harbors

Harbors Environmental Projects was initiated in 1971 at the University of
Southern California to develop baseline information in response to such needs for
EIA, EIR, and EIS documents. Literature search revealed a lack of physical,
chemical, and oceanographical information and an absence of coordinated
biological studies. In addition to baseline data, it soon became apparent that
experimental research was needed on tolerances to thermal effluents, fish cannery wastes, domestic wastes, oil, and proposed major dredging and filling
projects.

Table I shows the various monitoring procedures used in monitoring eight
stations in 1971–1972, 44 stations in the harbors and east of the harbors in 1973
and 1974, and 42 stations in 1978. Settling racks were deployed at 24 of the
stations. Some of the results from monitoring and research have been published
in Marine Studies of San Pedro Bay, California (Soule and Oguri, 1972–1978)
and as Allan Hancock Foundation (1976). Abbott *et al.* (1973) presented histori-

TABLE I

PARAMETERS MEASURED, LOS ANGELES–LONG BEACH HARBORS

Monitoring	Method
	A. Monthly
Abiotic parameters	
Temperature	
Salinity	Martek electronic remote probe, at 1-m intervals through the water
Dissolved oxygen	column
pH	
Light transmittance	Hydroproducts transmissometer, remote probe with self-contained
	light path, at 1-m intervals through depth
Ammonia	Solorzano (1969)
Nitrite	Strickland and Parsons (1968)

TABLE I (*Continued*)

Monitoring	Method
Abiotic parameters (*continued*)	
Nitrate	Modified Strickland and Parsons
Phosphate	(Allan Hancock Foundation, 1976)
Biotic parameters	
Biological oxygen	Standard Methods (American Public Health, 1971) modified by Juge
demand (BOD)	and Greist (1975), surface samples
Total coliforms	
Fecal coliforms	Millepore (1972), Allan Hancock Foundation (1976)
Fecal streptococcus	
Bacterial standard	American Society for Microbiology (1957), Allan Hancock Founda-
plate count	tion (1976)
Primary productivity,	Modified Steeman-Neilsen (1952) ^{14}C light and dark bottles, standard
phytoplankton	light source incubator with ambient water temperature
Chlorophyll	Spectrophotometry, Strickland and Parsons (1968) equations
Assimilation ratio	
Zooplankton species	253-μm net surface tow with flow meter
Water column fouling	Glass microscope slides in wood frame rack, plastic screened,
fauna, larvae and	suspended at 3-m depth
juveniles	

B. Quarterly

Abiotic parameters	
Sediment grain size	Pettijohn (1957), Felix (1969), Gibbs (1971), Allan Hancock Founda-
	tion (1976)
Trace metals	American Public Health Association (1971), Allan Hancock Founda-
pesticides	tion (1976)
Biotic parameters	
Benthic fauna	Campbell grab or Reinecke box corer, 0.5-mm screen
Fish species	Otter trawl, gill netting

C. Biweekly (Outer Los Angeles Harbor Only)

Abiotic parameters	
Temperature	
Salinity	Martek electronic remote probe, at 1-m intervals through the water
Dissolved oxygen	column
pH	
Oil and grease	American Public Health Association (1971)
Biotic parameters	
BOD	Standard Methods (American Public Health Association, 1971),
	modified by Juge and Greist (1975), surface samples

D. Weekly (Outer Los Angeles–Long Beach Harbors) 1973–1974

Biotic parameters	
Bird census	Observations of nesting, resting, feeding, and transit

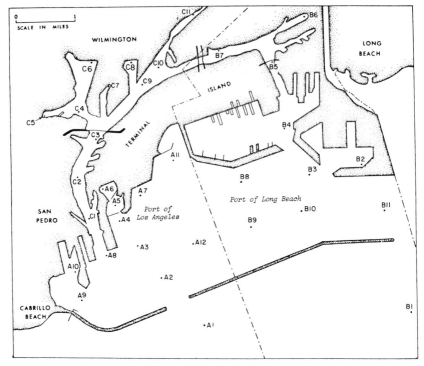

Fig. 1. Monitoring in the ports.

cal background for the harbors and preliminary results of settling rack studies at the Helgoland Symposium.

In the major studies for the ports all of the parameters listed in Table I were analyzed for 1973 and 1974 using a newly developed multiple discriminant analysis package—Ecosystems Analysis (Smith, 1976; Allan Hancock Foundation, 1976).

Classification and analysis methods were used to group the stations according to their biological characteristics (species). It became evident that the harbor could be categorized as divided into inner slip (polluted) areas, main channel (semipolluted) areas, and outer harbor (clean) areas—except for a limited area at a sewer outfall in the outer harbor. Seasonal effects were not great, but certain single stations switched from one group to another according to the season.

Zooplankton characteristics were not as well defined, but computer mapping techniques identified the harbor areas with greater diversity or population density (Fig. 1).

The settling rack technique, by which bryozoan data were gathered monthly, demonstrated the biological potential of the water column and the influx of adults on tides or reproduction in the harbor for the periods. Over 4 million separate animal occurrences were recorded of more than 1600 species in 15 phyla. Settling racks collected 70% of the taxa, while plankton and benthic sampling collected about 40% of the taxa, without regard to the number of individuals.

2. *Multiple Discriminant Analysis*

Because bryozoans were not quantified, as were the plankton and benthic organisms, they were represented by presence–absence data and new methods were required for analysis. The techniques utilized include classification of data, using the Bray–Curtis Index and hierarchical clustering strategy to group the sites and species into dendrograms. Multiple discriminant analysis then determines the relationships between species and parameters and plots them on axes in multidimensional space. Methods were discussed and results of initial analyses were presented at the Fourth International Conference of the International Bryozoology Association, Woods Hole, Massachusetts, in September, 1977 (Soule *et al.*, (1979).

Figures 2–8 show the orientation along axes 1 and 2 or 1 and 3 and the directions of parameters in space. The numbers beneath vectors represent coefficients of separate determination; the higher the value, the more important the

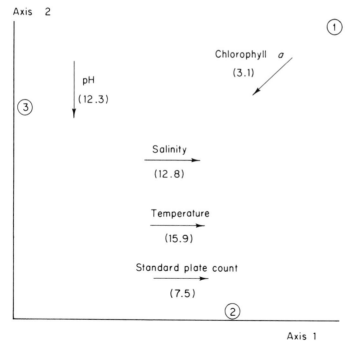

Fig. 2. Weighted discriminant analysis, Winter 1973–1974. 1, *Membranipora* sp.; 2, *Bugula neritina*; 3, *Bugula californica*.

Dorothy F. Soule and John D. Soule

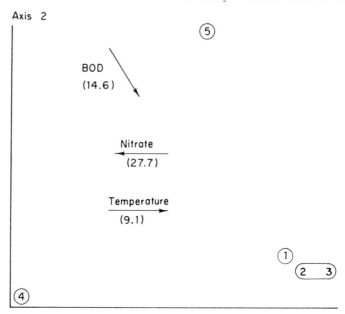

Fig. 3. Weighted discriminant analysis, Spring 1974. 1, *Membranipora* sp.; 2, *Bugula califor-nica;* 3, *Bugula neritina;* 4, *Cryptosula pallasiana;* 5, *Celleporaria brunnea.*

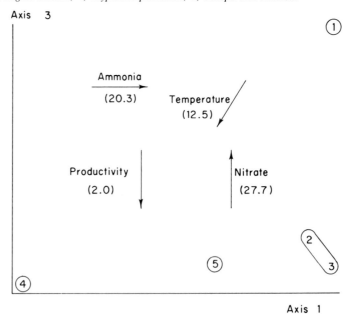

Fig. 4. Weighted discriminant analysis, Spring 1974. 1, *Membranipora* sp.; 2, *Bugula califor-nica;* 3, *Bugula neritina;* 4, *Cryptosula pallasiana;* 5, *Celleporaria brunnea.*

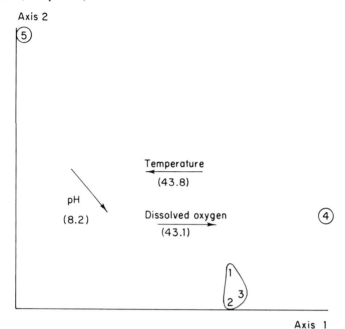

Fig. 5. Weighted discriminant analysis, Summer 1974. 1, *Membranipora* sp.; 2, *Bugula neritina;* 3, *Bugula californica;* 4, *Cryptosula pallasiana;* 5, *Scruparia ambigua.*

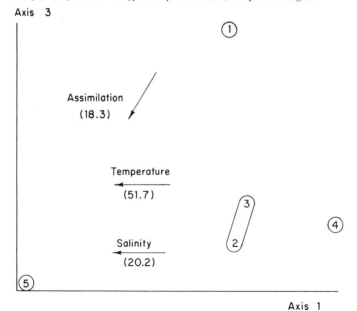

Fig. 6. Weighted discriminant analysis. Summer 1974. 1, *Membranipora* sp.; 2, *Bugula neritina;* 3, *Bugula californica;* 4, *Cryptosula pallasiana;* 5, *Scruparia ambigua.*

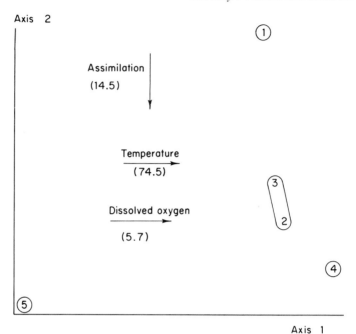

Fig. 7. Weighted discriminant analysis, Autumn 1974. 1, *Membranipora* sp., 2, *Bugula californica;* 3, *Bugula neritina;* 4, *Cryptosula* pallasiana; 5, *Celleporaria brunnea.*

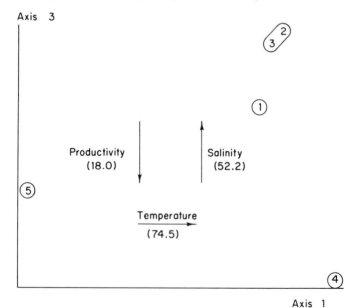

Fig. 8. Weighted discriminant analysis, Autumn 1974. 1, *Membranipora* sp.; 2, *Bugula californica;* 3, *Bugula neritina;* 4, *Cryptosula* pallasiana; 5, *Celleporaria brunnea.*

parameter is supposed to be. Statistical validity has not been determined for these techniques, and the management of presence–absence data presents some problems. It must always be remembered that coincidence in time and space does not imply causality.

The vectors for salinity, temperature, oxygen, and pH may in fact represent only slight differences between species and localities, as can be seen in the histograms of salinity (Fig. 9). Temperature and dissolved oxygen show wider ranges (Figs. 10 and 11).

The intriguing implication of the multiple discriminant results was that it suggested possible correlations between phytoplankton (chlorophyll *a* and productivity) or between bacteria (standard plate count). This may indicate various feeding strategies of the species for part or all of the year, which offers new directions for experimental research. Further analyses will be carried out on 1978 data and it will be interesting to see if results are at all consistent. If not, the plots then may only indicate coincidence in time and space of physical and biological entities.

3. Principal Component Analysis of Bryozoan Occurrence in Los Angeles Harbor (1973 and 1974)

Bryozoan occurrence data and corresponding abiotic and nutrient data for 1973 and 1974 were collected (151 sets of observations, 11 species, 13 variables). Incomplete observation sets were eliminated, reducing the total of 91 sets of observations, 8 species, and 11 variables. The four most common species (*M. villosa,* 50 occurrences; *B. neritina,* 30; *B. californica,* 20; and *C. pallasiana,* 7) were distributed over each of the 11 variables. This resulted in 44 histograms. (The other four species occurred only once each.)

No statistical tests were performed to determine the significance of apparent differences in species distribution. Principal Component Analysis (PCA) was performed on the 11 variables to determine the factors of environmental variation in the harbor. The variables were)1) temperature, (2) salinity, (3) O_2, (4) DO, (5) BOD, (6) bacteria count, (7) nitrite, (8) nitrate, (9) ammonia, (10) productivity, and (11) chlorophyll *a*. All eight species were plotted on the plane of the first two factors resulting from the PCA.

The PCA yielded five factors accounting for 74% of the variation in the supplied data. The other 26% is spread over six probably meaningless factors. The first two factors account for 40% of the variation and correlate highly with seven of the variables (temperature, BOD, bacteria, nitrate, ammonia, productivity, and chlorophyll *a*) and less highly with one factor (oxygen). Factors 3, 4, and 5 account for 14, 11, and 10%, respectively, and each correlates highly with one variable (nitrite, salinity, and oxygen, respectively). This conveniently takes care of all 11 input variables.

All eight species are plotted on the plane formed by factors 1 and 2. (It was felt that the best way to deal with factors 3, 4, and 5 was to refer to the histograms of

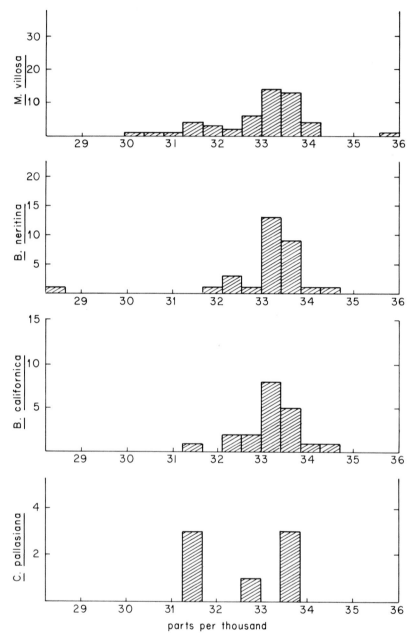

Fig. 9. Distribution of common bryozoans in Los Angeles Harbor with salinity.

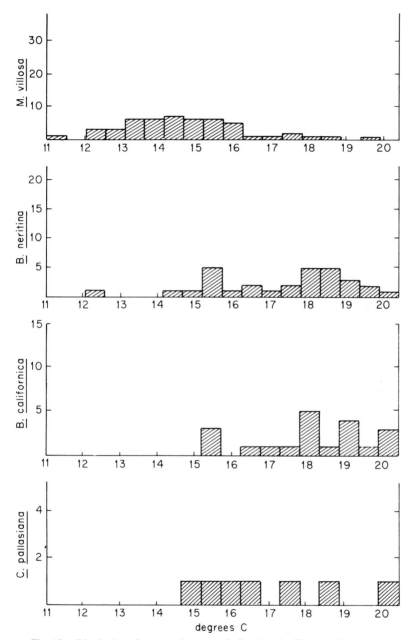

Fig. 10.   Distribution of common bryozoans in Los Angeles Harbor with temperature.

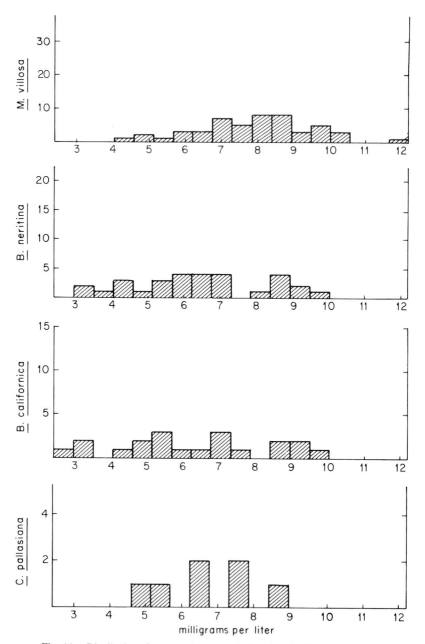

Fig. 11. Distribution of common bryozoans in Los Angeles Harbor with oxygen.

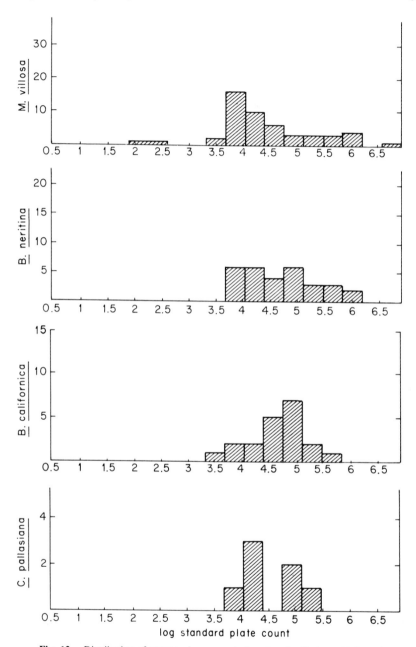

Fig. 12. Distribution of common bryozoans in Los Angeles Harbor with bacteria.

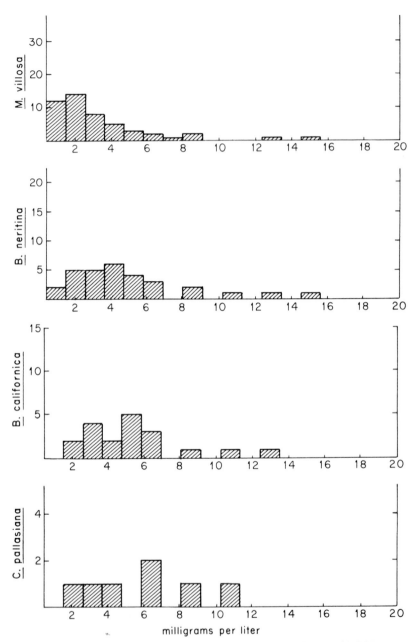

Fig. 13. Distribution of common bryozoans in Los Angeles Harbor with BOD.

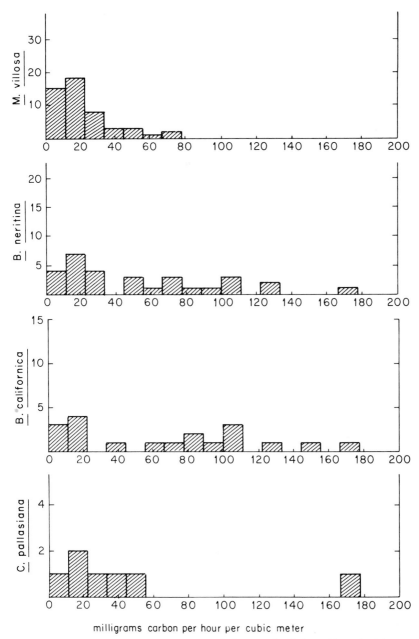

milligrams carbon per hour per cubic meter

Fig. 14. Distribution of common bryozoans in Los Angeles Harbor with productivity.

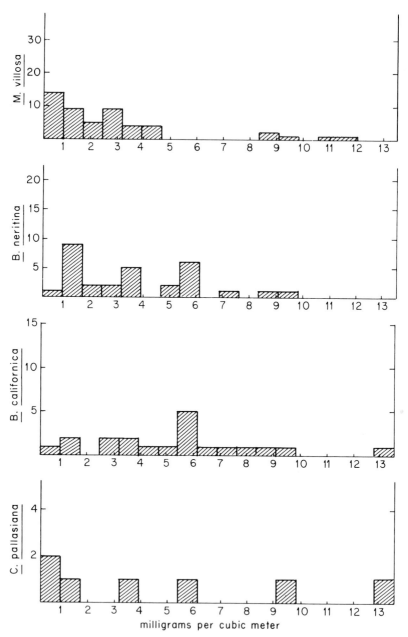

Fig. 15.   Distribution of common bryozoans in Los Angeles Harbor with chlorophyll *a*.

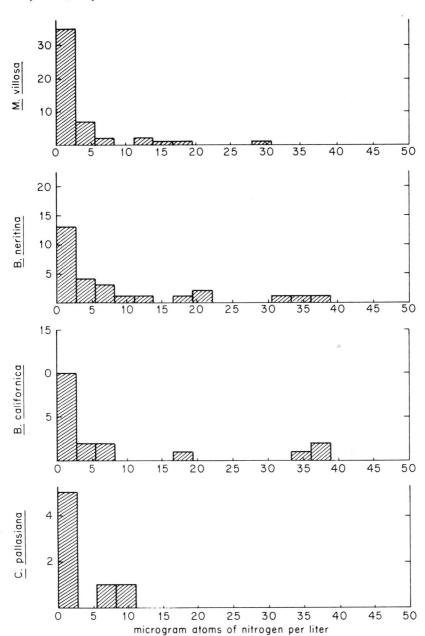

Fig. 16. Distribution of common bryozoans in Los Angeles Harbor with ammonia.

Dorothy F. Soule and John D. Soule

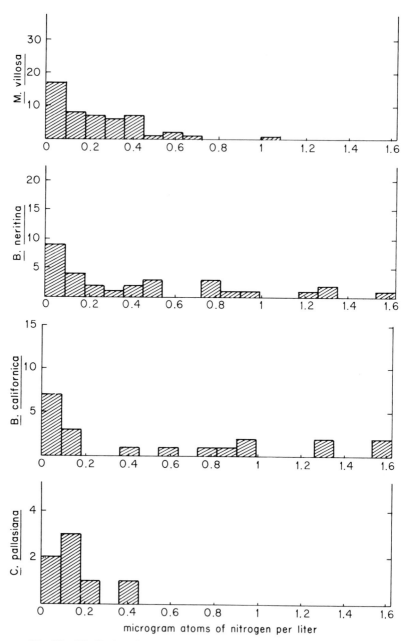

Fig. 17. Distribution of common bryozoans in Los Angeles Harbor with nitrite.

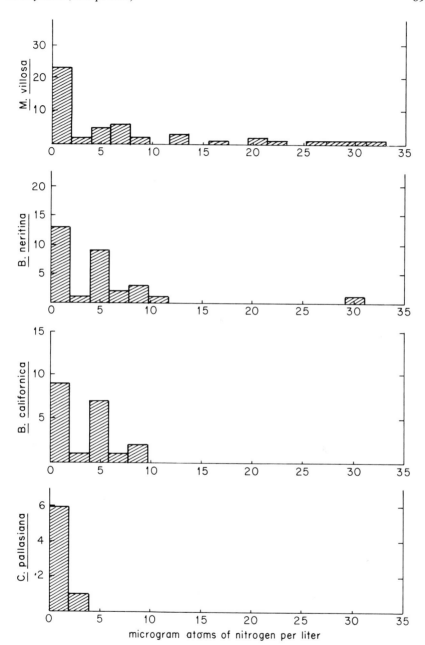

Fig. 18. Distribution of common bryozoans in Los Angeles Harbor with nitrate.

the variables with which they correlate most highly.) All the species exist in a wide range of temperature, productivity, and nitrate, with *M. villosa* preferring slightly lower temperatures and getting along at higher nitrate levels than the other three commonly occurring species. All of the species exist in a relatively narrow range of a vector characterized by BOD, ammonia, and bacteria. This range appears a bit wider at medium temperatures than at extreme temperatures.

PCA based on the data shown for each separate parameter presented different pictures (Figs. 9–18) of the one-to-one relationship of species to parameters. PCA (Fig. 19) displays the interactions in a two-dimensional manner for all the variables, whereas Discriminant Analysis (DA) portrayed them in multiple dimensions. In effect, discriminant analysis takes two variables and creates a ''new'' variable and repeats this procedure. Distance matrices are used to calculate coefficients of separate determination.

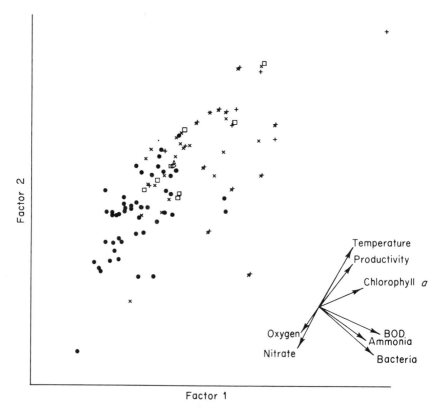

Fig. 19. Principal component analyses distribution of bryozoans with factors 1 and 2 (factors are groups of variable correlations). ●, *M. villosa;* ×, *B. neritina;* +, *B. california;* □, *C. pallasiana;* ◇, *S. unicornis;* △, *C. reticulum;* *, *C. brunnea;* ·, *S. ambigua.*

The comparison can be made with *Membranipora* (*villosa*), which appeared to be at the higher end of the vector for temperature in DA, based on coefficients of separation. Yet in the histograms (PCA) *M. villosa* actually occurred during lower temperature periods. The DA placement is the summation of parameters, and temperature might be coincidental, or higher, in placing *M. villosa* at that given period in comparison with the other species present at that time.

References

Abbott, B. C., Soule, D. F., Oguri, M., and Soule, J. D. (1973). *In situ* studies of the interface of natural and man-made systems in a metropolitan harbor. *Helgol. Wiss. Meeresunters.* **24,** 455–464.

Allan Hancock Foundation (1976). "Environmental Investigations and Analyses, Los Angeles–Long Beach Harbors," Final Report to the U.S. Army Corps of Engineers, Los Angeles District. Harbors Environ. Proj., University of Southern California, Los Angeles.

Allman, G. (1856). "Monograph of the Fresh-water Polyzoa." Ray Society, London.

Al-Ogily, S. M. (1977). Anti-fouling roles of antibodies produced by marine algae and bryozoans. *Nature (London)* **165,** 728–729.

American Public Health Association (1971). "Standard Methods for Examination of Water and Waste Water," 13th ed. APHA, New York.

American Society for Microbiology (1957). "Manual of Microbiological Methods." McGraw-Hill, New York.

Andrews, J. D. (1973). Effects of tropical storm *Agnes* on epifaunal invertebrates in Virginia estuaries. *Chesapeake Sci.* **14,** 223–234.

Barrois, J. (1877). Mémoire sur l'embryologie des bryozoaires. *Trav. Inst. Zool. Lille* **1,** 1–305.

Barroso, M. G. (1912). Briozoos de la estacion de biologia maritima de Santander. *Trab. Mus. Cienc. Nat.* **5,** 3–63.

Beatty, J. A., and Blackwelder, R. E. (1974). Names of invertebrate phyla. *Syst. Zool.* **23,** 545–547.

Boardman, R. S., Cheetham, A. H., and Oliver, W. H., eds. (1973). "Animal Colonies." Dowden, Hutchinson & Ross, Stroudsburg, Pennsylvania.

Brown, D. A. (1958). The relative merits of the class names "Polyzoa" and "Bryozoa." *Bull. Zool. Nomencl.* **15,** Doc. 25/5, 540–542.

Bushnell, J. H. (1974). Bryozoans (Ectoprocta). "Pollution Ecology of Freshwater Invertebrates" (C. W. Hart, Jr. and S. L. H. Fuller, eds.), pp. 157–194. Academic Press, New York.

Busk, G. (1855). Class Bryozoa. *In* "Catalogue of the Reigen Collection of Mazatlán Mollusca in the British Museum" (P. P. Carpenter, ed.), pp. 1–6. Warrington.

Carrada, G. C. (1973). Briozoi litorali della Ria di Vigo (Spagna Nord-Occidentale). *Invest. Pesq.* **37,** 9–15.

Carriker, M. R. (1976). The crucial role of systematics in assessing pollution effects on the biological utilization of estuaries. *In* "Estuarine Pollution Control and Assessment," pp. 486–506. U.S. Environ. Protect. Agency, Office of Water Planning and Standards, U.S. Govt. Printing Office, Washington, D.C.

Chesher, R. H. (1975). Biological impact of a large-scale desalination plant at Key West, Florida. *Elsevier Oceanogr. Ser. (Amsterdam)* **12,** 99–153.

Cook, P. L. (1968). Polyzoa from West Africa. The Malacostega. Part I. *Bull. Br. Mus. (Nat. Hist.), Zool.* **16,** 115–160.

Crisp, D. J., and Ryland, J. S. (1960). Influence of filming and of surface texture on the settlement of marine organisms. *Nature (London)* **185,** 119.

Cuffey, R. J. (1969). Bryozoa versus Ectoprocta—the necessity for precision. *Syst. Zool.* **18,** 250-251.

Cuffey, R. J. (1974). Delineation of bryozoan construction roles in reefs from comparison of fossil bioherms and living reefs. *Proc. Int. Coral Reef Symp. 2nd, 1973,* Vol. I, pp. 357-364.

Daniel, A. (1960). The primary film as a factor in the settlement of marine foulers. *J. Madras Univ., Sect. B* **30,** 189-200.

Dudley, J. E. (1973). A note on the taxonomy of three membraniporine ectoprocts from Chesapeake Bay. *Chesapeake Sci.* **14,** 282-285.

Dyer, K. R. (1972). "Estuaries: A Physical Introduction." Wiley, New York.

Eiban, A. (1976). Einfluss von Benetzungsspannung und Ionen auf die Substratbesiedlung und das Einsetzen der Metamorphose bei Bryozoenlarven (*Bowerbankia gracilis*). *Mar. Biol.* **37,** 249-254.

Ekman, S. (1953). "Zoogeography of the Sea." Sidgwick & Johnson, London.

Evans, E. C., III, Peeling, T. J., Murchison, A. E., and Stephen-Hassard, Q. D. (1972). "A Proximate Biological Survey of Pearl Harbor, Oahu." Nav. Undersea Res. and Dev., San Diego, California.

Fiala-Medioni, A. (1973). Bryozoaires du benthos Rocheux de Banyuls-sur-mer Inventaire faunistique et notes écologiques. *Vie Milieu, Ser. A* **23,** 273-308.

Geraci, S., and Relini, G. (1971). Fouling di zone inquinati. Observazioni del Porto di Genoa. I. Bryozoi. *Pubbl. Stn. Zool. Napoli* **38,** Suppl., 21-32.

Ghiselin, M. T. (1977). On changing the names of higher taxa. *Syst. Zool* **26,** 346-349.

Gornitz, V. (1972). Chelation. *In* "Encyclopedia of Geochemistry and Environmental Sciences" (R. W. Fairbridge, ed.), pp. 149-152. Van Nostrand-Reinhold, Princeton, New Jersey.

Harmer, S. F. (1911). The terms Polyzoa and Bryozoa. *Proc. Linn. Soc. London* **123,** 70-71.

Harmer, S. F. (1947). On the relative merits of the names Bryozoa and Polyzoa as the name for the class in the Animal Kingdom now known by one or another of these names. *Bull. Zool. Nomencl.* **1,** 230-231.

Hastings, A. B. (1927). Zoological results of the Cambridge Expedition to the Suez Canal, 1924. XX. Report on the Polyzoa. *Trans. Zool. Soc. London* **22,** 331-353.

Hesse, R., Allee, W. C., and Schmidt, K. P. (1937). "Ecological Animal Geography." Wiley, New York.

Hincks, T. (1880). On the terms Polyzoa and Bryozoa. *Ann. Mag. Nat. Hist.* [5] **5,** 127-129.

Howard, J. D., and Frey, R. (1975). Estuaries of the Georgia coast, USA. Sedimentology and biology. I. *Senckenbergiama Marit.* **7,** 1-31 and 33-101.

Hutchins, L. W. (1952). Bryozoa. *In* "Marine Fouling and its Prevention," pp. 140-143, 163, 167, 190-192, and 204-207. U. S. Nav. Inst., Annapolis, Maryland.

Hyman, L. H. (1959). "The Invertebrates: Smaller Coelomate Groups," Vol. 5. McGraw-Hill, New York.

Jebram, D. (1977). Experimental techniques and culture methods. *In* "Biology of Bryozoans" (R. W. Woollacott and R. L. Zimmer, eds.), pp. 273-306. Academic Press, New York.

Jones, T. R. (1880). On the nomenclature of Polyzoa, Busk. *Ann. Mag. Nat. Hist.* [5] **5,** 220.

Juge, D. M., and Greist, G. C. (1975). A modification of BOD methods for use in the marine environment. *In* "Marine Studies of San Pedro Bay, California" (D. F. Soule and M. Oguri, eds.), Part 8, pp. 46-55. Allan Hancock Found. and Sea Grant Program, University of Southern California, Los Angeles.

Leatham, W., Kinner, P., Maurer, D., Biggs, R., and Treasure, W. (1973). Effect of spoil disposal on benthic invertebrates. *Bull. Mar. Pollut.* **4,** 122-125.

Lee, S. W., and Trott, L. B. (1973). Marine succession of fouling organisms in Hong Kong, with a comparison of woody substrates and common, locally available antifouling paints. *Mar. Biol.* **20,** 101-108.

Lee, W. L. (1978). Resources in invertebrate systematics. *Am. Zool.* **18**, 167-185.

Long, E. R. (1970). "Second Year of Marine Biofouling Studies off Oahu, Hawaii. Oceanographic Cruise Summary," Informal Rep. No. 70-48, pp. 1-12. Nav. Oceanogr. Office, Washington, D.C.

Loosanoff, V. L. (1969). Maturation of gonads of oysters *Crassostrea virginica,* of different geographical areas subjected to relatively low temperatures. *Veliger* **11**, 153-163.

Lutaud, G. (1977). The bryozoan nervous system. *In* "Biology of Bryozoans" (R. W. Woollacott and R. L. Zimmer, eds.), pp. 377-410. Academic Press, New York.

Lynch, W. F. (1949). Modification of the responses of two species of *Bugula* larvae from Woods Hole to light and gravity: Ecological aspects of the behavior of *Bugula* larvae. *Biol. Bull. (Woods Hole, Mass.)* **97**, 302-310.

Lynch, W. F. (1958a). The effects of certain organic compounds and antimitotic agents on metamorphosis of *Bugula* and *Amaroecium* larvae. *J. Exp. Zool.* **137**, 117-152.

Lynch, W. F. (1958b). The effects of X-rays, irradiated sea water, and oxidizing agents on the rate of attachment of *Bugula* larvae. *Biol. Bull. (Woods Hole, Mass.)* **114**, 215-225.

Lynch, W. F. (1959). Factors influencing metamorphosis of larvae of some of the sessile organisms. *Proc. Int. Congr. Zool., 15th, 1958* Sect. III, Paper 9, pp. 1-3.

Marcus, E. (1938). Bryozoarios marinhos brasilieros. II. *Bol. Fac. Philos. Sci. Let. Zool.* **2**, 1-137.

Marcus, E. (1940). Mosdyr. (Bryozoaeller Polyzoa). *Dan. Naturh. Foren.* **46**, 1-401.

Maturo, F. J. (1959). Seasonal distribution and settling rates of estuarine Bryozoa. *Ecology* **40**, 116-127.

Mawatari, S. (1951). The natural history of a common fouling bryozoan *Bugula neritina* (Linnaeus). *Misc. Rep. Res. Inst. Nat. Resour. (Tokyo)* **19-21**, 47-54.

Mayr, E. (1968). Bryozoa versus Ectoprocta. *Syst. Zool.* **17**, 213-216.

Mayr, E. (1969). "Principles of Systematic Zoology." McGraw-Hill, New York.

Menon, N. R. (1972). Heat tolerance. *Mar. Biol.* **15**, 1-11.

Menon, N. R., and Nair, N. B. (1971). Ecology of fouling bryozoans from the Cochin waters. *Mar. Biol.* **8**, 280-307.

Menon, N. R., and Nair, N. B. (1974a). The growth rates of four species of intertidal bryozoans in Cochin backwaters. *Proc. Indian Natl. Sci. Acad., Sect. B* **38**, 397-402.

Menon, N. R., and Nair, N. B. (1974b). On the nature of tolerance to salinity in two euryhaline intertidal bryozoans *Victorella pavida* Kent and *Electra crustulenta* Pallas. *Proc. Indian Natl. Sci. Acad., Sect. B* **38**, 414-424.

Menon, N. R., Katti, R. J., and Shetty, H. P. C. (1977). Biology of marine fouling in Mangalore waters. *Mar. Biol.* **41**, 127-140.

Millepore Corporation (1972). "Biological Analysis of Water and Wastewater," Manual AM 302. Millepore Corp.

Nielsen, C. (1977a). The relationships of Entoprocta, Ectoprocta, and Phoronda. *Am. Zool.* **17**, 149-150.

Nielsen, C. (1977b). Phylogenetic considerations: The protostomian relationships. *In* "Biology of Bryozoans" (R. M. Woollacott and R. L. Zimmer, eds.), pp. 519-534. Academic Press, New York.

Nitsche, H. (1869). Beitrage zur Kenntnis der Bryozoen. *Z. Wiss. Zool.* **20**, 1-36.

Osburn, R. C. (1944). Bryozoa of Chesapeake Bay. *Chesapeake Biol. Lab. Publ.* **63**, 1-55.

Osburn, R. C. (1950). Bryozoa of the Pacific coast of America. Part 1. Cheilostomata—Anasca. *Allan Hancock Pac. Exped.* **14**(1), 1-270.

Osburn, R. C. (1952). Bryozoa of the Pacific coast of America. Part 2. Cheilostomata—Ascophora. *Allan Hancock Pac. Exped.* **14**(2), 271-612.

Osburn, R. C. (1953). Bryozoa of the Pacific coast of America. Part 3. Cyclostomata, Ctenostomata, Entoprocta and addenda. *Allan Hancock Pac. Exped.* **14**(3), 613-841.

Pinter, P. (1969). Bryozoan-algal associations in southern California waters. *Bull. South. Calif. Acad. Sci.* **68**, 199-218.

Powell, N. A., Sayce, C. S., and Tufts, D. F. (1970). Hyperplasia in an estuarine bryozoan attributable to coal tar derivatives. *J. Fish. Res. Board Can.* **27**, 2095-2096.

Prenant, M., and Bobin, G. (1956). "Faune de France. No. 60. Bryozoaires 1. Entoproctes, Phylactolémes, Cténostomes." Fed. Fr. Soc. Sci. Nat., Paris.

Prenant, M., and Bobin, G. (1966). "Faune de France. No. 68. Bryozoaires 2. Chelostomes Anasca." Fed. Fr. Soc. Sci. Nat., Paris.

Prosser, C. L. (1969). Molecular mechanisms of temperature adaptation. *Publ. Am. Assoc. Adv. Sci.* **84**, 1-390.

Ross, J. R. P., and McCain, K. W. (1976). *Schizoporella unicornis* (Ectoprocta) in coastal waters of Northwestern United States and Canada. *Northwest Sci.* **50**, 160-171.

Ryland, J. S. (1960). Experiments on the influence of light on the behavior of polyzoan larvae. *J. Exp. Biol.* **37**, 783-800.

Ryland, J. S. (1962). The effect of temperature on the photic responses of polyzoan larvae. *Sarsia* **6**, 4-48.

Ryland, J. S. (1965). "Catalogue of Main Marine Fouling Organisms," Vol. 2. Org. Econ. Coop. Dev.,

Ryland, J. S. (1970). "Bryozoans." Hutchinson, London.

Scheer, B. T. (1945). The development of marine fouling communities. *Biol. Bull. (Woods Hole, Mass.)* **89**, 102-121.

Scheltema, R. S. (1968). Dispersal of larvae by equatorial ocean currents and its importance to the zoogeography of shoal-water tropical species. *Nature (London)* **217**, 1159-1162.

Schopf, T. J. M. (1967). Names of phyla: Ectoprocta, Entoprocta, and Bryozoa. *Syst. Zool.* **16**, 276-278.

Schopf, T. J. M. (1968). Ectoprocta, Entoprocta, and Bryozoa. *Syst. Zool.* **17**, 470-472.

Schopf, T. J. M. (1974). Long-term (3-5 yr) records of gene frequencies in natural populations of an abundant, subtidal species (the bryozoan *Schizoporella errata*). *Biol. Bull. (Woods Hole, Mass.)* **147**, 498.

Schopf, T. J. M. (1977). Population genetics of bryozoans. *In* "Biology of Bryozoans" (R. M. Woollacott and R. L. Zimmer, eds.), pp. 459-486. Academic Press, New York.

Schopf, T. J. M., and Gooch, J. L. (1971). Genetic variation in the marine ectoproct *Schizoporella errata*. *Biol. Bull. (Woods Hole, Mass.)* **141**, 235-246.

Silen, L. (1942). Origin and development of the Cheilo-Ctenostomatous stem of Bryozoa. *Zool. Bidr. Uppsala* **22**, 1-59.

Smith, R. W. (1976). Numerical analysis of ecological data. Ph.D. Dissertation, University of Southern California, Los Angeles.

Soule, D. F. (1973). Morphogenesis of giant avicularia and ovicells in some Pacific Smittinidae. *In* "Living and Fossil Bryozoa" (G. P. Larwood, ed.), pp. 485-495. Academic Press, New York.

Soule, D. F. (1974). Thermal effects and San Pedro Bay. *In* "Marine Studies of San Pedro Bay, California" (D. F. Soule and M. Oguri, eds.), Part 3, pp. 1-20. University of Southern California, Los Angeles.

Soule, D. F., and Oguri, M., eds. (1972-1978). "Marine Studies of San Pedro Bay, California," Parts 1-15. Harbors Environ. Proj., Allan Hancock Found., Sea Grant Program, & Inst. Mar. Coastal Stud., University of Southern California, Los Angeles.

Soule, D. F., and Soule, J. D. (1964). The Ectoprocta (Bryozoa) of Scammon's Lagoon, Baja California, Mexico. *Am. Mus. Novit.* **2199**, 1-56.

Soule, D. F., and Soule, J. D. (1968). Bryozoan fouling organisms from Oahu, Hawaii with a new species of *Watersipora*. *Bull. South. Calif. Acad. Sci.* **67**, 203-218.

Soule, D. F., and Soule, J. D. (1972). Ancestrulae and body wall morphogenesis of some Hawaiian and eastern Pacific Smittinidae (Bryozoa, Ectoprocta). *Trans. Am. Microsc. Soc.* **91**, 251–260.

Soule, D. F., and Soule, J. D. (1975). Species groups in Watersiporidae. *In* "Bryozoa 1973," pp. 299–309. Doc. Lab. Geol. Fac. Sci., Lyon, France.

Soule, D. F., and Soule, J. D. (1979). Marine zoogeography and evolution of Bryozoa in the southern hemisphere. *Proc. Int. Symp. Mar. Biog. Evol., South. Hemisphere, 1978* (in press).

Soule, D. F., Morris, P. A., and Soule, J. D. (1977). Comparisons of estuarine management needs of a tropical Pacific Mexican harbour and a temperate California harbour. *Mar. Res. Indonesia* **19**, 21–38.

Soule, D. F., Soule, J. D., and Henry, C. A. (1979). The influence of environmental factors on the distribution of estuarine bryozoans, as determined by multivariate analysis. *Proc. Int. Conf., Int. Bryozool. Assoc., 1977* (in press).

Soule, D. F., Oguri, M., and Soule, J. D. (1978). The impact of the *Sansinena* explosion and Bunker C spill on the marine environment of outer Los Angeles Harbor. *In* "Marine Studies of San Pedro Bay, California" (D. F. Soule and M. Oguri, eds.), Part 15, pp. 1–226. Harbors Environ. Proj., University of Southern California, Los Angeles.

Soule, J. D. (1958). The case for Bryozoa and Ectoprocta. *Bull. Zool. Nomencl.* **15**(B), Doc. 25/17, 1070–1072.

Soule, J. D. (1959). Results of the *Puritan*-American Museum of Natural History Expedition to western Mexico. 6. Anascan Cheilostomata (Bryozoa) of the Gulf of California. *Am. Mus. Novit.* **1969**, 1–54.

Soule, J. D. (1961). Results of the *Puritan*-American Museum of Natural History Expedition to western Mexico. 13. Ascophoran Cheilostomata (Bryozoa) of the Gulf of California. *Am. Mus. Novit.* **2053**, 1–66.

Soule, J. D. (1963). Results of the *Puritan*-American Museum of Natural History Expedition to western Mexico. 16. Cyclostomata (Ectoprocta), and Entoprocta of the Gulf of California. *Am. Mus. Novit.* **2144**, 1–34.

Soule, J. D., and Soule, D. F. (1974). The bryozoan-coral interface on coral and coral reefs. *Proc. Int. Coral Reef Symp., 2nd, 1973,* Vol. I, pp. 335–340.

Soule, J. D., and Soule, D. F. (1977). Fouling and bioadhesion: Life strategies of bryozoans. *In* "Biology of Bryozoans" (R. M. Woollacott and R. L. Zimmer, eds.), pp. 437–457. Academic Press, New York.

Stebbing, T. R. R. (1911). The terms Polyzoa and Bryozoa. *Proc. Linn. Soc. London* **123**, 61–72.

Steemann, Nielsen, E. (1952). The use of radioactive carbon for measuring organic production in the sea. *Rapp. P.-V. Reun., Const. Int. Explor. Mer.* **144**, 92–95.

Straughan, D. (1975). Investigation of ovicell hyperphasia in bryozoans chronically exposed to natural oil seepage. *Water, Air, Soil Pollut.* **5**, 39–45.

Ström, R. (1977). Brooding patterns of bryozoans. *In* "Biology of Bryozoans" (R. M. Woollacott and R. L. Zimmer, eds.), pp. 23–55. Academic Press, New York.

Thompson, J. V. (1830). "Zoological Researches and Illustrations," Vol. V.

Vernberg, F. J., and Vernberg, W. B. (1969). Thermal influence on invertebrate respiration. *Chesapeake Sci.* **10**, 234–240.

Vernberg, F. J., and Vernberg, W. B., eds. (1974). "Pollution and Physiology of Marine Organisms." Academic Press, New York.

Vernberg, F. J., Calabrese, A., Thurberg, F. P., and Vernberg, W. B., eds. (1977). "Physiological Responses of Marine Biota to Pollutants." Academic Press, New York.

Waters, A. W. (1880). On the terms Bryozoa and Polyzoa. *Ann. Mag. Nat. Hist.* [5] **5**, 34–36.

Wilson, D. P. (1951). Larval metamorphosis and substratum. *Ann. Biol.* **27**, 491–501.

Winston, J. E. (1977). Distribution and ecology of estuarine ectoprocts: A critical review. *Chesapeake Sci.* **18**, 34–57.

Woollacott, R. M., and Zimmer, R. L. (1972). Origin and structure of the brood chamber in *Bugula neritina* (Bryozoa). *Mar. Biol.* **16,** 165–170.

Woollacott, R. M., and Zimmer, R. L., eds. (1977). "Biology of Bryozoans." Academic Press, New York.

Zimmer, R. L. (1973). Morphological and developmental affinities of the lophophorates. *In* "Living and Fossil Bryozoa" (G. P. Larwood, ed.), pp. 593–599. Academic Press, New York.

ZoBell, C. E., and Allan, E. C. (1935). The significance of marine bacteria in the fouling of submerged surfaces. *J. Bacteriol.* **29,** 239–251.

CHAPTER 3
Bristle Worms (Annelida: Polychaeta)

DONALD J. REISH

I. Introduction

Polychaetous annelids are segmented worms that generally bear many setae ("bristles") which usually arise from paired lateral extensions of the body (parapodia). Polychaetes comprise a class of phylum Annelida, which also includes the Oligochaeta (earthworms) and Hirudinea (leeches). Adult polychaetes measure from less than 1 mm to over 1 m in length. The external morphology is extremely varied and often specific for a family: many species appear much like earthworms (lumbrinerids and capitellids), while others may bear various types of appendages originating from the anterior end and/or along the sides of the body. The color of living specimens is generally tan or brown, but some, especially the feather dusters (sabellids and serpulids), may be brightly colored. Setae are extremely variable and may be species specific.

Polychaetes are conveniently separated into the errantiates (free moving) and sedentarians (tube dwellers); however, this distinction is superficial since some errantiates are tube dwellers and some sedentarians move about freely. Tubes are constructed from mucus, to which sediment may or may not adhere, or from calcium carbonate secretions from the collar region (serpulids). Tubes may in some cases be specific for the species.

A. Ecological Occurrence

Polychaetes are present in all oceanic regions of the world and in all habitats. A few species have been reported from fresh water, the sabellid *Manayunkia speciosa* being the best known in North America. Intertidal species occur along rocky shores where they are present within algal holdfasts, under rocks, or wherever some protection can be afforded.

Polychaetes are richly abundant within the sediment of intertidal mud flats, and to a much lesser extent within sandy beaches. The subtidal benthos, especially the soft bottom, is where polychaetes are the most numerous in terms of number of species and specimens. They occur in the pelagic environment either as larvae or as adults, and in the latter case, the species are adapted morphologically for swimming.

B. Importance of Polychaetes in the Marine Environment

Regardless of the type of measurement employed, i.e., number of species, number of specimens, standing crop, or productivity, polychaetes of the soft bottom community are either the dominant group or an important contributor to the particular parameter used (Knox, 1977). Wherever a sample is taken of the subtidal sediments, polychaetes are certain to be present.

Table I summarizes the findings by Sanders *et al.* (1965, as reported by Knox, 1977) of a transect off the North American coast from a depth of 97 to 5001 m. From these data the maximum number of polychaete specimens, nearly 5000 per square meter, were taken from depths of 200 to 400 m—with the number falling off rapidly to 14 per square meter at 5001 m depth. However, the percentage of polychaete species present within a sample did not follow this trend. While the percentage occurrence of polychaetes was highest also at the 200–400 m depth, the percentage of poly'·haete species present was usually 33 to 50 regardless of depth. The importance of polychaetes within the benthic community is established, therefore, as considerable, regardless of depth.

The relationship of the number of species and specimens to the total number of species and specimens in the community is summarized in Table II according to the geographical region. Except for the Arctic Basin all data presented are for relatively shallow water (0–157 m). When the variability due to screen size is discounted, polychaetes are important components of the benthic community in terms of both number of species and specimens, regardless of the geographical region.

The occurrence of polychaete specimens as a fraction of the total population varied from 0.9 to 98.9 with an average of 44.3% for 61 examples cited by Knox (1977). The lower figures were measured from intertidal or subtidal sandy beaches. These values ranged from 25.5 to 83.6% for the data included in Table II.

TABLE I

RELATIONSHIP OF THE NUMBER AND PERCENTAGE OF POLYCHAETE SPECIMENS PER SQUARE METER TO DEPTH[a]

Depth (m)	Number of polychaete specimens	Polychaetes (%)
97	1260	40.9
200	4982	77.2
400	3494	78.7
824	2201	73.1
1500	698	70.0
2086	737	65.8
2500	265	72.6
2873	358	56.3
3752	144	36.8
4540	137	57.9
4977	32	47.8
5001	14	51.9

[a] Screen size, 0.42 mm; data condensed from Knox, 1977 (after Sanders *et al.*, 1965).

TABLE II

COMPARISONS OF THE TOTAL NUMBERS AND TOTAL SPECIES WITH POLYCHAETE NUMBERS AND POLYCHAETE SPECIES WITH GEOGRAPHICAL REGION[a]

Locality	Depth (m)	Screen size (mm)	Faunal density (No./m²)	Polychaete density (No./m²)	Polychaete (%)	Total number of species	Number of polychaete species	Polychaete species (%)
Arctic Basin, Scotland		0.15	634	250	39.4	67	19	28.4
Buzzards Bay, Massachusetts	146	0.5	4435	1133	25.5	79	38	48.1
Beaufort, North Carolina	12–30	0.2	15,622	4139	26.5	79	25	31.5
San Juan, Washington	0	1.0	1407	412	39.4	6	2	33.3
Kingston, Jamaica	6–16	—	240	123	51.1	154	64	41.6
Northumberland, England	80	0.5	1080	903	83.6	66	40	60.6
Catalina Coast, Spain	157	1.0	258	170	65.9	449	251	50.3
Antarctic Peninsula	100	1.0	6987	2822	40.3	—	30	—

[a]Data condensed from Knox (1977).

A similar percentage was noted when the number of polychaete species was compared to the total number present. This percentage ranged from 28.4 to 60.6 in the data included in Table II; Knox (1977) reported an average of 44.1% for all 61 studies reported. Additional measures of the importance of polychaetes, while limited, show that these organisms comprise an average of 42.9% of the standing crop (biomass of some authors) and they accounted for over 70% of the productivity in benthic communities over a period of 1 year (Knox, 1977).

Through their feeding activities polychaetes play an important role in the movement of sediments. They may be deposit detrital feeders either at the surface layer or at various depths within the sediment. Suspension feeders entrap fine particulate matter on ciliated appendages. Both feeding types have an influence on sediment characteristics, especially the distribution of grain size, the entrapment of moving particles, and the aeration of soils beneath the surface layer. How important a role the polychaetes play in sediment transport is difficult to ascertain because of the limited data available. Knox (1977) summarized that a single deposit feeding polychaete passes from 96 to 400 ml of sediment through its gut per year, as compared to a clam passing 257–365 ml per year.

Perhaps the most neglected area of the ecological importance of polychaetes in the marine benthos is the part these organisms play as food for fish and birds. All too often the polychaetes present within a gut of these vertebrates are listed as unidentified polychaete material. Such material can be identified, but, for the most part, such analyses are made by the vertebrate zoologist who lacks the skills for identifying polychaetes.

Recently, a study was initiated (Reish and Ware, 1976) to determine what species of fish were feeding upon benthic polychaetes. Thus far the record stands at nearly 1000 specimens of *Armandia bioculata* taken from the gut of one white croaker collected in Los Angeles Harbor. Shore birds were found to feed upon the larger polychaetes during low tide in Bahia de San Quintin, Baja California (Reish, unpublished data). A total of 77 specimens of *Neanthes arenaceodentata* was taken from the crop of a dunlin.

In summary, the importance of the polychaetes in the marine benthos has been demonstrated in many ways. Regardless of the measure, be it number of species, specimens, or standing crop, polychaetes are consistently the most important group of organisms in this environment. With this conclusion one must insist that all marine benthic studies, whether academic pursuits or environmental studies, should include polychaetes and that they must be adequately identified.

C. IDENTIFICATION OF POLYCHAETES

The knowledge of polychaete systematics is still primarily in the "alpha" stage. New species are still being described from all parts of the world; for example, 234 new species of polychaetes were described in 1973 (Zoological Record, 1976). As a further indication of the state of polychaete systematics, Hartman (1969) recorded 701 species of polychaetes from California, yet during the first year of operation of the Bureau of Land Management's Outer Continental Shelf study of southern California at least 100 new species of polychaetes have been encountered (G. F. Jones, personal communication). Hartman (1969, p. 3) stated, "It is common to find unknown polychaetes in incidental samples, such as those taken in fisheries research, in studies concerned with population densities and pollution studies." Anyone initiating benthic pollution studies, especially in offshore waters, should be cognizant that new species will be encountered.

The ability to learn how to identify polychaetes is not difficult in itself. The primary attribute is the willingness to be patient. One must possess the ability to seek out and interpret minute morphological details. The procedures are similar to those used in other lesser known marine invertebrates such as sponges, hydroids, isopods, amphipods, and bryozoans.

Polychaetes are extremely varied in their external morphology, but usually the members of a family resemble each other. It is possible for the uninitiated to learn recognition of the most commonly occurring families within a short period of time.

Division of polychaetes into families generally relies on the larger, distinct morphological characters such as possession of branchiae, tentacles, or elytrae. Separation of genera within a family depends upon the particular family, but frequently the number and distribution of prostomial branchiae and tentacles, the distribution of unique structures such as elytrae, the shape of parapodial lobes, and the type and distribution of setae are used. Specific identification is also dependent upon the family, but the details of setae are the most widely used characteristic.

Identification of polychaetes requires the use of both dissecting and compound microscopes of excellent quality. The purchase of infrared microscopes, in the interest of economy, is false economy since it will be more difficult and thus more time consuming to identify the polychaetes to species. In a monitoring program in which more than one person is identifying polychaetes it is possible for two people to share one compound microscope. Microscope lights are equally important and in many instances two lights are helped to use with one dissecting microscope, especially if the specimens are less than 1 cm in length. Jeweler's forceps have proven to be the most useful of such instruments for manipulating and removing parapodia from the body for microscopic examination.

The recent appearance of Fauchald's key (1977a) to the polychaete worms will help alleviate much of the difficulty associated with identifying polychaetes. His key is a dichotomous one to the 85 families and subfamilies and is followed by keys to the genera within each group. Keys to species are not included (see below for further discussion). Each genus is defined, the genotype named, and the number of known species recorded. The invalid genera are named at the end of each family, with their deposition. The introductory material is especially useful to the novice, and includes description of techniques such as collection, fixation, preservation, and dissection. There are ample figures of polychaetes including an illustrated glossary of morphological characters.

Since Fauchald's key does not include keys to species, which would be a herculean task, the identifier should proceed to local keys. Unfortunately, local keys do not always exist, and some of those that do may be of limited value or out of date. Table III summarizes the regional references to polychaetes. Anyone undertaking to identify polychaetes from a particular geographical area should obtain copies of the pertinent references plus those for nearby geographical areas. In addition, copies of the literature on the polychaetous annelids, including the catalog to the species of polychaetes by Hartman (1951a, 1959, 1965), are essential references to augment regional keys. The acquisition of these references does not necessarily mean that the novice will soon be an accomplished polychaetologist; he or she will still need to learn anterior from posterior (it can be tricky with the maldanids) as well as the bits and pieces of worms.

If a large-scale monitoring program is contemplated by a staff which has not had previous experience in the identification of polychaetes, there are certain pitfalls that can be avoided to facilitate the acquisition of scientific competency in identifying polychaetes to species. Once the samples are collected, screened, preserved, and brought to the laboratory, the sample should be washed through a fine screen, i.e., a 0.246-mm mesh sieve, to remove formaldehyde and excess sediment. A portion of the sample should be placed under a dissecting micro- scope and the polychaetes removed. Some laboratories favor the use of a 1% solution of rose bengal, which stains polychaetes and other living organisms a pink to rose color and facilitates soriting; however, other laboratories frown on this practice because they believe that it does not facilitate sorting and that it discolors the specimens and masks any naturally occurring pigmentation. Ini- tially it is better to remove all polychaetes and fragments and place them in a separate dish or vial. Once the sample has been sorted to animal group, examine the polychaetes under the dissecting microscope and attempt to separate similar appearing ones together. Once this is accomplished, take one species and attempt to identify it first to family, then genus, and finally species. Do not attempt to identify a poorly preserved or fragmented specimen; it may be possible later as experience is gained. The initial goal of a novice identifier should be the im- mediate recognition of the approximately 40 principal families. Once the iden- tifier has mastered this task, the major hurdle to identifying polychaetes is over. Every identifier will eventually discover that he had been attempting to identify a palp, a tentacle, a fragment, a posterior end, etc., and experience, in this case, is the best teacher.

TABLE III

SUMMARY OF THE REFERENCES TO POLYCHAETOUS ANNELIDS BY GEOGRAPHICAL REGION

Geographical region	Key	Notes	Reference
Alaska	+	Keys to some species	Hartman (1948)
	+	Includes North Atlantic	Pettibone (1954)
British Columbia	+		Berkeley and Berkeley (1948, 1952)
British Columbia and Washington	+	Errantiate only	Banse and Hobson (1974)
Oregon	+		Hartman and Reish (1950)
California	+		Hartman (1968, 1969)
Hawaiian Islands	+		Hartman (1966a)
Gulf of California	−	List all known species	Reish (1968a); Kudinov (1975a,b)
Canadian Atlantic	−		Treadwell (1948)
New England	+	Errantiate only	Pettibone (1963)
North Carolina	+		Day (1973)
Gulf of Mexico	+	Errantiate only	Gardiner (1976)
Panama	+		Hartman (1951b)
France	+		Fauchald (1977b)
Germany	+	Classic; in French	Fauvel (1923, 1927)
South Africa	+	In German	Hartmann-Schröder (1971)
U.S.S.R.	+	Translated into English	Day (1967)
Japan	+		Ushakov (1965)
Vietnam	+		Imajima and Hartman (1964)
India	+		Gallardo (1967)
Marshall Islands	+		Fauvel (1953)
Antarctica	+		Reish (1968b)
	+		Hartman (1964, 1966b)

II. The Relationship between Ecological Factors and Distribution of Polychaetes

Many environmental factors play roles in determining the distribution of natural populations of polychaetes. Undoubtedly, many of these factors are inter-related and may have synergistic effects, but thus far the study of such relationships has been minimal. The principal abiotic factors include sediment characters, salinity, temperature, depth, and dissolved oxygen concentration of the overlying water. Biotic factors which have been studied include food availability and intra- and interspecies compositions. Each species is capable of withstanding a certain amount of variation of a particular ecological factor. Stress on the species is exerted at the extremes of the factor, where the continual existence of species is precarious at best. Populations of the species at the extremes may be transitory, moving in or out of an area, or they may be vegetative populations which are unable to reproduce under the stressed conditions. In contrast to the extremes, there is some region where conditions are ideal for the species; it is here that the maximum number of individuals is found and where reproduction occurs.

In the absence of man-made changes, ideal conditions do not always necessarily remain ideal. Long-term changes in one or more natural environmental factor(s) may occur which could determine the future distribution of the species. Just because a particular polychaete association does not occur at a specific locality year after year does not necessarily mean that some man-made change caused this change. All too often man's activity is cited as the cause of an alteration of the environment, without first having a sound scientific basis for such a conclusion. The ability to make a valid scientific assessment of a situation is the primary reason for the importance of monitoring programs, as discussed below.

A. SEDIMENT

The particle size of the sediment is perhaps the single most important ecological factor influencing the distribution of polychaete annelids. Earlier studies characterized sediment characteristics subjectively, for example, sand, muddy sand, sandy mud, and fine muds. More recent studies have characterized sediments according to the diameter of the particle as measured by standardized sieves; data are reported as percentages according to the Udden–Wentworth size classification (Barnes, 1959). Since most of the sediments in the subtidal environments are within the sand, silt, and clay groups, these data are represented diagrammatically as an equilateral triangle with each sediment type at a point. Data are placed on the diagram on the basis of the percentage composition of these sediments (Emery, 1960). Precise characterization of sediments is especially important in estuaries and shallow offshore waters since this is the region where the greatest sediment variability occurs. While the distribution of any species is dependent upon many environmental factors, unifactorial analysis, such as particle size, gives us correlations which are useful in interpolating data.

Earlier studies on the relationship of polychaete species to particle size were generally analyzed visually by the researcher, who considered the optimum particle size for a species to be that which has the greatest number of individuals present. Computers now analyze these data in much the same manner. The relationship of benthic polychaetes to sediment particle size was studied in Bahiá de San Quentín, Baja California, by Reish (1963a). Some of these findings are summarized in Table IV. The majority of the benthic species preferred silty sediments to the very fine or fine sands. In most instances, for example *Prionospio malmgreni,* trends were noted; that is, an average of 107 specimens was taken from stations characterized by silts, 57 specimens from stations with very fine sands, and only 8 from stations with fine sands. A few species such as *Capitita ambiseta* and *Scoloplos acmeceps,* apparently preferred very fine sands. No numerically dominant species preferred fine sands to the other sediment types.

Sanders (1960) found that the percentage of clay present within a sediment sample was important in determining the distribution of infauna such as *Nephtys incisa.* The greatest number of individuals of this species were present when the percentage clay present ranged from 10 to 20. Since clays are smaller than silt particles and therefore have a larger surface area per unit weight, they would tend

TABLE IV

Average Number of Specimens per Sample Polychaetes According to Sediment Type in Bahiá de San Quentín, Baja, California[a]

Species	Silt	Very fine sand	Fine sand
Brania clavata	8	7	3
Exogone verugera	205	36	104
Neanthes candata	14	31	8
Lumbrineris minima	10	5	1
Scoloplos ohlini	6	13	2
S. acmeceps	18	33	25
Prionospio malmgreni	107	57	8
Cossura candida	101	21	4
Capitita ambiseta	85	164	1
Axiothella rubrocincta	3	10	1
Pista alata	42	5	1
Fabricia limnicola	101	42	34
Megalomma pigmentum	17	5	9

[a]Data condensed from Reish (1963a).

to bind a greater quantity of organic matter, which is the food for deposit feeders such as *Nephtys*.

Sediment characteristics, especially within estuaries or nearshore environments, and hence benthic polychaete populations, can be altered by natural and man-made events. Heavy river runoffs from rains or spring thaws can alter the sediment characteristics by deposition of new sediments or resuspension of existing ones. Discharge of industrial or domestic effluents can alter the physical as well as chemical characteristics of the sediments. Construction of breakwaters, jetties, etc., can alter the existing current patterns, which in turn may alter the sediment characteristics.

The importance of particle size in settlement of polychaete larvae, as well as other larvae, was demonstrated by Wilson (1952, 1955) working with *Ophelia bicornis*. The larvae of this species, in order to settle and metamorphose, required clean sands with an optimum particle size between 0.2 and 0.45 mm, to which a moderate population of microorganisms adhere. If such an environment is provided, then larvae undergo settlement and metamorphosis. However, in the absence of such conditions, the larvae are able to postpone settlement for an extended period of time. Obviously, however, the longer the period of time that metamorphosis and settlement are delayed, the greater the possibility that the larvae can be eaten.

B. SALINITY

Decrease in salinity can become a factor in determining the distribution of polychaetes, especially within estuaries. The salinity concentration of an estuary can be extremely variable and is dependent upon the size and configuration of the estuary, the tidal amplitude, and the amount and frequency of freshwater runoff. Salinity measurements are typically made of the overlaying water, which may or may not reflect the actual salinity of the sediment (Sanders *et al.*, 1965). In general, as one proceeds upstream in an estuary the number of marine species, including polychaetes, decreases, and these in turn are replaced by an increasing number of freshwater species (Alexander *et al.*, 1935).

Smith (1955) found the intertidal distribution of *Nereis diversicolor* to be restricted to a brackish water zone produced by underground seepage of fresh water. The polychaete was absent from similar nearby areas that lacked this seepage. The influence of salinity on benthic populations was shown by Wolff (1971) in a Dutch estuary. The river entering this estuary was dammed, which resulted in an increase in the salinity. Marine faunal components, including polychaetes, were able to inhabit and survive in areas where they were previously absent. Rainfall in excess of 1.25 cm either reduced the population or killed one or more intertidal species of polychaetes. Greater amounts of rainfall over a longer period of time completely eliminated *Capitella capitata, Strebispio bene-*

dicti, and *Polydora muchalis.* However, repopulation was rapid since these species also occur in subtidal waters and are able to reproduce throughout the year in southern California waters (Stone and Reish, 1965).

Salinity values at any given point can be more or less the same, with resulting stable populations of polychaetes and other infaunal groups—or it can vary considerably depending upon the freshwater input, with a resultant widely fluctuating numbers of species and specimens. It is therefore imperative to know whether the particular estuary is stable or fluctuates before an environmental assessment can be made. If the estuary is found to have a widely fluctuating population, it is important to know the biological conditions for at least a period of 1 year.

C. TEMPERATURE

The temperature of the water can affect polychaetes in a variety of ways from the cellular to the community level. The influence of temperature upon marine invertebrates has been extensively reviewed by Kinne (1970). Less is known about the influence of temperature on polychaetes, as compared with such organisms as pelecypods, decapods, or echinoderms. The geographical distribution of polychaetes is influenced by temperature as it is for other marine invertebrates. Many distributional patterns in polychaetes, such as the faunal break at Point Conception, California, can be related to geography of the land mass and ocean current patterns, which in turn influence the temperature of the water.

Experimental studies of various types have been conducted with temperature in hopes of elucidating its relationship to polychaetes. The heat tolerance of polychaetes, generally measured as a TL_m value, has been measured with respect to vertical and geographical distribution. De Silva (1967) found the tolerance to extremes in temperature to be greatest in *Spirorbis pagenstecheri,* which extended highest in the tide horizon, and least in *S. corallinae,* which was always subtidal. Two additional species of *Spirorbis,* which had distributions intermediate of the two, had corresponding tolerance ranges midway between the two. The heat tolerance was higher for *Diopatra cuprea* and *Clymenella torquata* collected from North Carolina than from Massachusetts; however, the Massachusetts populations could tolerate lower temperatures than those from North Carolina.

The influence of temperature on reproduction and spawning has been documented in polychaetes and the findings follow what has been observed more thoroughly in other marine invertebrates. A laboratory population of *Neanthes arenaceodentata* which is maintained at 20°C will complete its life cycle in 5 months, but requires 8 months for reproduction at 15°C. Furthermore, fewer specimens produce fewer number of eggs at the lower temperature. *Neanthes* did not survive experimental temperatures of 10°C (Reish and Gerlinger, 1977).

Epitokal swarming in *Neanthes succinea* occurred at temperatures between 16° and 21°C and rarely below (Kinne, 1954).

Generally, spawning in polychaetes is related to the spring rise in temperature. Spawning in *Nereis grubei* occurs throughout the year in southern California (Reish, 1954) but only between February and June in the central California population of the same species (Schroeder, 1968). Undoubtedly, heated effluents from electrical generating stations could influence polychaete distribution and survival, but specific studies, especially those involving predischarge characteristics, have not been studied with this group in mind.

D. Depth

Little is known of the relationship between water depth and polychaetes other than the range at which it occurs. Some species of polychaetes may be restricted to a specific zone in the intertidal environments, while others may extend from the intertidal zone to a depth of 1000 m or more. Since depth, temperature, and sediment characteristics are closely related, studies are usually focused on the latter two parameters. Since pressure is related to depth, studies on the effect of pressure on polychaetes have been limited (Knight-Jones and Qasim, 1955; Naroska, 1968).

E. Dissolved Oxygen

The amount of dissolved oxygen present within the water is perhaps the single most important environmental parameter governing not only the distribution of polychaete species, but also whether or not any species can live at a particular locality. A concentration of 5.0 mg/liter dissolved oxygen present within the water column is considered the minimal amount necessary to sustain fish, and is probably sufficient for most polychaetes. The dissolved oxygen content of the water mass can be easily measured by chemical or electronic methods. It is possible to measure the amount of dissolved oxygen present in subtidal quiet waters within 3.0 cm of the substrate–water interface, but measurement at the interface or within the interstitial water where the polychaetes live has not been done.

Generally, the dissolved oxygen content of the water mass has not been considered to be a limiting factor to polychaetes because, in most instances, an adequate supply is present. Whenever oxygen does become a limiting factor, it is associated with protected waters and pollution. However, oxygen depletion and mass mortality of benthic organisms in natural areas has been reported (Reish, 1963b; Tulkki, 1965).

The minimum amount of dissolved oxygen necessary for survival, as determined by the 28-day LC_{50}, was measured under laboratory conditions for four species of polychaetes which were indicative of varying ecological conditions (Reish, 1967). The 28-day LC_{50} was 2.95 mg/liter dissolved oxygen for *Nereis grubei,* an indicator of the healthy zone; 0.9 and 0.65 mg/liter for *Neanthes arenaceodentata* and *Dorvillea articulata,* respectively, both indicators of the semihealthy zone; and 1.5 mg/liter for *Capitella capitata,* the polluted zone indicator. Apparently, judging from these data, some factor other than dissolved oxygen accounts for the exclusion of the semihealthy zone indicators from the polluted areas.

Field experiments were conducted to measure different criteria of survival (Reish and Barnard, 1960). Male and female *Capitella* were suspended together in plastic traps and placed in various field localities of different dissolved oxygen concentrations. Various degrees of responses were noted: in a level with little or no oxygen the animals did not feed and died within a few days; in an intermediate level, with a median oxygen level of 2.9 mg/liter, the animals fed but did not reproduce; and in an upper level, in which the median oxygen was 3.5 mg/liter and the animals were able to reproduce.

Laboratory experiments were conducted on an inbred strain of *Neanthes arenaecodentata* under experimentally reduced dissolved oxygen concentrations to determine whether or not sublethal effects could be induced and measured (Raps and Reish, 1971; Abati and Reish, 1972; Cripps and Reish, 1973; Davis

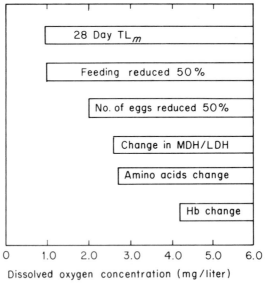

Fig. 1. The effect of reduced dissolved oxygen concentration on various biological activities in *Neanthes arenaceodentata.* (After Reish, 1974a.)

and Reish, 1975). Since it had been determined earlier that the 28-day LC_{50} of dissolved oxygen to *Neanthes* was 0.9 mg/liter (Reish, 1967), the basis for these experimental studies was the working hypothesis that subtle changes must be occurring within the organism prior to its actual death. Subtle changes were induced by experimentally reducing the amount of dissolved oxygen present in its environment. These findings are summarized in Fig. 1 (Reish, 1974a). It can be seen from this summary figure that different physiological processes are affected by different amounts of dissolved oxygen present, with feeding the least sensitive of the measured processes and an increase in the production of hemoglobin being the most sensitive. It is readily apparent in these laboratory studies that a determination of the LC_{50} to reduced dissolved oxygen conditions does not adequately indicate what is actually occurring within the organism.

III. Polychaetes as Indicators of Pollution

A. HISTORICAL

The use of organisms as indicators of various degrees of marine water quality is based on the ecological understanding that a natural environment is characterized by balanced biological conditions and contains a great diversity of plant and animal life with no one species dominating. This conclusion is particularly true within the marine photic zone whenever a variety of habitats are present. If a polluting substance is emptied into the region, then the most sensitive organisms are killed, which eliminates the enemies of the more tolerant species. In the absence of predators, the population of tolerant species increases in numbers of specimens. If the amount of polluting substance is increased, a further reduction in the number of species occurs, with the surviving ones increasing further in number. If the quantity of polluting substance is great, all surviving species may be killed, leaving an area totally devoid of macroscopic life. Therefore, a knowledge of what species are present is of paramount importance in evaluating the effect of the polluting substance.

The use of benthic organisms in evaluating the effects of pollution has been favored by many biologists because the organisms on or within the substratum will reflect the environmental conditions of the water prior to the time of sampling, whereas analysis of the water mass for planktonic organisms will reflect the conditions only at the time of sampling. The importance of benthic organisms in assessing the environmental condition of an area was well established in the freshwater studies by Kolkwitz and Mansson (1908), Forbes and Richardson (1913), Patrick (1949), and by Gaufin and Tarzwell (1952). The same generalities have been found to hold true for the marine environment, as will be discussed below.

The use of polychaete annelids as indicators of various degrees of marine pollution is a recent development, largely within the past 25 years. It is only logical that such studies followed and used similar criteria as had been developed in the freshwater environment. Wilhelmi (1916) made first reference to the use of polychates as indicators. He stated that the polychaete *Capitella capitata* played a similar role in marine waters as the oligochaete *Tubifex* does in the fresh waters of Germany. He followed the saprobic system developed earlier by Kolkwitz and Mansson (1908). *Capitella* was the principle member of polysaprobic polluted zone community, following the concept of Kolkwitz and Mansson. No further studies were undertaken by Wilhelmi on this topic, and his early observations were apparently not cited in the literature for some 40 years.

As an offshoot from the early Danish benthic studies by Petersen and his students, Blegvad (1932) sampled the bottom-dwelling communities around the Copenhagen domestic waste discharge within the harbor. He was able to divide the region surrounding the outfall into three zones: an inner zone lacking benthic animals, with the substratum characterized by a sulfide odor; an intermediate one containing a few animal species, most of which were polychaetes; and a third zone showing no measurable effects of the discharge, and possessing natural benthic populations.

B. Protected Bays

The modern day use of the indicator organism concept in marine waters began with the concurrent studies of Filice (1954a,b) working in San Francisco Bay and Reish (1955) working in Los Angeles–Long Beach (Anonymous, 1952) and nearby areas. Both of these studies were initially funded by the then State Water Pollution Control Boards. Filice separated the Castro Creek region of San Francisco Bay into ecological areas similar to Blegvad: an inner zone largely lacking benthic animals, and an intermediate one characterized by a few tolerant species (notably the polychaetes *Capitella capitata, Neanthes succinea,* and *Streblospio benedicti* and the mollusks *Mya arenaria* and *Macoma inconspicua*).

Reish (1955, 1959) based his indicator concept of various degrees of pollution entirely upon the use of polychaetes. This emphasis on the use of polychaetes was based in part by the primary research interest of the investigator and in part because of the dominance of polychaetes in the benthos of Los Angeles–Long Beach Harbors. The terminology used by Reish follows that employed by Patrick (1949) in her studies of the streams in the Conestoga Basin, Pennsylvania. On the basis of benthic polychaete populations it was possible to divide Los Angeles–Long Beach Harbor into five ecological areas, characterized as healthy bottom, polluted bottom, and very polluted bottom (Fig. 2). In addition, these areas were characterized by chemical and physical conditions of the water and sediment (Table V).

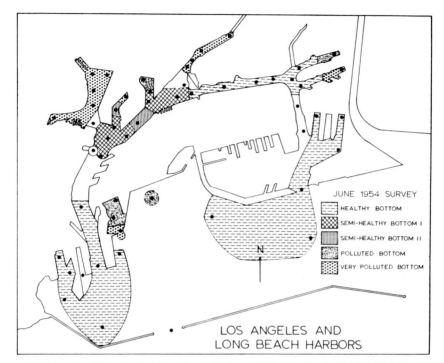

Fig. 2. Map of Los Angeles–Long Beach Harbors, California, showing the benthic biological conditions in June, 1954. (After Reish, 1959.)

The healthy bottom was characterized by a diversity of marine invertebrates, with three polychaete species, *Tharyx parvus, Cossura candida,* and *Nereis procera* being the most common forms present. An adequate supply of dissolved oxygen was present in the deeper waters, and the sediment was generally low in organic carbon content and lacked noticeable odors.

Semihealthy bottom Zone I was distinguished by the presence of the *Polydora (C.) paucibranchiata* and *Dorvillea articulata* (= *Stauronereis rudolphi* or *Schistomerigos longicornis* of recent authors). The dissolved oxygen content was suppressed, with an average reading of 3.2 mg/liter near the bottom. Some of the sediment samples possessed sulfide odors but the organic carbon content was low.

Semihealthy Zone II was similar to semihealthy Zone I, with the exception of the dominance by the large polychaete *Cirriformia luxuriosa* and the general absence of *Polydora* and *Dorvillea*. The polluted bottom was characterized by the polychaete *Capitella capitata,* frequently in large numbers. Of particular importance was that the presence of *Capitella* was correlated with areas receiv-

TABLE V

SUMMARY OF BIOLOGICAL, CHEMICAL, AND PHYSICAL CHARACTERISTICS OF FIVE ECOLOGICAL AREAS OF LOS ANGELES–LONG BEACH HARBORS[a,b]

Characteristic	Healthy bottom, *Tharyx parvus, Cossura candida, Nereis procera*	Semihealthy bottom I, *Polydora paucibranchiata, Dorvillea articulata*	Semihealthy bottom II, *Cirriformia luxuriosa*	Polluted bottom, *Capitella capitata*	Very polluted bottom, no animals
Number of animal species (average)					
Polychaetes	7	5	5	3	0
Nonpolychaetes	3	2	2	2	0
Dissolved oxygen (ppm) (median)					
Surface	6.0	2.5	2.5	3.5	1.6
20-ft depth	6.0	3.2	3.2	3.5	2.2
pH (median)					
Surface	7.8	7.3	7.4	7.6	7.5
20-ft depth	7.8	7.4	7.6	7.6	7.5
Substrate	7.2	7.2	7.2	7.3	7.1
Nature of substrate (in order of importance)	Gray mud, black mud, black sulfide mud	Black sulfide mud, gray clay, sand, and mud, black mud	Black sulfide mud, gray clay, black mud	Black sulfide mud	Black sulfide mud
Organic carbon of substrate (%) (median)	2.5	2.0	2.7	2.7	3.4

[a]Data from Reish (1959).
[b]Dominant species of polychaete.

ing waste discharges of biological origin such as fish cannery wastes or domestic sewage. The sediment always possessed a strong sulfide odor, but the average deep-water dissolved oxygen reading of 3.5 mg/liter was higher than observed within the semihealthy zones.

No macroscopic animal life was present within the benthos of the very polluted zones. The dissolved oxygen in the deeper waters averaged 2.2 mg/liter but at some stations was absent. The organic content of the sediment measured as high as 10.7%, with an average of 4%. The sediment at these stations always possessed a strong sulfide odor.

Summarizing these generalities set down by the author (1959), which in turn have been the basis for subsequent studies dealing with polychaetes as indicators of pollution, we find that as one proceeds from the healthy zone to the very polluted zone there is a change in the polychaete species composition, a reduction in the number of polychaete species present until finally there is an elimination of all species, a reduction in the amount of dissolved oxygen present within the water near the bottom, an increase in the amount of organic carbon content of the sediment, and an increase in the frequency of sediment possessing sulfide odors. While the selection and distribution of these polychaetes into these five zones was based on nonmathematical analyses, the author was particularly gratified to see Boesch (1977) draw similar conclusions using the same data (Reish, 1959, Tables 16–21) and subjecting it to computer analysis. Furthermore, it points out the usefulness of publishing such raw data.

C. OFFSHORE WATERS

A different ecological response is noted in offshore waters, where you typically find only three zones based on the distance from a domestic outfall sewer (Turner et al., 1968). An inner zone with impoverished benthic fauna and Capitella capitata dominating, an intermediate zone with an enriched benthic fauna, and an outer zone with a natural benthic population.

The previous use of polychaetes as indicators of various degrees of marine pollution is summarized in Table VI according to geographic region, the type of benthic environmental response, and whether or not Capitella capitata was present. The author realizes that this list is by no means exhaustive; undoubtedly many reports from governmental agencies and private consulting firms as well as published studies were overlooked. Initial perusal of Table VI indicates that most of the studies have been done in the Northern Hemisphere, especially in North America and Europe. In most instances, different degrees of pollution, or zones, were observed. The number of zones, which was based on either a subjective or computer analysis of data, ranged from two to six different associations of benthic animals. Variation in the number of zones was dependent on whether or not an abiotic zone was present, the type of environment studied, and the mag-

TABLE VI

THE USE OF POLYCHAETOUS ANNELIDS AS INDICATORS OF MARINE POLLUTION

Geographical region	Benthic environmental response[a]	Presence of *Capitella capitata*	Reference
British Columbia	3	+	Balch *et al.* (1976)
California	3	+	Filice (1954a,b)
	5	+	Reish (1955, 1959, 1960, 1973)
	3	+	Reish (1961, 1973)
	6	+	Hartman (1956)
	—	+	Turner *et al.* (1968)
	3	+	Crippen and Reish (1969)
	4	+	Hill and Reish (1975)
	2	+	Greene (1976)
Baja California	5	+	Lizarrage-Partida (1974)
Chile	3	+	Gallardo *et al.* (1972)
Japan	4	+	Kitamori (1972)
Massachusetts	—	+	Grassle and Grassle (1974)
New York	3	+	Smith (1971)
	3	+	Ristich *et al.* (1977)
Jamaica	3	+	Wade (1972, 1976)
England	4	+	Pearson and Rosenberg (1976)
	2	+	Gray (1976)
	3	+	Wharfe (1977)
	2	+	Halcrow *et al.* (1973)
Denmark	3	−	Blegvad (1932)
	—	+	Bonde (1967)
Denmark–Sweden	4	+	Hendriksson (1969)
Sweden	3	+	Bagge (1969)
	—	−	Olsson *et al.* (1973)
	5	+	Leppäkoski (1975)
	4	+	Pearson and Rosenberg (1976)
	3	+	Rosenberg (1974, 1976)
Finland	3	−	Tulkki (1965)
Baltic Sea			Melvessalo *et al.* (1975)
Germany	4	+	Wilhelmi (1916)
France	2	+	Gilet (1960)
	2	+	Bellan (1964)
	3	+	Bellan and Peres (1970)
	3	+	Bellan *et al.* (1975)
	4	+	Bellan (1967)
	3	+	Arnoux *et al.* (1973)
Italy	4	+	Cognetti (1972)
South Africa	—	−	Oliff *et al.* (1967)
	3	−	Christie and Moldan (1977)
Texas	3	−	Holland *et al.* (1973)

[a] Number of ecological zones.

nitude of the study. Therefore, if an abiotic zone or very polluted zone is present, four different ecological zones are noted: (1) an abiotic zone; (2) a polluted zone with *Capitella capitata* dominating; (3) semihealthy assemblage or polychaetes that have generally some spionids and dorvilleids present; and (4) a healthy or natural population unaffected by pollution.

Capitella capitata was found to be present in the majority of the reports summarized in Table VI. It has been observed in both areas polluted by domestic as well as industrial wastes. However, *Capitella* has been reported to thrive in areas receiving wastes of biological origin, i.e., domestic sewage or fish cannery wastes (Reish, 1957). *Capitella* is most certainly the most commonly used and commonly referred to polychaete as an indicator of the polluted marine environment; it is also probably the most commonly used marine organism as an indicator of marine pollution.

D. VALIDITY OF POLYCHAETES AS INDICATORS

The usefulness of organisms as indicators of pollution has been questioned since the concept was introduced by Kolkwitz and Mansson (1908) for freshwater organisms. Certainly there has been a considerable amount of discussion on the usefulness of marine organisms as indicators of pollution with much of the attention directed toward *Capitella capitata*. Actually, if the various arguments against the use of *Capitella* as an indicator of the polluted environment are adequately addressed, the validity of the entire indicator concept might be resolved. This question is obviously one that cannot be answered simply in an unqualified yes, *Capitella* is an indicator of marine pollution or, no it is not. *Capitella* is a definite indicator under certain conditions and definitely not under others. The ability to tell when it is and when it is not requires the scientific judgment of the investigator. It is possible for an experienced pollution biologist to make such an assessment. In an attempt to impart this ability to make this assessment to others, it is necessary to discuss some of the arguments for and against the use of *Capitella* as an indicator.

One of the underlying reasons for the use of the indicator species concept is the desire for a quick answer to the question of whether or not a particular area is polluted, and if so, how severely. It is hoped by some that this assessment could be made in the field at the time of sampling. It is quite possible to make such an evaluation, with reservations, particularly if the area has been well studied. If such an evaluation could not be made in the field, then perhaps its presence or absence could be noted by examination of the sample in the laboratory. The primary reason here is to eliminate the expense of time and money of sorting and identifying all the organisms in the sample.

Perhaps the most frequently cited argument against the use of *Capitella* as an indicator is its occurrence, often in large numbers, in a estuarine area far from any source of pollution. *Capitella* is a normally occurring inhabitant of a natural unpolluted estuary, and it is logical to assume, in the absence of any fossil record, that it occurred on earth long before man evolved. The particular niche which *Capitella* occupies within the estuary is that region where the water is neither marine nor fresh water. This region is characterized by a reduced number of species because the water is too fresh for marine species and too saline for freshwater species (Alexander *et al.*, 1935). Many authors refer to *Capitella* as an opportunistic species (e.g., Grassle and Grassle, 1974). Since *Capitella* has a short life history and is capable of reproducing throughout the year, at least in some localities (Reish, 1961, 1974b), it is capable of colonizing an area rapidly whenever some change occurs in the environment be it a natural phenomenon (Reish, 1973) or man-made disaster (Grassle and Grassle, 1974). In the former case *Capitella* was the first benthic organism to settle on the subtidal benthos following a die-off of the existing population caused by a depletion of the dissolved oxygen as a result of an extensive red-tide bloom (Reish, 1973). Within 8 months the *Capitella* population had been reduced to its previous levels and the normal association of benthic polychaetes had been reestablished. The mechanism by which this population of *Capitella* was reduced is unknown. The majority of species which replaced *Capitella* are either filter or detrital feeders (i.e., *Capitita ambiseta, Prionospio cirrifera, Euchone limnicola,* and *Haploscoloplos elongatus*); however, it is possible, but not known, that species such as *Glycea americana, Goniada littorea, Nephytys caecoides,* and a nemertean are carnivorous and feed upon *Capitella* (Reish, 1963b).

In the case of a man-made disaster, a large population of *Capitella* settled in Wild Harbor, Massachusetts, following an oil spill (Grassle and Grassle, 1974). In both the natural and man-related changes in the benthic environment, an environmental disturbance occurred which killed or reduced the existing population significantly. Since *Capitella* has a short life history, in the absence of competition it was able to inhabit these disturbed environments and build up a population rapidly.

As discussed above, *Capitella* is an inhabitant of an unpolluted estuary but not a normal inhabitant of the offshore benthos. Therefore, if *Capitella* is taken in large numbers from the benthos within an estuary or a harbor, then it is important to determine why such a population occurs at that locality. Is it because it is the expected species in this particular environment or is it the result of some change caused by man? If it is initially ascertained to be the result of some man-related activity, then further analyses or additional sampling are required in order to answer this question with adequate scientific authority. If *Capitella* occurs in offshore benthic sampling, then it is important to investigate the possible cause of the occurrence of this population. Is it the result of an underground freshwater spring which caused a reduced salinity (Hartman, 1960), or a buildup of plant debris at the head of a canyon (Hartman, 1963)? Both of these could be considered natural phenomena. These are the type of questions which arise with the occurrence of a large population of *Capitella* in the marine environment and these are the type of questions which must be answered cautiously and with scientific judgment. Once those situations in which the occurrence of *Capitella* in a particular area has been established as a natural or a man-related phenomenon, then the usefulness and validity of *Capitella* as an indicator of marine pollution can be employed with reasonable assurance. Certainly the occurrence of *Capitella* in previous marine pollution studies, as summarized in Table VI, indicates its usefulness and validity as an indicator of the polluted environment. I am well aware that the validity of *Capitella* as an indicator will not be resolved by this discussion, especially since the usefulness of certain freshwater organisms as indicators in that environment has been questioned throughout much of this century and since *Capitella* has really only been employed as a marine pollution indicator for 20 years.

E. Use of Polychaetes in Monitoring Programs

The primary purpose of benthic monitoring programs is to determine whether or not the waste discharge is adversely affecting the environmental conditions in the area. Monitoring programs are long-range studies which include, in addition to a determination of the biological conditions, the chemical, geological, and physical characteristics of both the water mass and the sediment. The magnitude of the monitoring program depends upon the size and nature of the discharge(s), the oceanographic conditions of the area, and the pressure exerted by governmental agencies as well as by an enlightened population.

Results from long-term monitoring programs are limited to a few studies, primarily because of limited interest in such programs in the past. However, such programs are becoming more common in recent years because of the heightened general interest of the population in environmental matters. A knowledge of the benthos plays an important role in any monitoring program because this area reflects what the ecological conditions have been for some time previous to the time of sampling.

1. Monitoring a New Outfall Sewer

If a new waste discharge system is planned for an area that previously received waste water, it is important that a monitoring program be established. It is also important to have a knowledge of the predischarge conditions, or that the baseline be measured prior to emptying wastes into the area. The magnitude of the baseline study depends upon the volume and nature of the discharge, as well as the local oceanographic conditions. Benthic analyses should be made for the polychaetes as well as other marine invertebrates. Again, depending upon the size of the new discharge, it is advantageous to have a minimum of 1 year of data prior to initiation of operations. Careful identifications should be made of the polychaetes in these initial samplings since, especially in the case of a contemplated large discharge; it is quite possible that the biological staff is a new one which only has limited experience. All specimens should be retained, since it is possible that as the staff gains experience in the identification of the polychaetes of the area, earlier mistakes may be discovered. As a further aid to ensure accurate identifications, it may be advantageous to employ a competent polychaete systematist to verify identifications made by the staff. The use of a consultant will hasten the acquiring of staff competency. After the accumulation of 1 year of data, the information should be analyzed and benthic communities determined, preferably by computer program. Indicator organisms of each community, such as advocated by Thorson (1956), can be established as a convenience in distinguishing the various communities.

After discharge operations commence, the same stations should be sampled at the same time interval. Depending upon the nature and amount of the discharge as well as local oceanographic conditions, sampling should continue at this rate for 2 years. Data analyses should be on a continual basis, looking especially for any changes in species composition or community structure. If changes are observed, then it is important to determine, if possible, whether or not the changes are the result of the waste discharge or because of a natural change in the environment. Additional samples should be taken from a sufficient distance from the discharge to serve as "control" stations. After about 2 years of postdischarge data gathering, a thorough analysis of all data should be made to determine if the sampling program is adequate and to ascertain what samples are necessary and if any can be eliminated. A monitoring program should not be merely an exercise in collecting data; the data must be meaningful. It is possible that some stations and/or some parameters should be eliminated or the frequency of sampling be reduced. Special attention should be noted if polychaetes such as *Capitella capitata* or *Stauronereis rudolphi* appear in the postdischarge benthic samples.

The monitoring program conducted at Pt. Loma, California by the City of San Diego is perhaps the best known study of a large-scale operation in which conditions were measured prior to the commencement of operation of a primary treatment plant in 1963. A benthic survey was made in 1959 and 1962 prior to the construction of the sewer. The survey was repeated in 1965 and more or less continuously since that time. Unfortunately, many of these data have not been published, nor have the benthic organisms been stressed (Turner *et al.*, 1968; Chen, 1970). *Capitella capitata*, which was not present during the predischarge period, was present at the discharge point 2 years following commencement of discharge operations. The presence of this worm may have indicated the beginning of deterioration of the benthic environment.

2. *Monitoring an Existing Discharge*

In this case no data are available on the predischarge environment. Unfortunately, this is the rule rather than the exception. Los Angeles County discharges approximately 351 million gallons per day of primary effluent and 1.9 million gallons per day of digested sludge concentrate off the Palos Verdes coast (Shafer, 1977). The benthic monitoring program was started in 1971, over 35 years after the first wastes were discharged into the area. Three distinct benthic communities (Fig. 3) were noted, with *Capitella capitata, Nereis procera,* and *Ophiodromus pugettensis* located at the terminus of the different sewers. A second group was distributed near the discharge as well as up and down the coast; this group was characterized by *Caulleriella hamata* and *Telepsavus costarum.* The third group, dominated by many species of polychaetes, did not occur near the discharges. In the past 2 or 3 years the sanitation district has been upgrading the nature of their disposal, and the benthos has reflected this change. The community with *Capitella* has been reduced in area as well as in number of specimens, and the other two groups have moved in closer to the discharge point. Although, as Greene (1976) stated, the chemical nature of the sediments is still far from natural, the observed biological changes indicate that conditions are improving. It can be seen that the fruits of many years of monitoring this marine environment are beginning to be meaningful. Changes in the benthic communities were rapid

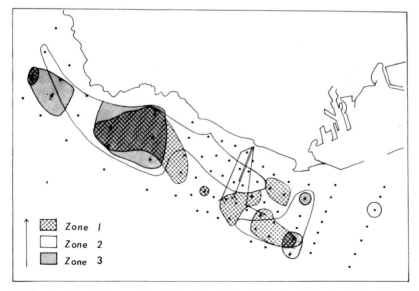

Fig. 3.   Benthic animal population off the Palos Verdes Peninsula, California, in the vicinity of Whites Point outfall sewer. (After Reish, 1973.)

and reflective of the clean up of the wastes discharged into the area. By knowing what benthic polychaetes are present, biological proof may be offered that it is possible to improve the environment by upgrading the discharge water quality.

IV. The Effect of a Disturbed Environment on Polychaetes

As discussed previously, changes in the benthic polychaete communities can be the result of natural environmental fluctuations, natural but sudden ecological alterations in the environment, or be anthropogenic in origin. It is essential to be able to determine, if at all possible, the cause of the change in the polycaete distribution, especially from a water quality management point of view.

Some of the natural causes were discussed under Section II, which dealt with the factors which influence the occurrence and distribution of polychaetes. This section deals with those events which are caused by man's activity which in turn brings about changes in polychaete distribution. Under each cause of an environmental disturbance, both field and laboratory studies are discussed whenever such data are available. The majority of the laboratory studies have been undertaken in this decade with the primary purpose of understanding what has happened or what will happen in the field. In this short period of time laboratory studies have yielded meaningful data but difficulties arise when attempts are made to apply these data to field conditions. It is difficult at best to consider all the variables that occur in the marine environment. We have not yet achieved the ability to simulate a complex benthic polychaete community in experimental laboratory studies (with or without the other components, let alone the pelagic element).

A. Organic Enrichment

Organic enrichment generally refers to domestic sewage but other wastes could be included here such as fish cannery or food processing discharges. The amount of organic enrichment depends upon the type of sewage treatment, whether or not storm and industrial wastes are included, as well as the quantity of discharge and the oceanographic conditions. Much of the work concerned with indicator organisms and benthic polychaete communities described above was based on the effects of domestic wastes.

The initial effect of a new domestic sewer discharging primary wastes is to increase the productivity of the benthos as measured by the size and number of specimens of the polychaetes as well as other macroinvertebrates. This increase in productivity will plateau providing the discharge remains small, i.e., 1.3 million gallons per day (Turner *et al.,* 1965). If, however, the amount of waste discharged is increased with time then the benthic fauna begins to be impoverished with a reduction in the number of species and specimens of both polychaetes and other invertebrates. *Capitella capitata,* and perhaps other polychaete indicators such as *Stauroneris rudolphi,* begin to appear near the point of discharge. Such was the condition about 2 years after the commencement of operations of the sewer for the city of San Diego (Turner *et al.,* 1968).

Continual increase in the amount of waste discharge into open oceanic waters, which frequently is the case in urban areas, leads to the expansion of the impoverished zone as a result of an increase in organic matter in which the environment can no longer assimilate. This description more or less fits the conditions at Palos Verdes Peninsula, California, described earlier. Domestic discharges into open oceanic waters do not usually produce the azoic or very polluted zones which can develop in protected waters. Any significant change in the nature or amount of discharge can bring about a change in the benthic polychaete population, as evidenced by the benthic change at Palos Verdes following the upgrading of that discharge.

Bellan and his colleagues have been monitoring the City of Marseille, France, domestic discharge at Cortiou for over a decade (Bellan *et al.,* 1975). An azoic area was present in the immediate vicinity of the discharge. Beyond the azoic zone they found a *Capitella capitata* community which increased slightly in area during the 1963–1973 decade. The spionid polychaete *Scolelepis fuliginosa* was associated with *Capitella* in this polluted zone. The most dramatic environmental change during these years was noted in what Bellan referred to as the subnormal zones which, in addition to *Capitella* and *Scolelepis,* also was characterized by *Neanthes caudata* and *Cirriformia tentaculata.* This zone increased in size at least 10 times and extends outward to include the offshore islands off the Marseille Peninsula. Thus far, pollution from Marseille Harbor area has not yet overlapped with the area affected by the Cortiou outfall, but the combined effect of these two major sources of pollution could possibly overlap in the near future in the absence of the initiation of any pollution abatement program.

No laboratory studies have been undertaken insofar as is known to measure the effect of domestic wastes on polychaetes.

B. Oil Pollution

The sea can be polluted by oil in many ways: (1) by the wreck of a ship, (2) by the cleaning of ballast tanks, (3) by waste discharges from oil refineries, (4) by spilling oil during transfer from dock to ship, or vice versa, or (5) by natural oil seeps. If any single event can be cited as the one which triggered the worldwide ecological movement of the 1960s, it was the wreck and the aftermath of the "Torrey Canyon" disaster off Land's End, England, in March 1967. Detergents used in clean up of the oil caused more damage than the oil itself.

1. Field Studies

After the extensive oil spill in West Falmouth, Massachusetts, Grassle and Grassle (1974) followed the settlement of polychaetes and other invertebrates in this environmentally disturbed area. Species with the most opportunistic life histories were able to settle in the area affected by the oil pollution, and included the polychaetes *Capitella capitata, Polydora ligni, Syllides verrilli, Microphthalmus aberrans, Steblospio benedicti,* and *Capitita ambiseta.* All these species have some type of brood protection, so that the newly released larvae settle almost immediately in the nearby area and lead to a rapid increase in the populations of those species. Furthermore, the first three species listed above reproduce throughout the year in these waters, which greatly facilitates the buildup of populations. *Capitella* increased most rapidly at the affected stations, followed by *Polydora* and *Syllides.* The rapid buildup of the population of these species was followed by an equally sharp decline in the same order. Other polychaetes replaced those as the population of the initial inhabitants declined. The speed of recovery of a normal population depended upon the amount of oil on the bottom. It was suggested that the large population of *Capitella,* which rose to about $200,000/m^2$ within the 3 months following the spill, may be the result of genetic flexibility rather than a higher tolerance to the pollutant. By electrophoretic studies of the malate dehydrogenase loci, the Grassles were able to document the short-term selection for a single genotype. Their hypothesis suggests that genetic flexibility may account for the ability of an opportunistic species such as *Capitella* to take advantage, so to speak, of a situation.

The Consolidated Slip, East Basin of Los Angeles Harbor, has been known to be azoic since the initial benthic samples were taken in 1951 (Anonymous, 1952). This region received 16 million gallons per day of oil refinery waste; the sediment consisted of over 10% organic material at the inner most stations. Two changes, one temporary and one permanent, have occurred since 1951. In 1953, much of the accumulated pollutants were dredged from the bottom to enable larger ships to enter this area of the harbor. With the elimination of the sludge beds from the bottom, marine animals settled rapidly in 1954 but, because the area was still receiving oil refinery wastes, these organisms were killed and further settlement was prevented by these accumulated wastes by late 1954. Conditions remained much the same until the 1968–1970 period, when the pollution abatement program was initiated in Los Angeles Harbor. Oil refinery wastes were no longer allowed to be emptied into the Consolidated Slip. Since pollution abatement was initiated, the changes in benthic conditions have been striking. This benthic environment partially recovered from a very polluted zone to either a polluted or a semipolluted zone within 2 years (Reish, 1973). There are indications that this recovery will continue to extend the semipolluted zone and, perhaps, even the healthy zone, into the Consolidated Slip permanently.

2. *Laboratory Studies*

The toxicity of oils vary, not only according to the type of oil but also to the individual components within the oil (Rice *et al.,* 1977). However, because it remains longer within the marine environment, laboratory studies have focused largely on the water-soluble fraction (WSF) of the oils. Acute, chronic, and sublethal effects of four oils have been measured using polychaetes—five of which were laboratory inbred strains (Carr and Reish, 1977; Rossi and Anderson, 1976; Rossi, 1976; Rossi *et al.,* 1976). The techniques for preparing the various oils (No. 2 fuel oil, South Louisiana crude, Bunker "C," and Kuwait crude) have been outlined by Anderson *et al.* (1974) and were followed in all of these studies. Studies have generally focused on the WSF fraction of No. 2 fuel oil and South Louisiana crude since these are the most and least toxic, respectively, of the four oils (Rossi *et al.,* 1976). The effects of these oils are summarized in Table VII for these two oils. The WSF of No. 2 fuel oil was more toxic than South Louisiana crude to the six species tested. *Neanthes* and the two species of *Ophryotrocha* were more sensitive to both oils than the other species. It is of interest to note that *Capitella,* which was the first species to inhabit the benthos following an oil spill (Grassle and Grassle, 1974), and *Cirriformia* were the most tolerant of the species tested.

The effects of the WSF of the two oils on reproduction of three species have been measured. Data consisted of counting the number of larvae produced and noting the concentration at which a significant suppression occurred (Table VIII). The relative toxicity of the WSF fraction of these two oils to *Neanthes, Ophryotrocha,* and *Ctenodrilus* follows the same pattern as noted in the 96-hour toxicity test. The level of reproductive suppression was nearly one order of magnitude less for *Neanthes,* but only slightly less for the two species. Such observations should lead one to be cautious in drawing environmental conclusions based on the experiments conducted with one species.

Rossi (1976) conducted additional experiments on growth and reproduction of *Neanthes* with the WSF of No. 2 fuel oil. The growth rate of larval *Neanthes* was the same regardless of the concentration of oil until feeding commenced at segments 18–21 at which time growth was inversely related to the concentration. Rossi suggested that the oils may be sequestered in the yolk during the nonfeeding stage. Rossi also reared *Neanthes* through three generations in a particular concentration of WSF of No. 2 fuel oil to determine whether or not there were any accumulative effects. Fecundity was inversely related to concentration of oil. There were some indications of a reduction in the number of young produced in each successive generation within a concentration.

These laboratory studies measuring the effects of various WSF of oils on polychaetes, while limited in number, do serve to indicate the value of growth and reproductive studies in assessing the long-term effects of a pollutant.

TABLE VII

THE 96-HOUR LC_{50} AS A PERCENTAGE OF THE WATER-SOLUBLE FRACTION OF OILS TO POLYCHAETES

Species	Type of oil			
	No. 2 fuel oil	South Louisiana crude	Bunker C	Kuwait crude
Neanthes arenaceodentata[a]	2.7	12.5	3.6	>10.4
Ophryotrocha diadema[b]	2.9	12.9	—	—
Ophryotrocha puerilis[b]	2.2	17.2	—	—
Cirriformia luxuriosa[b]	>8.7	>19.8	—	—
Crenodrilus serratus[b]	4.1	>19.8	—	—
Capitella capitata[a]	2.3	12.0	0.9	9.8
Capitella capitata[b]	>8.7	>19.8	—	—

[a]Data from Rossi *et al.* (1976).
[b]Data from Carr and Reish (1977).

TABLE VIII

EFFECT OF WATER-SOLUBLE FRACTION OF TWO OILS ON SURVIVAL AND REPRODUCTION IN POLYCHAETES[a,b]

Species	96-Hour LC_{50}		Significant reproductive suppression	
	No. 2 fuel oil	South Louisiana crude	No. 2 fuel oil	South Louisiana crude
Neanthes arenaceodentata	2.7	12.5	0.43	—
Ophryotrocha diadema	2.9	12.9	1.7	9.9
Ophryotrocha puerilis	2.2	17.2	—	—
Cirriformia spirabranchia	>8.7	>19.8	—	—
Ctenodrilus serratus	4.1	>19.8	2.2	9.9
Capitella capitata	>8.7	>19.8	—	—

[a] Data condensed from Carr and Reish (1977) and Rossi (1976).
[b] Data given in milligrams per liter.

107

C. HEAVY METALS

The discharge of heavy metals into the marine environment is considered by many to be a serious pollution problem. The discharge of mercury into Minimata Bay, Japan, and the movement of this element through the food chain to man dramatized such effects in a very tragic way. As an aftermath of the discovery of the cause of Minimata disease, many studies were undertaken to measure the effects of various metals on marine organisms, including polychaetes. Studies on the effects of metals include bioassays (Oshida *et al.*, 1976), sublethal effects (Reish, 1978), body burden levels (Table IX), and the attempt to relate body burden levels to reproduction (Oshida, 1977; Reish and Gerlinger, 1977). Heavy metals enter the marine environment by a variety of routes: (1) by surface runoff from rain, (2) by direct fallout from air into the ocean, (3) by resuspension of sediments by boat activities or dredging, and (4) by waste discharges from sewage and industrial plants.

1. Body Burden

Table IX summarizes the known data on the body burden levels of 11 metals to 24 species of polychaetes. Many variables must be taken into consideration in the interpretation of these data, such as dry versus wet weight, purged versus non-purged (which is generally not stated), field versus laboratory uptake studies, and whether or not there is any known biological function of the metal.

The highest burden levels in polychaetes have been measured in laboratory studies dealing with cadmium, lead, iron, and zinc. Too few data are available to determine whether certain species accumulate one or more metals more rapidly than others or that high values are associated with a particular feeding type. The most extensive studies undertaken thus far has been those with *Nereis diversicolor* conducted by Bryan (1971) and Bryan and Hummerstone (1971, 1973a,b, 1977) in English estuaries. They found a direct relationship between the concentration in the tissues of *Nereis* and the level within the sediments. In most instances the concentration of a particular metal in the sediments was higher than in the worm. Specimens of *Nereis* collected from polluted estuaries were more resistant to $ZnSO_4$ than those taken from cleaner areas (Bryan and Hummerston, 1973a). Oshida (1977) attempted to relate body burden levels of chromium to reproduction through two life cycles in *Neanthes arenaceodentata*. Tissue levels in excess of 40.5 mg/gm dry weight in the P_1 generation did not significantly suppress reproduction in terms of number of offspring produced; however, in the F_1 generation of worms exposed to the same concentration of chromium, he found a significant decrease in the number of offspring produced when the tissue levels exceeded 30.3 mg/liter. A meaningful area of research with heavy metals seems to be relating body burden levels to reproductive success and to determine whether or not a population becomes more tolerant or sensitive to a metal with each succeeding generation.

2. Bioassays

The toxicity of metals to polychaetes has been reviewed by Reish *et al.* (1976a, 1978). The effect of six metals on polychaetes has been measured under laboratory conditions including cadmium, chromium, copper, lead, mercury, and zinc (Table X). The four species of laboratory populations of polychaetes that have been used in the majority of these studies are *Neanthes arenaceodentata, Ophryotrocha diadema, Ctenodrilus serratus,* and *Capitella capitata* (Reish, 1978). The most toxic elements were mercury followed by copper. The 96-hour LC_{50} varied from <0.1 to 0.09 mg/liter for mercury and 0.16 to 0.3 mg/liter for copper. The toxicity of chromium to these species varied from 3.12 to >5.0 mg/liter. The toxicity of zinc was 1.8 to 10.7 mg/liter. Concentrations of cadmium and lead, the least toxic of the six metals tested, were 4.2 to >20.00 and 7.7 to >20.0 mg/liter, respectively. The effects of these six metals on reproduction were measured for *Ophryotrocha* and *Ctenodrilus*. There was a sublethal concentration at which the reproductive rate was significantly reduced and a still lower concentration at which reproduction was not affected. Interestingly, the magnitude of difference between the 96-hour LC_{50} and that level at which there was no significant suppression of reproduction was less than one for animals exposed to mercury and copper, but was greater than one for the remaining four metals. Perhaps mercury and copper are so toxic that some animals are killed rapidly and those that are not are largely unaffected and can continue to live and complete their life cycle. The action of cadmium, chromium, lead, and zinc may take a longer period of time to show an effect or the animal may be able to excrete at least some of the metal. It is obvious from these data that the uptake and fate of a metal in a polychaete is a fruitful area for future research.

D. DETERGENTS

The effect of detergents on polychaetes has been limited, and these studies were stimulated largely because of the use of a detergent following the wreck of the *Torrey Canyon*. Intertidal specimens of *Cirriformia tentaculata* and *Cirratulus cirratus* were unaffected by an English oil terminal spill, but were killed by dispersants used to clean up the oil. Recovery was evident 2 years later at the locality where dispersants were used (George, 1971).

Bellan and his colleagues have conducted bioassays with *Capitella capitata* and *Scolelepis fuliginosa,* using many different types of French detergents (Bellan *et al.*, 1969, 1972; Foret-Montardo, 1970; Foret, 1972, 1974). Nonionic detergents were more toxic than anionic ones; *Capitella* was slightly more resistant than *Scolelepis* to the same detergent. Foret (1972) observed abnormal larvae of *Capitella* in the second generation of specimens reared in higher concentrations of sublethal amounts of a nonionic detergent (an ester of polyethylene glycol). These larvae possessed two posterior ends and failed to develop beyond

TABLE IX

Trace Metal Concentrations in Body Tissues of Polychaetes

Species	Geographic region	As	Cd	Cr	Co	C
Polynoid	Puerto Rico					
Phyllodoce maculata	Scotland					2.44.
Sigambra (nr.) *tentaculata*	Puerto Rico					
Nereis diversicolor	England					9–22
						22–1
			<0.1–3.4	>0.1–2.4	1.6–7.9	22–7
			0.08–3.6			
			140–4010[a]			
			0.2–0.4[b]		1.6–5.1[b]	
	Finland					
Nereis sp.	North Carolina					
Neanthes arenaceodentata	California		0.5–81.7[a,b]			
			<02–40.6[a]			
		<0.7[a,b]	0.5–290[a,b]			4.9–1
Neanthes succinea	Hawaii					
Neanthes virens	England					19–2
Ceratonereis sp.	Puerto Rico					
Aglaophamus sp.						
Glycera americana	North Carolina					
Diopatra ornata						
Arabella iricolor						
Protaricia sp.						
Paraprionospio sp.						
Magelona sp.						
Chaetopterus variopedatus						
Cirriformia luxuriosa	California					6.8–6
Cirratulid	Puerto Rico					
Capitella capitata	California		0.33–1300[a,b]			9.3–3
Capitella sp.	Puerto Rico					
Armandia maculata						
Ampharetid						
Amphitrite ornata	North Carolina					

[a]Laboratory study.
[b]Wet weight.

			(μg/gm dry weight)			
Fe	Pb	Mn	Hg	Ni	Zn	Reference
					55-59	Phelps (1967)
						McLusky and Phillips (1975)
					2490	Phelps (1967)
	0.7-1.1[b]				34-36[b]	Bryan (1971)
	0.7-5.9				155-199	Bryan and Hummerstone (1971)
-462		7.6-12.5		2.1-5.2	170-258	Bryan and Hummerstone (1977)
					130-350	Bryan and Hummerstone (1973a)
					198-3630[b]	Bryan and Hummerstone (1973a)
		3.7-20.1				Bryan and Hummerstone (1973b)
	0.1-4.8[b]				2.2-37.5[b]	Wharfe and van de Brock (1977)
			0.01			Nuorteva and Häsänen (1975)
					82	Cross et al. (1970)
						Oshida et al. (1976)
						Oshida (1977)
27[a,b]	4.2-3300				28-91[a,b]	Reish and Gerlinger (1977)
			0.08-1.44[a,b]			Luoma (1977)
						Raymont and Shields (1963)
					845	Phelps (1967)
					2961	Phelps (1967)
					164	Cross et al. (1970)
					87	Cross et al. (1970)
					143	Cross et al. (1970)
					1239	Phelps (1967)
-24.6						Phelps (1967)
						Phelps (1967)
					100	Cross et al. (1970)
						Milanovich et al. (1976)
8						Phelps (1967)
3100[a,b]	9-1200[a,b]				9-3200[a,b]	Reish and Gerlinger (1977)
1						Phelps (1967)
6						Phelps (1967)
0						Phelps (1967)
					78	Cross et al. (1970)

TABLE X

COMPARISON OF THE TOXICITY OF HEAVY METALS TO POLYCHAETES TO THE CONCENTRATION IN SEWAGE EFFLUENTS OFF SOUTHERN CALIFORNIA[a,b]

Metal	96-Hour LC$_{50}$				Significant reproduction suppression				Sewage effluent[c]
	Neanthes arenaceodentata	*Capitella capitata*	*Ctenodrilus serratus*	*Ophryotrocha diadema*	*Neanthes arenaceodentata*	*Capitella capitita*	*Ctenodrilus serratus*	*Ophryotrocha diadema*	
Cadmium	12.0	7.5	4.3	4.2	1.0	0.56	2.5	1.0	0.02–0.06
Chromium	>1.0	5.0	4.3	7.5	No data	0.1	0.05	1.0	0.21–0.86
Copper	0.3	0.2	0.3	0.16	No data	No data	0.1	0.25	0.19–0.6
Lead	>10.0	6.8	7.2	14.0	3.1	0.2	1.0	5.0	0.04–0.26
Mercury	0.022	<0.1	0.042	0.09	No data	No data	0.05	0.1	0.001–0.003
Zinc	1.8	3.5	7.1	1.4	0.32	0.56	0.5	0.5	0.24–1.79

[a]Data from Reish et al. (1976b); Reish and Gerlinger (1977); Reish and Carr (1978).
[b]Data given in milligrams per liter.
[c]Data from Mitchell and McDermott (1975) for three largest discharges.

the eight-segmented stage. The incidence of abnormal larvae was 11.3 and 16.3% in the two instances where they were observed. In another study Bellan *et al.* (1972) measured the effects of a polyethylene glycol fatty acid on development, reproduction, and subsequent development of the F_1 generation of *Capitella*. The results which demonstrate the importance of long-term bioassays conducted through one complete life cycle are summarized in Table XI. Not only was there an inverse relationship between the concentration of detergent and reproductive success, as measured by the number of benthic worms produced, but also the detergent affected each stage of its life cycle. While the number of benthic worms produced in the lowest concentration was not statistically significant as compared to the control, a trend is indicated. Foret (1974) concluded that development of ovarian tissue in *Capitella* was the most sensitive stage to the action of the various detergents. The search has continued to create new, but less toxic, detergents or dispersants to be used in cleaning oil spills, but it is apparent from the data reported by Bellan *et al.* (1972) that long-term bioassays involving reproduction should be conducted prior to any approval of the use of these chemicals.

E. RADIONUCLIDES

Radioactive materials can reach the ocean via a variety of routes such as the testing of nuclear devices, the release of liquid waste effluents from nuclear fuel processing plants, and the deposit of nuclear waste materials in cannisters at deep sea dumpsites. Unfortunately, baseline studies were not made of polychaetes and other invertebrates prior to the detonation of nuclear devices at oceanic island sites during the late 1940s and 1950s.

Nothing is known on whether or not the disposal of radioactive wastes in cannisters at dumpsites has caused an adverse effect on the polychaetes or other components of the benthic community. Again, unfortunately, no baseline studies were taken prior to the initial use of these dumpsites some 25 years ago. The Environmental Protection Agency has recently initiated studies at an Atlantic and Pacific radioactive dumpsite to determine, among other things, whether or not these cannisters are leaking radioactive materials, and, if so, if such release of these wastes is causing any adverse effect on the environment.

Comparisons of the polychaete populations were made between "control" sites and previous studies conducted outside the area; the author (unpublished data) concluded that the reported small amount of leakage of radioactive wastes which has occurred has not caused any discernible change in the polychaete fauna. However, since most of the specimens are minute, it is impossible to measure whether or not any of these organisms have taken up any of the radioactive wastes.

Beasley and Fowler (1976) exposed *Nereis diverscolor* to sediment contaminated by testing of nuclear devices or by liquid wastes from a nuclear fuel plant.

TABLE XI

RELATIONSHIP OF THE CONCENTRATION OF A DETERGENT TO REPRODUCTION, DEVELOPMENT, AND LARVAL SURVIVAL IN *Capitella capitata*[a]

Concentration (mg/liter)	No. of females developing ovarian tissue	No. of females laying eggs	No. of eggs (mean)	No. of trochophores (mean)	No. of free trochophores (mean)	No. of metatrochophores (mean)	No. of benthic (mean)
Control	21	19	285	250	183	167	153
0.01	21	17	285	227	176	158	133
0.1	19	14	288	214	165	144	114
1.0	19	13	315	217	159	108	77
10.0	17	10	278	161	133	78	47
100.0	11	6	238	124	102	51	25

[a]Data reduced from Bellan *et al.* (1972) based on an initial number of 25 males and females each.

The uptake of plutonium and americium was small, with the uptake of plutonium similar from both sources but greater for americium from nuclear testing contaminated sediment. The authors concluded that water may be the predominant pathway for accumulation rather than via the sediments.

F. DREDGING ACTIVITIES

Dredging can effect polychaetes and other benthic organisms in four ways: by physical removal of the sediment in which they live; by physical burial by sediments at the disposal site; by an increase in water turbidity; and by a release of toxic chemicals which previously have been bound to the sediment. The increase in turbidity of the water has either little or no effect on polychaetes, but the other environmental changes caused by dredging can have an effect on polychaetes.

Since dredging is largely associated with maintaining or constructing shipping channels or docking facilities, most of this activity takes place within protected waters such as bays, harbors, and estuaries. If dredging removes at least the upper 1 m of sediments, then all benthic animals will be physically removed. Repopulation of the benthos follows, with animals moving from nearby unaffected areas or by way of larval stages. McCauley *et al.* (1977) showed that the infauna readjusted to predredging conditions in 28 days in the dredged area and 14 days in the nearby areas disturbed by deposition. Recovery was within 2 weeks at the spoil site. Since both sites had similar estuarine faunas composed of the polychaetes *Streblospio benedicti, Pseudopolydora kempi, Polydora ligni, Eteone lighti, Capitella capitata, Notomastus (C.) tenuis,* and *Glycinde armigra,* it was impossible to ascertain whether the specimens collected at the spoil site were the original ones or transported ones. Polychaetes are able to burrow to the surface of sediment after burial in light of the laboratory study by Saila *et al.* (1972). *Nephtys incica* was able to surface in less than 24 hours after being buried by 21 cm of sediment. *Streblospio benedicti* regained the surface within 24 hours after being buried 6 cm. In light of these data it seems quite likely that the reestablishment of the benthic fauna in Coos Bay (McCauley *et al.,* 1977) was from the buried specimens, although the depth of sediment deposited was not given. In contrast, Rosenberg (1977) found that it took 1.5 years for a Swedish estuary to recover from dredging and that recovery was largely through larval settlement by such organisms as the polychaetes *Pholoe minuta, Pectinaria koreni,* and *Nephtys hombergi.*

Major concern is expressed that dredging activities will release toxicants to the water mass which may result in mass mortalities of marine organisms. Very few data are available on the effects of such actions on polychaetes. The discharge of mine tailings, which acts more or less as continuous deposition of dredge spoils, caused a decrease in density, weight, and lengths of *Ammotrypane aulogaster* in areas where deposition was greatest (Jones and Ellis, 1977).

It is reasonable to assume that in the near future a considerable amount of data will be accumulated on the effects of dredge spoils on marine animals, including polychaetes, in light of the recent requirements established by the Environmental Protection Agency and the U.S. Army Corps of Engineers (1977). Bioassay procedures were outlined and a list of recommended appropriate sensitive benthic infaunal species prepared. The following genera were included under polychaetes: *Neanthes, Nereis, Nephtys, Glycera, Urechis* (not a polychaete), *Magelona, Owenia, Diopatra,* and *Glycinde.* Members of these genera can be important elements of the benthic fauna, but unfortunately, with the exception of *Neanthes* and *Nereis,* experience in working with living species of the remaining genera is limited or nonexistent. It is doubtful, at the present state of knowledge of these latter genera, that any results from bioassay using dredge spoils will yield any meaningful data. If such organisms are going to be used it is imperative that explorative tests be run in order to develop standardized procedures with each species within the framework of the guidelines (Environmental Protection Agency, 1977).

V. Conclusions

In the foregoing discussion data were presented to substantiate the importance of polychaetes in the marine benthic environment and their role in assessing various degrees of water quality. Polychaetes are the most important macroinvertebrate benthic animal, and it is therefore essential that this group be included in any assessment of benthic conditions. It is necessary that identifications be made carefully and accurately to ensure correct interpretation of data. Benthic studies now include monitoring as well as initial survey work, and it is necessary to rely on a qualified technical staff to do the work of identifying these organisms to species. Less and less of this work is being done by the professional polychaetologist, and more by the technician. Therefore, it becomes increasingly more important for the professional to work closely with the technicians to assist them in identifications and verifications, and to provide them with aids that will make their work more efficient. It is equally important for the technician to have contact with the professional polychaetologist, both for assistance and because they come in contact with a greater number of potentially unique species than the professional.

Arguments have been presented which demonstrate the validity of the use of polychaetes as indicators of various degrees of pollution. The universal indicator, *Capitella capitata* is indeed useful on a worldwide scale; however, in some instances, its presence does not indicate a polluted environment, and so points to the importance of careful scientific judgment. Other species, to a lesser extent, have been useful as indicators of semipolluted conditions, while still others are known only in natural environments. Because particular species may indicate a particular environmental condition, it is important, in the interest of economy, that we do not record the data only as numbers. Numbers of species and specimens can be used to augment data on the names and numbers of specimens, but they do not replace the importance of knowing the species present.

We have seen the development of two major lines of research dealing with the subject of polychaetes and pollution in the past decade or so. One has been the expansion of survey work and the subsequent development of monitoring programs, some of which have become increasingly sophisticated. The second major endeavor has been the growth of laboratory experimentation in the attempts to learn what field data mean. Initial laboratory experiments were concerned with short-term toxic effects; gradually long-term experiments studying chronic effects were initiated, followed by those dealing with the entire life cycle, and the most recent type of study has been concerned with biochemical changes. However sophisticated these experiments may become, relating the results to field conditions is important and must always be considered. It is in the field that we must interprete our results.

Generally, larvae are considered to be the most sensitive stage in the organism's life cycle. Very few laboratory studies have been concerned with larvae, and the results of some of these have indicated that the generality does not always hold true. Why does it hold true in some cases and not in others? Is it dependent upon the type of toxicant or the species? In another vein, we need to see greater use of the analytical chemist to make our increasingly more sophisticated experiments yield more meaningful information. We essentially know nothing about the uptake, metabolism, and release, if any, of any of the toxicants that affect polychaetes.

In the next decade or two, I look for the continual expansion of surveys and monitoring programs with greater use of computer analyses of data. As pollution abatement programs are initiated, I think that there will be increasing evidence of the usefulness and validity of the indicator concept. The 96-hour bioassay using polychaetes will be expanded as an aid to solve immediate environmental problems. Long-term studies, involving a reproductive period, will be more prevalent as a means of measuring long-term sublethal effects. Specific studies will be initiated to determine if a particular element or compound is toxic and, if so, can it be removed from an effluent to render the the discharge less toxic. There will be an increasing awareness of the role of polychaetes in the marine food chain, and with it the possible transfer and magnification of different toxicants as the material moves up the chain. Since polychaetes are the dominant benthic macroinvertebrate, I look for their use in both field and laboratory situations to help us solve many of our problems in the marine environment.

Donald J. Reish

References

Abati, J. L., and Reish, D. J. (1972). The effects of lower dissolved oxygen concentrations and salinity on the free amino acid pool of the polychaetous annelid *Neanthes arenaceodentata*. *Bull. South. Calif. Acad. Sci.* **71**, 32–39.

Alexander, W. B., Southgate, B. A., and Bassindale, K. (1935). "A Survey of the River Trees," Part II, Pap. No. 5, pp. 1–171. Water Pollut. Res. Board, London.

Anderson, J. W., Neff, J. M., Cox, B. A., Tatem, H. E., and Hightower, H. E. (1974). Characteristics of dispersions and water-soluble extracts of crude and refined oils and their toxicity to estuarine crustaceans and fish. *Mar. Biol.* **27**, 75–88.

Anonymous (1952). "Los Angeles-Long Beach Pollution Survey," Report to Los Angeles Regional Water Pollution Control Board, No. 4, pp. 1–43.

Arnoux, A., Auclair, D., and Bellan, G. (1973). Etude de la pollution chemique des sédiments marins du secteur de Cortiou (Marseille): Relations avec le peuplements macrobenthiques. *Tethys* **5**, 115–123.

Bagge, P. (1969). Effects of pollution estuarine ecosystems. I. Effects of effluents from wood processing industries on the hydrography, bottom and fauna of saltkällefjord (W. Sweden). *Merentutkimuslait. Julk./Havsforskningsinst. Skr.* **228**, 3–118.

Balch, N., Ellis, D., Littlepage, J., Marles, E., and Pym, R. (1976). Monitoring a deep marine wastewater outfall. *J. Water Pollut. Control Fed.* **48**, 429–457.

Banse, K., and Hobson, K. D. (1974). Benthic errantiate polychaetes of British Columbia and Washington. *Bull. Fish. Res. Board Can.* **185**, 1–111.

Barnes, H. (1959). "Apparatus and Methods of Oceanography," Part One. Wiley (Interscience), New York.

Beasley, T. M., and Fowler, S. W. (1976). Plutonium isotope ratios in polychaete worms. *Nature (London)* **262**, 813–814.

Bellan, A. (1967). Pollution et peuplements benthiques sur substrat meuble dans la région de Marseille. I. Le secteur de Cortiou. *Rev. Int. Oceanogr. Med.* **6–7**, 53–87.

Bellan, A., Carvelle, F., Foret-Montardo, P., Kaim-Malka, R. A., and Leung Tack, K. (1969). Contribution à l'étude de différence facteurs physiochimiques pollutants sur les organisms marins. I. Action des détergents sur la polychète *Scolelepis fuliginosa* (note préliminaire). *Tethys* **1**, 367–374.

Bellan, G. (1964). Influence de la pollution sur la faune annélidienne des substrats meubles. *Commun. Int. Explor. Sci. Mer Medit., Symp. Pollut. Mar. Microorgan. Prod. Pet.* pp. 123–126.

Bellan, G., and Peres, J. M. (1970). Etat général des pollutions sur les cates méditerranéenes de France. *Quad. Civ. Stn. Idrobiol. Milano* No. 1, pp. 35–65.

Bellan, G., Reish, D. J., and Feret, J. P. (1972). The sublethal effects of a detergent on the reproduction development, and settlement in the polychaetous annelid *Capitella capitata*. *Mar. Biol.* **14**, 183–188.

Bellan, G., Kaim-Malka, R. A., and Picard, J. (1975). Evolution récente des différentes aurioles de pollution marin des substrats meubles liées au grand collector de Marseille-Cortiou. *Bull. Ecol.* **6**, 57–66.

Berkeley, E., and Berkeley, C. (1948). "Canadian Pacific Fauna. 9. Annelida. 9b (1). Polychaeta Errantia," pp. 1–100. Can. Fish. Res. Board, Toronto.

Berkeley, E., and Berkeley, C. (1952). "Canadian Pacific Fauna. 9. Annelida. 9b (2). Polychaeta Sedentaria," pp. 1–139. Can. Fish. Res. Board, Toronto.

Blegvad, H. (1932). Investigations of the bottom fauna at the outfalls of drains in the Sound. *Dan. Biol. Stn.* **37**, 1–20.

Boesch, D. F. (1977). "Application of Numerical Classification in Ecological Investigations of Water Pollution," EPA-600/3-77-033. Corvallis Environ. Res. Lab., U.S. Environ. Prot. Agency, Washington, D.C.

Bonde, G. J. (1967). Pollution of a marine environment. *J. Water Pollut. Control Fed.* **39**, R45–R63.

Bryan, G. W. (1971). The effects of heavy metals (other than mercury) on marine and estuarine organisms. *Proc. R. Soc. London, Ser. B* **177**, 389–410.

Bryan, G. W., and Hummerstone, L. G. (1971). Adaption of the polychaete *Nereis diversicolor* to estuarine sediments containing high concentrations of heavy metals. Part I. General observations and adaption to copper. *J. Mar. Biol. Assoc. U. K.* **51**, 845–863.

Bryan, G. W., and Hummerstone, L. G. (1973a). Adaptation of the polychaete *Nereis diversicolor* to estuarine sediments containing high concentrations of zinc and cadmium. *J. Mar. Biol. Assoc. U. K.* **53**, 839–857.

Bryan, G. W., and Hummerstone, L. G. (1973b). Adaptation of the polychaete *Nereis diversicolor* to manganese in estuarine sediments. *J. Mar. Biol. Assoc. U.K.* **53**, 859–872.

Bryan, G. W., and Hummerstone, L. G. (1977). Indicators of heavy-metal contamination in the Looe Estuary (Cornwall) with particular regard to silver and lead. *J. Mar. Biol. Assoc. U.K.* **57**, 75–92.

Carr, R. S., and Reish, D. J. (1977). The effect of petroleum hydrocarbons on the survival and life history of polychaetous annelids. *In* "Fate and Effects of Petroleum Hydrocarbons in Marine Organisms and Ecosystems" (D. A. Wolfe, ed.), pp. 168–173. Pergamon, Oxford.

Chen, C. W. (1970). Effects of San Diego's wastewater discharges on the ocean environment. *J. Water Pollut. Control Fed.* **42**, 1458–1467.

Christie, N. D., and Moldan, H. (1977). Effects of fish factory effluents on benthic macrofauna of Saldanha Bay. *Mar. Pollut. Bull.* **8**, 41–45.

Cognetti, G. (1972). Distribution of Polychaeta in polluted waters. *Rev. Int. Oceanogr. Med.* **25**, 23–24.

Crippen, R. W., and Reish, D. J. (1969). An ecological study of the polychateous annelids associated with fouling material in Los Angeles Harbor with special reference to pollution. *Bull. South. Calif. Acad. Sci.* **68**, 170–187.

Cripps, R., and Reish, D. J. (1973). The effect of environmental stress on the activity of malate dehydrogenase and lactate dehydrogenase in *Neanthes arenaceodentata* (Annelida: Polychaeta). *Comp. Biochem. Physicol. B* **46**, 123–133.

Cross, F. A., Duke, T. W., and Willis, J. N. (1970). Biogeochemistry of trace elements in a coastal plain estuary: Distribution of manganese, iron, zinc in sediments, water and polychaetous worms. *Chesapeake Sci.* **11**, 221–234.

Davis, W. R., and Reish, D. J. (1975). The effect of reduced dissolved oxygen concentration on the growth and production of oocytes in the polychaetous annelid *Neanthes arenaceodentata. Rev. Int. Oceanogr. Med.* **37–38**, 3–16.

Day, J. H. (1967). "A Monograph on the Polychaeta of Southern Africa." British Museum (*N.H.*), London.

Day, J. H. (1973). New Polychaeta from Beaufort with a key to all species recorded from North Carolina, U.S. Department of Commerce. *NOAA Tech. Rep., NMFS Circ* **375**, 1–140.

De Silva, P. H. D. H. (1967). Studies on the biology of spirorbinae (Polychaeta). *J. Zool. Soc. London* **15-2**, 269–279.

Emery, K. O. (1960). "The Sea off Southern California." Wiley, New York.

Environmental Protection Agency and the U.S. Army Corps of Engineers (1977). "Ecological Evaluation of Proposed Discharge of Dredged Material into Ocean Waters." 81 pp. EPA, Washington, D.C.

Fauchald, K. (1977a). The polychaete worms, definitions and keys to the orders, families and genera. *Sci. Bull., Nat. Hist. Mus., Los Angeles Cty.* **28**, 1–188.

Fauchald, K. (1977b). Polychaetes from intertidal areas in Panama, with a review of previous shallow-water records. *Smithson. Contrib. Zool.* **221**, 1–81.

Fauvel, P. (1923). Polychètes errantes. *Faune Fr.* **5**, 1–488.

Fauvel, P. (1927). Polychètes sédentaires. *Faune Fr.* **16**, 1–494.

Fauvel, P. (1953). Annelida Polychaeta. "The Fauna of India including Pakistan, Ceylon, Berma, and Malaya," pp. 1–507.

Filice, F. P. (1954a). An ecological survey of the Castro Creek area in the San Pablo Bay. *Wasmann J. Biol.* **12**, 1–24.

Filice, F. P. (1954b). A study of some factors affecting the bottom fauna of a portion of the San Francisco Bay estuary. *Wasmann J. Biol.* **12**, 257–292.

Forbes, S. A., and Richardson, R. E. (1913). Studies on the biology of the upper Illinois River. *Bull., Ill. State Lab. Nat. Hist.* **9**, 481–574.

Foret, J.-P. (1972). Etude des effets à longterme de quelques détergents (issues de la pétroléochemie) sur la séquence du deux espèces de Polychètes sédentaire: *Scolepis fuliginosa* (Claparede) et *Capitella capitata* (Fabricius). Thèse del Universite de Marseille-Luminy.

Foret, J.-P. (1974). Etude des effets a longterme de quelques détergents sur la séquence du développement de la polychète sédentaire *Capitella capitata* (Fabricius). *Tethys* **6**, 751–778.

Foret-Montardo, P. (1970). Etude de l'action desprocuits de base entrant dans la composition des détergents issus de la pétroléochemie vis-à-vis de quelques invertibratés benthique marins. *Tethys* **2**, 567–614.

Gallardo, V. A. (1967). Polychaeta from the Bay of Nha Trang, South Viet Nam. *Naga Rep., Scripps Inst. Oceanogr.* **4**, 35–279.

Gallardo, V. A., Castillo, J. G., and Yañez, L. A. (1972). Algunas consideraciones preliminares sabre la ecologia benthonica de los fondas sublitorales blandos en la Bahia de Concepcion. *Bol. Soc. Biol. Concepcion* **44**, 169–190.

Gardiner, S. L. (1976). Errant polychaete annelids from North Carolina. *J. Elisha Mitchell Sci. Soc.* **91**, 77–220.

Gaufin, A. R., and Tarzwell, C. M. (1952). Aquatic invertibrates of stream pollution. *Public Health Rep.* **67**, 57–64.

George, J. D. (1971). The effects of pollution by oil and oil-dispersants on the common intertidal polychaetes, *Cirriformia tentaculata* and *Cirratulus cirratus*. *J. Appl. Ecol.* **8**, 411–420.

Gilet, R. (1960). Water pollution in Marseilles and its relation with flora and fauna. *In* "Waste Disposal in the Marine Environment" (E. A. Pearson, ed.), pp. 39–56. Pergamon, Oxford.

Grassle, J. F., and Grassle, J. P. (1974). Opportunistic life histories and genetic systems in marine benthic polychaetes. *J. Mar. Res.* **32**, 253–284.

Gray, J. S. (1976). The fauna of the polluted River Tees Estuary. *Estuarine Coastal Mar. Sci.* **4**, 653.

Greene, C. S. (1976). Partial recovery of the benthos at Palos Verdes. *Annu. Rep., Coastal Water Res. Proj., South. Calif.* pp. 205–210.

Halcrow, M., MacKay, D. W., and Thornton, I. (1973). The distribution of trace metals and fauna in the Firth of Clyde in relation to the disposal of sewage sludge. *J. Mar. Biol. Assoc. U.K.* **53**, 74–739.

Hartman, O. (1948). The polychaetous annelids of Alaska. *Pac. Sci.* **2**, 1–58.

Hartman, O. (1951a). "Literature of the Polychaetous Annelids." Los Angeles.

Hartman, O. (1951b). The littoral marine annelids of the Gulf of Mexico. *Inst. Mar. Sci.* **9**, 7–124.

Hartman, O. (1956). "Contributions to a Biological Survey of Santa Monica Bay, California." Univ. of Southern California Press, Los Angeles.

Hartman, O. (1959). Catalogue of the polychaetous annelids of the world. *Occas. Pap. Allan Hancock Found.* No. 23, pp. 1–628.

Hartman, O. (1960). Systematic account of some marine invertebrate animals from the deep basins off Southern California. *Allan Hancock Pac. Exped.* **22**, 69–215.

Hartman, O. (1963). Submarine canyons of Southern California. Part 2. Biology. *Allan Hancock Pac. Exped.* **27**, 1–424.

Hartman, O. (1964). Polychaeta Errantia of Antarctica. *Antarct. Res. Ser.* **3**, 1–131.

Hartman, O. (1965). Catalogue of the polychaetous annelids of the world. Supplement and index (1960–1965). *Occas. Pap., Allan Hancock Found.* No. 23, pp. 1–197.

Hartman, O. (1966a). Polychaetous annelids of the Hawaiian Islands. *Bishop Mus., Occas. Pap.* **23**, 163–252.

Hartman, O. (1966b). Polychaeta Myzostomidae and Sedentaria of Anarctica. *Antarct. Res. Ser.* **7**, 1–158.

Hartman, O. (1968). "Atlas of the Errantiate Polychaetous Annelids from California." Allan Hancock Found., Los Angeles, California.

Hartman, O. (1969). "Atlas of the Sedentariate Polychaetous Annelids from California." Allan Hancock Found., Los Angeles, California.

Hartman, O., and Reish, D. J. (1950). The marine annelids of Oregon. *Oreg. State Monogr. Zool.* No. 6, pp. 1–64.

Hartmann-Schröder, G. (1971). Annelida, Borstenwürmer, Polychaeta. "Die Tierwelt Deutschlands und der angrenzenden Meeresteile nach ihren Merkmalen und nach ihren Lebenweise." Jena.

Hendriksson, R. (1969). Influence of pollution on the bottom fauna of the Sound (Öresund). *Oikos* **19**, 111–125.

Hill, L. R., and Reish, D. J. (1975). Seasonal occurence and distribution of benthic and fouling species of polychaetes in Long Beach Naval Station and Shipyard, California. *In* "Marine Studies of San Pedro Bay, California" (D. F. Soule and M. Oguri, eds.), Part 8, pp. 57–58. Harbors Environ. Proj., University of Southern California, Los Angeles.

Holland, J. S., Maciolek, N. J., and Oppenheimer, C. H. (1973). Galveston Bay benthic community structure as an indicator of water quality. *Contrib. Mar. Sci.* **17**, 169–188.

Imajima, M., and Hartman, O. (1964). The polychaetous annelids of Japan. Parts I and II. *Occas. Pap., Allan Hancock, Found.* No. 26, pp. 1–452.

Jones, A. P., and Ellis, D. V. (1977). Sub-obliterative effects of mine-tailing on marine infaunal benthos. II. Effect on the productivity of *Ammotrypane aulogaster. Water, Air, Soil Pollut.* **5**, 299–307.

Kinne, O. (1954). Über das Schwarmen und die Larvelentwicklung von *Nereis succinea* Leuckart (Polycnaeta). *Zool. Anz.* **153**, 114–126.

Kinne, O. (1970). Temperature, invertebrates. *Mar. Ecol.* **1**, No. 1, 407–514.

Kitamori, R. (1972). Faunal and floral changes by pollution in the coastal waters of Japan. *Int. Ocean Dev. Conf., 2nd, 19??* Vol. 1, pp. 71–77.

Knight-Jones, E. W., and Qasim, S. Z. (1955). Responses of some marine plankton animals to changes in hydrostatic pressure. *Nature (London)* **175**, 941.

Knox, C. A. (1977). The role of polychaetes in benthic soft-bottom communities. *In* "Essays on Polychaetous Annelids in Memory of Dr. Olga Hartman" (D. J. Reish and K. Fauchald, eds.), pp. 547–604. Allan Hancock Found.

Kolkwitz, R., and Mansson, M. (1908). Oekologie der pflanzlicken Saprabein. *Ber. Dtsch. Bot. Ges.* **26**, 505–519.

Kudinov, J. D. (1975a). Errant polychaetes from the Gulf of California, Mexico. *J. Nat. Hist.* **9**, 65–91.

Kudinov, J. D. (1975b). Sedentary polychaetes from the Gulf of California, Mexico. *J. Nat. Hist.* **9**, 205–231.

Leppäkoski, E. (1975). Assessment of degree of pollution on the basis of macrozoobenthos in marine and brackish-water environments. *Acta Acad. Abo., Ser. B* **35**, 1–90.

Lizarrage-Partida, M. L. (1974). Organic pollution in Ensenada Bay, Mexico. *Mar. Pollut. Bull.* **5**, 109–112.

Luome, S. N. (1977). Physiological characteristics of mercury uptake by two estuarine species. *Mar. Biol.* **41**, 269–273.

122   *Donald J. Reish*

McCauley, J. E., Parr, R. A., and Hancock, D. R. (1977). Benthic infauna and maintenance dredging. *Water Res.* **11**, 233–242.

McLusky, D. S., and Phillips, C. N. K. (1975). Some effects of copper on the polychaete *Phyllodoce maculata* (Firth of Forth). *Estuarine Coastal Mar. Sci.* **3**, 103–108.

Melvessalo, T., Pesonen, L., Varmo, R., and Viljamaa, H. (1975). Inshore effects of pollution in the biata of the Baltic, Southern Finland. *Verh., Int. Ver. Theor. Angew. Limnol.* **19**, 2340–2353.

Milanovich, F. P., Spies, K., Guram, M. S., and Sykes, E. E. (1976). Uptake of copper by the polychaete *Cirriformia spirabrancha* in the presence of dissolved yellow organic matter of natural origin. *Estuarine Coastal Mar. Sci.* **4**, 585–588.

Mithcell, F. K., and McDermott, D. J. (1975). Characteristics of municipal wastewater discharges, 1974. *Annu. Rep., Coastal Water Res. Proj., South. Calif.* pp. 163–165.

Naroska, V. (1968). Vergleichende Untersuchungen über den Einfluss des hydrostatischen Druckes auf Überlebensfähigkeit und Staffwichselintensität mariner Evertebraten und Teleasteen. *Kiel. Meeresforsch.* **24**, 95–123.

Nuorteva, P., and Häsäneu, E. (1975). Bioaccumulation of mercury in *Myoxcephalus quadricornis* (L.), (Teleostei, Cottidae) in an unpolluted area of the Baltic. *Ann. Zool. Fenn.* **12**, 247–254.

Oliff, W. D., Berrisford, C. D., Turner, W. D., Ballard, J. A., and McWilliam, D. C. (1967). The ecology and chemistry of sandy beaches and nearshore submarine sediments of Natal. II. Pollution criteria for nearshore sediments of the Natal Coast. *Water Res.* **1**, 131–146.

Olsson, I., Rosenberg, R., and Ölundh, E. (1973). Benthic fauna and zooplankton in some polluted Swedish estuaries. *Ambio* **2**, 158–163.

Oshida, P. S. (1977). A safe level of hexavalent chromium for a marine polychaete. *Annu. Rep., South. Calif. Coastal Water Res. Proj.*, pp. 169–180.

Oshida, P. S., Mearns, A. J., Reish, D. J., and Word, C. J. (1976). The effects of hexavalent and trivalent chromium on *Neanthes arenaeodentata* (Polychaeta: Annelida). *Tech. Memo, Coastal Water Res. Proj., South. Calif.* **TM 224**, 1–58.

Patrick, R. (1949). A proposed biological measure of stream conditions based on a survey of the Cosestoga Basin, Lancaster County, Pennsylvania. *Proc. Acad. Nat. Sci. Philadelphia* **101**, 227–341.

Pearson, T. H., and Rosenberg, R. (1976). A comparative study of the effects on the marine environment of wastes from cellulose industries in Scotland and Sweden. *Ambio* **5**, 77–79.

Pettibone, M. H. (1954). Marine polychaete worms from Point Barrow, Alaska, with additional records from the North Atlantic and North Pacific. *Proc. U. S. Natl. Mus.* **103**, 203–356.

Pettibone, M. H. (1963). Marine polychaete worms of the New England region. Part I. Aphroditidae through Trochochaetidae. *U.S., Natl. Mus., Bull.* **227**, 1–356.

Phelps, D. K. (1967). Partitioning of the stable elements Fe, Zn, Sc, and Sm within the benthic community, Añasco Bay, Puerto Rico. *Proc., Int. Symp.: Radioecol. Appl. Concentration Processes, 1966*, pp. 721–734.

Raps, M. E., and Reish, D. J. (1971). The effect of varying dissolved oxygen concentrations on the hemoglobin levels of the polychaetous annelid *Neanthes arenaceodentata*. *Mar. Biol.* **11**, 363–368.

Raymont, J. E. G., and Shields, J. (1963). Toxicity of copper and chromium in the marine environment. *Int. J. Air Water Pollut.* **7**, 435–443.

Reish, D. J. (1954). The life history and ecology of the polychaetous annelid *Nereis grubei* (Kinberg). *Occas. Pap., Allan Hancock Found.* No. 14, pp. 1–74.

Reish, D. J. (1955). The relation of polychaetous annelids to harbor pollution. *Public Health Rep.* **70**, 1168–1174.

Reish, D. J. (1957). The relationship of the polychaetous annelid *Capitella capitata* (Fabricius) to waste discharges of biological origin. *In* "Biological Problems in Water Pollution" (C. M. Tarzwell, ed.), pp. 195–200. U.S. Public Health Serv., Washington, D.C.

Reish, D. J. (1959). An ecological study of pollution in Los Angeles-Long Beach Harbors, California. *Occas. Pap., Allan Hancock Found.* No. 22, pp. 1–119.

Reish, D. J. (1960). The use of marine invertebrates as indicators of water quality. *In* "Waste Disposal in the Marine Environment" (E. A. Pearson, ed.), pp. 92–103. Pergamon, Oxford.

Reish, D. J. (1961). The use of the sediment bottle collector for monitoring polluted marine waters. *Calif. Fish Game* **47**, 261–272.

Reish, D. J. (1963a). A quantitative study of the benthic polychaetous annelids of Bahia de San Quintin, Baja, California. *Pac. Nat.* **3**, 399–436.

Reish, D. J. (1963b). Further studies on the benthic fauna in a recently constructed boat harbor in southern California. *Bull. South. Calif. Acad. Sci.* **62**, 23–32.

Reish, D. J. (1967). Relationship of polychaetes to varying dissolved oxygen concentrations. *Proc. Int. Conf. Water Pollut. Res., 3rd, 1966* Vol. 3, No. 10, pp. 1–10.

Reish, D. J. (1968a). A biological survey of Bahia de Los Angeles, Gulf of California, Mexico. II. Benthic polychaetous annelids. *Trans. San Diego Soc. Nat. Hist.* **15**, 67–106.

Reish, D. J. (1968b). The polychaetous annelids of the Marshall Islands. *Pac. Sci.* **22**, 208–231.

Reish, D. J. (1973). The use of benthic animals in monitoring the marine environment. *J. Environ. Plann. Pollut. Control* **1**, 32–38.

Reish, D. J. (1974a). The sublethal effects of environmental variables on polychaetous annelids. *Rev. Int. Oceanogr. Med.* **33**, 1–8.

Reish, D. J. (1974b). The establishment of laboratory colonies of polychaetous annelids. *Thalassia Jugosl.* **10**, 181–195.

Reish, D. J. (1978). The effects of heavy metals on polychaetous annelids. *Rev. Int. Oceanog. Med.* **49**, 99–104.

Reish, D. J., and Barnard, J. L. (1960). Field toxicity tests in marine waters utilizing the polychaetous annelid *Capitella capitata* (Fabricius). *Pac. Nat.* **1**, 1–8.

Reish, D. J., and Carr, K. S. (1978). The effect of heavy metals on the survival, reproduction, development, and life cycles for two species of polychaetous annelids. *Mar. Pollut. Bull.* **9**, 24–27.

Reish, D. J., and Gerlinger, T. V. (1977). "Toxicity of Formulated Mine Tailings on Marine Polychaeta." Mar. Biol. Consultants, Inc., Costa Mesa, California.

Reish, D. J., and Ware, R. (1976). The impact of waste effluents on the benthos and food habits of fish in outer Los Angeles Harbor. *In* "Marine Studies of San Bedro Bay, California" (D. F. Soule and M. Oguri, eds.), pp. 113–128. Harbors Environ. Proj., University of Southern California.

Reish, D. J., Kauwling, T. J., and Mearns, A. J. (1976a). Marine and estuarine pollution. *J. Water Pollut. Control Fed.* **48**, 1439–1455.

Reish, D. J., Martin, J. M., Piltz, F. M., and Word, J. Q. (1976b). The effect of heavy metals on laboratory populations of two polychaetes with comparison to the water quality standards in southern California marine waters. *Water Res.* **10**, 299–302.

Reish,'D. J., Kauwling, T. S., Mearns, A. J., Oshida, P. S., and Rossi, S. S. (1977). Marine and estuarine pollution. *J. Water Pollut. Control Fed.* **49**, 1316–1340.

Rice, S. D., Short, J. W., and Karineu, J. F. (1977). Comparative oil toxicity and general animal sensitivity. *In* "Fate and Effects of Petroleum Hydrocarbons in Marine Organisms and Ecosystems" (D. A. Wolfe, ed.), pp. 28–94. Pergamon, Oxford.

Ristich, S. S., Crandall, M., and Fortier, J. (1977). Benthic and epibenthic macroinvertebrates of the Hudson River. I. Distribution, natural history and community structure. *Estuarine Coastal Mar. Sci.* **5**, 255–266.

Rosenberg, R. (1974). Spatial dispersion of an estuarine benthic faunal community. *J. Exp. Mar. Biol. Ecol.* **15**, 69–80.

Rosenberg, R. (1976). Benthic faunal dynamics during succession following pollution abatement in a Swedish estuary. *Oikos* **27**, 414–427.

Rosenberg, R. (1977). Effects of dredging operations on estuarine benthic macrofauna. *Mar. Pollut. Bull.* **8**, 102–104.

Rossi, S. S. (1976). Interactions between petroleum hydrocarbons and the polychaetous annelid, *Neanthes arenaceodentata:* Effects on growth and reproduction; fate of diaromatic hydrocarbons accumulated from solution or sediments. Ph.D. Dissertation, Texas A&M University, College Station.

Rossi, S. S., and Anderson, J. W. (1976). Toxicity of water-soluble fractions of No. 2 fuel oil and South Louisiana crude oil to selected stages on the life history of the polychaete *Neanthes arenaceodentata. Bull. Environ. Contam. Toxic.* **16**, 18–23.

Rossi, S. S., Anderson, J. W., and Ward, G. S. (1976). Toxicity of water-soluble fractions of four test oils for the polychaetous annelids *Neanthes arenaceodentata* and *Capitella capitata.* Environ. Pollut. **10,**9–18.

Saila, S. B., Pratt, S. D., and Polgar, T. T. (1972). Dredge spoil disposal in Rhode Island Sound. *R. I., Univ., Mar. Tech. Rep.* No. 2, pp. 1–48.

Sanders, H. L. (1960). Benthic studies in Buzzards Bay. III. The structure of the soft-bottom community. *Limnol. Oceanogr.* **5**, 138–153.

Sanders, H. L., Hessler, R. R., and Hampson, G. R. (1965). An introduction to the study of deep-sea benthic faunal assemaloges along the Gay Head-Bermuda Transect. *Deep-Sea Res.* **12**, 845–867.

Schroeder, P. C. (1968). On the life history of *Nereis grubei* a polychaete annelid from California. *Pac. Sci.* **22**, 476–481.

Shafer, H. A. (1977). Characteristics of municipal wastewater discharges, 1976. *Annu. Rept., Coastal Water Res. Proj., South. Calif.* pp. 19–23.

Smith, R. J. (1955). Salinity variation in the interstitial water at Kames Bay, Millport, with reference to the distribution of *Nereis diversicolor. J. Mar. Biol. Assoc. U.K.* **34**, 33–46.

Smith, R. N. (1971). Reconnaissance studies of benthic organisms in New York Harbor. *N.Y. State Univ., Mar. Sci. Res. Cent., Tech. Rep. Ser.* No. 8, pp. 38–55.

Stone, A. N., and Reish, D. J. (1965). The effect of fresh-water run-offs on a population of estuarine polychaetous annelids. *Bull. South. Calif. Acad. Sci.* **64**, 111–119.

Thorson, G. (1956). Marine level bottom communities of recent seas, their temperature adaptation, and their "balance" between predators and food animals. *Trans. N.Y. Acad. Sci.* [2] **18**, 693–700.

Treadwell, A. L. (1948). "Polychaeta Canadian Atlantic Fauna," No. 96. Toronto.

Tulkki, P. (1965). Disappearance of the benthic fauna from the Basin of Bornholm (Southern Baltic) due to oxygen deficiency. *Cah. Biol. Mar.* **6**, 455–463.

Turner, C. H., Ebert, E. E., and Given, R. R. (1965). Survey of the marine environment offshore to San Elijo Lagoon, San Diego County. *Calif. Dep. Fish Game* **51**, 81–112.

Turner, C. H., Ebert, E. E., and Given, R. R. (1968). The marine environment offshore from Point Loma, San Diego County. *Calif. Dep. Fish Game, Fish Bull.* No. 140, pp. 1–85.

Ushakov, P. V. (1965). Polychaetous annelids of the Far Eastern Seas of the U.S.S.R. *Akad Nauk SSSR* **56**, 1–433 (translated into English by Isr. Program Sci. Transl., Jerusalem).

Wade, B. A. (1972). A description of a highly diverse soft-bottom community in Kingston Harbour, Jamacia. *Mar. Biol.* **13**, 57–69.

Wade, B. A. (1976). The pollution ecology of Kingston Harbour, Jamaica. Part 4. Benthic ecology. *Res. Rep. Zool. Dep., Univ. West Indies* **2**, No. 5, 1–104.

Wharfe, J. R. (1977). The intertidal sediment habitats of the lower Medway Estuary, Kent. *Environ. Pollut.* **13**, 79.

Wharfe, J. R., and van de Brock, W. L. F. (1977). Heavy metals in macroinvertebrates and fish from Lower Medway Estuary, Kent. *Mar. Pollut. Bull.* **8**, 31–40.

Wilhelmi, J. (1916). Übersicht über die biologische Beurteilung des Wasser. *Ges. Naturf. Freunde Berlin, Setzher.* pp. 297–306.

Wilson, D. P. (1952). The influence of the nature of the substratum on the metamorphosis of the larvae of marine animals, especially the larvae of *Ophelia bicornis* Saviguy. *Ann. Inst. Oceanogr. (Monaco)* **27**, 49–156.

Wilson, D. P. (1955). The role of microorganisms in the settlement of *Ophelia bicornis* Saviguy. *J. Mar. Biol. Assoc. U.K.* **34**, 531–543.

Wolff, W. J. (1971). The estuary as a habitat, an analysis of data on the soft-bottom macrofauna of the estuarine area of the Rivers Rhine, Meuse, and Scheldt. *Zool Verh. (Leiden)* **126**, 1–242.

Zoological Record (1976). Vermes. Part B. Annelida, Chaetognatha, Echiura, Gastrotricha, Gnathostomulita, Kinorhyncha, Priapulida, Rotifera, Sipuncula (literature for 1973). *Proc. Zool Soc. London* **110**, 1–118.

CHAPTER 4

Ostracods (Arthropoda: Crustacea: Ostracoda)

C. W. HART, JR., and DABNEY G. HART

I. Introduction

The Ostracoda belong to a well-defined order of the class Crustacea of the phylum Arthropoda and are among the most numerous of the microorganisms.

127

Recent ostracods (as opposed to fossil ostracods) are known to inhabit practically all types of aquatic environments. They occur in freshwater environments ranging from deep lakes to very shallow temporary roadside pools. Some are typically brackish water forms, and others may be encountered in marine environments from tidal pools to the abyssal depths. Many species show a great tolerance for fluctuations of environmental conditions such as temperature and salinity and are able to flourish in extreme environments such as hot water springs and saline pools. However, other groups are restricted to well-defined biotopes and are susceptible to slight changes in environmental factors.

Despite their being among the most ubiquitous and abundant of crustaceans, ostracods remain poorly investigated and comparatively little is known of their biology. The brackish water forms have been studied in more detail than the marine species, perhaps because they are procured more easily than marine ones. Studies of freshwater ostracods have been few, however, and those species are less well known than the brackish water or marine forms.

Both pelagic and benthic in their habits, almost all ostracods are directly or indirectly dependent on the nature of the bottom (Benson, 1961), and the responses of some species and species assemblages to certain outside stimuli have been documented. The most important environmental factors that have been investigated relative to ostracods are salinity, temperature, substrate, and depth; more limited data are available on phototaxis and pH tolerance. As was pointed out by Benson (1959), with changes in species concepts those species formerly considered to be eurythermic and eurybathic are now found to be more ecologically restricted than was previously thought, and as taxonomic work progresses more and more species are becoming useful ecological indicators.

The ecology of ostracods relevant to water pollution per se has received little attention, but ecological studies involving natural ostracod communities have been relatively numerous—primarily because ostracod assemblages have been found to be useful in studying geological horizons. Such studies are used to extrapolate backward in geological time in order to estimate previous environmental conditions and also to determine those assemblages most likely to indicate the presence of petroleum deposits.

While these studies have mainly involved marine and estuarine ostracods, they nevertheless represent a relatively large body of information, information that is usually centered around salinity, temperature, and substrate composition—but nonetheless information that could be of considerable importance to contemporary studies of aquatic pollution.

It is paradoxical that ostracod workers have been concerned with the effects of environmental conditions on ostracod assemblages, but have shown little interest in pollution ecology. At the same time, those biologists interested in assaying pollution have seldom considered ostracod assemblages.

Certain factors would appear to make ostracods worthy of considerable investigation as indicator organisms for assessing degrees and kinds of pollution:

1. Numerous species occur in relatively large numbers in virtually all aquatic habitats.
2. Their small size make them easy to collect and diminishes storage and handling problems.
3. Their numbers make their populations particularly amenable to satistical analysis.
4. Whether pelagic or benthic, they are directly or indirectly affected by the nature of the bottom, and their responses are therefore more likely to reflect long-term pollution effects than short-term natural environmental fluctuations.
5. Ostracod environmental studies have traditionally emphasized species assemblages rather than individual species, an approach that would strengthen interpretations of data regarding pollution effects.

II. Ostracod Assemblages

The research emphasis placed on ostracod assemblages rather than individual species, together with their interrelationship with the substrate, would appear to make them ideal organisms for use in pollution ecology studies.

Various investigators have defined ostracod assemblages according to a variety of criteria—ranging from considering only living individuals, to living plus dead, to only dead. Baker and Hulings (1966) felt that, in general, only live ostracods should be correlated with ecological data since dead valves are susceptible to current transport and may give a false picture of true distributions. In some cases, however, they feel that live and dead ostracods can be used if the work is carried out carefully. Kilenyi (1971) discussed this problem at length, and also felt that the transport of dead ostracods by currents is responsible for the lack of correlation between living and dead assemblages (bio- and thanatocoenoses).

In ecological assemblages of ostracods all members appear to have essentially the same environmental requirements, and there appear to be no implications of interrelationships between species. Some assemblages overlap, and within overlap areas assemblages occupy an environment in which particular sets of conditions are present.

Baker and Hulings (1966) divided their ecological "indicator" assemblages into primary and secondary types. The primary indicators include the most abundant species with both live and dead representatives; secondary indicators include species that are very abundant and occur in 55% or more of the samples, but have no living representatives.

A. MARINE ASSEMBLAGES

In a study of ecological assemblages of recent marine ostracods in Puerto Rico, Baker and Hulings (1966) recognized five assemblages based on water depth. In their study they concluded that 52 of the 230 ostracod species found off Puerto Rico were useful ecological indicators.

Rosenfeld (1977) studied 50 ostracod species from 220 surface sediment samples from the Baltic Sea. Analysis of these samples indicated four ostracod associations characterizing different environments from the upper water layer of low salinity to the lower water layer of higher salinity. He noted maximum diversity and population density in the western Baltic Sea within the euphotic zone, and minimum diversity and population densities in the lower layer of the eastern Baltic Sea.

Skaumal (1977), in studying the recent ostracods from the North Sea in the vacinity of Helgoland, noted that the horizontal distribution of ostracod abundance increased from rocky beaches through the progression of large rock pools, mussel banks, sandy bays, and cliffs (with the exception of the sandy bays). The cliff niche—described as having rock layers covered with *Enteromorpha,* cliffs covered with *Fucus,* rock channels connected with the sea and containing *Cladophora, Chaetomorpha,* etc., and accumulations of debris—was the most densely populated area. Skaumal considered the most important physical factors affecting the horizontal distributions to be the type of substrate and the extent of exposure to wave action.

Williams (1969), in studying the ostracods from marine intertidal localities on the coast of Anglesey, divided the ostracods into two assemblages: those populations occurring on beach sediment and those on the algae adjoining the beach. A number of species were found to be specific to those separate habitats. The populations were not specifically confined within those habitats, and there was a great deal of mixing of the two populations especially at the boundaries between the habitats.

In a study of the distributions of 65 ostracod species in the southern Irish Sea, Whatley and Wall (1969) noted four assemblages inhabiting four distinct environments. They concluded that ostracod numbers on that very exposed coast, subject to low fertility and high turbidity, were extremely low, and they postulated that the absence of one species of ostracod may be related to the absence of the alga *Ascophyllum nodosum,* an association previously pointed out by Colman (1940).

Preliminary laboratory data showing ecological preferences for an ostracod inhabiting rock pools that were often exposed to varying environmental conditions throughout the year and during the day and night cycle were carried out by Ganning (1967). In studies of the effect of light, dissolved oxygen, temperature, and salinity, Ganning noted diurnal vertical migration from the shallow parts of pools at night to the deep parts during the daytime. He found that one ostracod survived better in hypo- than in hyperoxygenated waters, and that it showed a well-defined reaction to different temperatures with a preference for 15°C. In addition he found that the ostracod exhibited a preference for the most common rock pool salinity, 6‰, and he discussed the possible influences of light, oxygen, CO_2, and temperature combinations and variations.

In the Gulf of Mexico near the western and southern coast of Florida, Benson and Coleman (1963) observed that water depth (and the correlated amount of light penetration), wave base, and proximity to shore appeared to be the major factors influencing the distribution of ostracod species in that area.

B. Estuarine Assemblages

In a pioneer study of the seasonal distribution of estuarine ostracods, Tressler and Smith (1948) determined the distribution and seasonal fluctuations of ostracods living in the littoral zone in front of the Chesapeake Biological Laboratory at Solomons, Maryland. In that study, temperature, transparency, density, dissolved oxygen, pH, and salinity data were correlated with numbers of individuals of each species in the area. Those forms inhabiting shallow water, aquatic plants, weeds in the deeper water, and at the deep end of the laboratory pier were recorded, as were those species occurring in significant numbers during various seasons. They concluded that temperature and dissolved oxygen seemed to be the most important factors influencing seasonal distribution at Solomons, but that salinity showed no clear influence and that pH apparently had no effect.

As a result of a study of the ostracods in canals and lagoons joining the sea in the Tramandai area, Rio Grande Do Sul, Brazil, Pinto de Ornellas (1974) concluded that associations of ostracods in that area are much more valuable for characterizing an ecological environment than are foraminifera, because of the consistency of associations of ostracods for each type of environment.

Benson and Maddocks (1964), in a study of intertidal sediment samples collected from the shores of the Knysna Estuary, Union of South Africa, noted a transition along the estuary from mesohaline to euhaline that was reflected by changes in the composition of the ostracod populations.

Puri *et al*. (1964), in a study of the ecology of the Gulf of Naples, identified no single environmental factor that appeared to control ostracod distribution in that area. They stated that the distribution patterns seemed to be due to the interrelationship of many factors, some of which were probably unknown. They recognized two major ostracod assemblages representing two vaguely defined bottom environments in the Gulf of Naples. These were the nearshore assemblage, inhabiting relatively shallow waters, and the offshore assemblage. As with other studies, they noticed that the boundaries between the assemblages were poorly defined. The major environmental factors that might have affected ostracod distributions in the Bay of Naples were depth, type of substrate, and possibly salinity.

Bate (1971), in examining the recent ostracods of the Abu Dhabi Lagoon, recognized six ostracod assemblages: the nearshore shelf, the back-reef, the tidal delta, the lagoon terrace, the lagoon channel, and the tidal pool. He considered the distribution of the ostracod species to be controlled by water depth, salinity, and turbulence—with food supply playing a subsidiary role because it controls the size of the populations.

Benson and Kaesler (1963) noted faunal assemblages in two slightly hyper-saline lagoons on the west coast of Sonora, Mexico, representing an open gulf biotope, a lower lagoon biotope, an upper lagoon biotope, and a tidal flat biotope. They felt that salinity, tidal current, and exposure to atmosphere were the physical factors that seemed to have the most effect on the local distribution of the ostracods. They considered nine species to be sufficiently restricted ecologically so that they could be used as biotope indicators.

Benson (1959), in studying the ostracods from an estuary, a saltwater lagoon, and a large open bay on the west coast of Baja California, showed that 28 species belonging to 19 genera had sufficient ecologic restriction and stratigraphic range to be used as indicators in ancient sediments, and he felt that the salinity and water depth were the two factors that seemed to affect the distribution of os-tracods most greatly. In addition, Benson observed that the distribution of plant types greatly influences the distribution of ostracods that live closely associated with them.

Krutak (1975) found that living populations of ostracods in coastal Mississippi generally decreased from estuarine to the gulf environment, and reached a maximum in the estuaries and a minimum in the open Gulf. His data suggest an inverse relationship between living foraminiferal populations and total popula-tions in the same environment, implying that marginal marine foraminifera may be poor paleoecologic indicators.

Kornicker and Wise (1962) studied three species of ostracods collected alive in the bays and lagoons along the coast of southern Texas, and correlated them with temperature, salinity, water depth, and bottom types.

King and Kornicker (1970), in a study of bays and lagoons of the central Texas coast, studied ostracod assemblages emphasizing salinity, temperature, seasonal changes, and substrate in relation to population sizes and compositions. They attributed varying ostracod abundance to varying food supply and in part to salinity. Good agreement was observed between ostracod abundance, shrimp abundance, and gross photosynthesis in the northern part of one of the bays. Although seasonal trends were observed in relative abundances of individual species in part of the area studied, the faunal assemblages were remarkably constant. Assemblages in some areas were deemed more variable because of the occasional influx of exotic species transported into the bay by rivers. Ostracods inhabiting oyster reef environments and muds in the deeper waters of the bays were more abundant than those in the well-sorted, shifting subtidal sands. Ostracods inhabiting the poorly sorted sands were abundant and seemingly were not influenced by variations in other sediment characteristics or by the kinds of plants living in the area.

In an ancillary study, Kornicker (1975) noted that ostracods were transported to exotic environments on transplanted oysters, and he emphasized that those species should be recognized to ensure correct ecological and zoogeographical interpretations.

Kornicker (1977), in studying the ostracods of the Indian River complex in Florida, noted that ostracod abundance increased during the spring, suggesting that a spring bloom of vegetation may be the controlling factor. Data on vegetation were not available, but comparisons of changes in salinity, temperature, and water depth with the abundance of ostracods suggested that increasing salinity during the spring may also be a factor. He also noted that predators have a major effect on macrobenthos density in the study area, and that possibly the meiobenthos is also affected.

Kilenyi (1969), in a study of the ostracods of the Thames estuary, quantitatively analyzed 250 samples. He discussed the problem of biocoenosis versus thanatocoenosis and suggested ways of differentiating the two. He recognized seven major ostracod biofaces in the Thames estuary and considered them to be controlled by salinity, temperature, and the nature of the sediment. Prohibiting factors were considered to be "liquid mud" and pollution in the inner estuary and "black mud" and a certain type of grain size in the outer estuary.

Elliott *et al.* (1966) studied the distribution of recent ostracods in the Rappahannock Estuary, Virginia. Using 69 bottom samples, they related ostracod distribution to the pattern of estuarine circulation and to the salinity of the waters. They recognized three ostracod assemblages on the basis of dominant species: a river assemblage, a shoal assemblage, and a basin assemblage. They suggested that the abundance of individuals and the variety of species may be the result of greater production associated with river-borne nutrients and with the growth of benthic plants. Generally, they found ostracods to be most abundant on the shoals, especially in the middle estuary, where diversity of the species was also greatest. They noted that a few species were distinctive of particular environments, but that an equal number are distributed throughout the estuary.

Keyser (1977) studied 183 samples from 120 stations collected in the brackish water region of southwest Florida and plotted the distribution of 36 ostracod species according to salinity, substrate, water temperature, pH, water depth, turbidity, and currents. He found that the most important factor controlling the distribution of ostracods in that region was salinity, and recognized five assemblages with respect to their salinity preferences: the limnic-oligohaline assemblage, the oligo-mesohaline assemblage, the meso-polyhaline assemblage, the poly-euhaline, and the euryhaline assemblage.

III. Freshwater Environments

Although freshwater ostracods are far less well known than estuarine or marine species, Bowen (1976) has broken ground in ostracod pollution work in a study of the effects of thermal effluents on ostracods in Par Pond, South Carolina. Par Pond and "Pond B" are on the grounds of the Energy Research and Development Administration's (now the Department of Energy) Savannah River Plant near Aiken, South Carolina, and receive, or did receive, heated effluents from nuclear reactors.

A thermal gradient is produced in the arm of Par Pond receiving the effluent, and although summer temperatures may reach nearly 50°C in this arm, the rest of the lake has ambient temperatures. The lake has been characterized as moderately productive, with the low nutrient levels characteristic of lakes in South Carolina. Pond B received heated effluent from 1961 to 1964, and now has temperatures normal for the region. It is less productive than Par Pond and has a lower diversity of aquatic plants and animals.

Bowen noted that ostracods were most abundant at the "Cold Dam" in Par Pond (mean minimum and mean maximum temperatures between June and August 1973, 27.4°C and 33.3°C, respectively; mean minimum and mean maximum temperatures from October 1972 through September 1973, 17.5°C and 22.7°C, respectively). At that site the numbers taken per tow were four times those at any other site. Total numbers were next highest at the "Hot Dam" (mean minimum and mean maximum temperatures for June through August 1973, 34.2°C and 40.1°C, respectively; mean minimum and mean maximum temperatures from October 1972 to September 1973, 26.6°C and 35.3°C, respectively). The lowest numbers of ostracods were found at the North Cove (mean minimum and mean maximum temperatures June through August 1973, 26.1°C and 30.9°C, respectively; mean minimum and mean maximum temperatures from October 1972 through September 1973, 16.1°C and 21.7°C, respectively). At all sites she found that the number of specimens collected in tow samples were much higher than those in core samples.

As for community structure, she found that of the seven ostracod species collected in Par Pond and Pond B, all species were collected from only one site, the Cold Dam. Statistical analyses indicated that minimum (but not maximum) temperature was nearly significant with respect to diversity, and the Cold Dam cores were significantly more diverse than all other samples, followed by the

Cold Dam tows and the North Cove cores. The Hot Dam and then Pond B samples (both cores and tows) showed the lowest diversities.

As to seasonal cycles, Bowen found that the number of species at the Hot Dam fell during the time that minimum temperatures drastically increased (by 15.5°C, from 20 to 35.5°C). Species not exposed to this coupling of high minimum and high maximum temperatures did not show fluctuating populations.

Artificial substrates were colonized by species that were rare in the surrounding areas near the Hot Dam. The substrates were installed coincident with a reactor shutdown, at which time all thermal effluent ceased. When the reactor operations resumed, the increased temperature coincided with a sharp drop in species diversity in waters outside the substrate, but a similar increase in diversity did not occur within the substrate. Bowen found that community structure, but not total number, was temperature related.

IV. Marine and Estuarine Environments

The large ostracod populations of endemic species found in estuaries and lagoons reflect the ability of these organisms to withstand relatively large fluctuations in environmental conditions.

Benson (1961) recognized three (or possibly four) intergrading ostracod assemblage types in water where a salinity gradient exists from freshwater to marine: (1) *freshwater,* with ostracods being scarce in waters more saline than 2‰; (2) *brackish waters,* which contain ostracod populations with their greatest abundance at about 10‰; and (3) *marine waters,* which contain ostracods that seldom survive to reproduce where the salinity is less than 17‰. He also recognized that the brackish water environment may be represented by two adjacent and overlapping faunas—an oligohaline assemblage at 0.2–2.0‰; a more saline brachyhaline assemblage at 2.0 to 17‰.

The patterns of distributions of ostracods are influenced by a complex of environmental factors, and these factors vary diurnally as well as seasonally. Thus, the restricted as well as the more ubiquitous species must be environmentally tolerant in order to survive the wide range of variations found in estuaries. These variations are of greatest magnitude in the middle estuary.

V. Specific Environmental Factors

The typical environments of a number of species have been described, and certain factors have been correlated with the presence or absence of some species. Yet there are comparatively few data from the field or the laboratory that quantitatively illustrate the effects of different environmental factors on ostracod populations. This is a field where much work needs to be done.

A. Temperature

Temperature is one of the more obvious environmental factors that would be expected to influence ostracod populations. These animals are unique in that some appear to have the highest temperature tolerances of any aquatic metazoan and have been found living in waters of thermal springs with temperatures as high as 50°C (Mason, 1939). Some investigators have concluded that temperatures have negligible effects (Hoff, 1942; McKenzie, 1964), but other studies have shown that ostracods are distributed according to temperature. Indeed, depending on the species, ostracods may be either extremely stenothermal or be able to tolerate wide temperature fluctuations. As with salinity (see below) marine species probably tolerate narrower thermal fluctuations than do estuarine and freshwater species, whose natural environments expose them to greater cyclical changes. Certain ostracod species have their optimum temperature in cold water, others in warm water. The range of tolerance above and below the optimum temperature varies in different species.

As far as known, the only structural feature of ostracods affected by temperature differences is size. Arctic representatives of certain species are sometimes as much as twice the size of specimens living in more temperate regions. In general, the number of marine ostracod species increases toward the tropics, where it reaches its maximum, and many genera do not occur outside tropical environments. Temperature has been shown to accelerate ostracod growth rate, and in certain cases the time required for the development from egg to sexual maturity may be reduced by 50% when cultures are kept at constant relatively warm temperatures.

B. Salinity

Because of the relative constancy of salinity and temperature in the marine environment, these factors are not considered as significant in controlling those ostracod distributions as are sediment types, organic carbon content, and depth. Marine species in general are probably less tolerant of wide salinity and temperature fluctuations than are estuarine and freshwater species, a factor that should increase their usefulness as indicators of environmental changes in the marine milieu.

Elofson (1941) and Wagner (1957) have shown that the behavior of ostracods with respect to salinity follows the observations of Remane (1934) on other groups of organisms, in that the number of marine species decreases regularly down to a salinity of about 10‰, and shows an abrupt drop in waters of lower salinity. On the other hand, the number of freshwater species diminishes very rapidly when salinity surpasses 3‰. In the meio-mesohaline range (3–10‰) ostracod species are not numerous, but large numbers of individuals often occur. These may be considered the true brackish water species. Many of these species, however, may tolerate a considerable variation in salinity.

In a study of the marine benthic invertebrate communities of Hadley Harbor, Massachusetts, Hulings (1969) suggested that in the marine salinity is not a factor in ostracod ecology since it is relatively constant throughout the year. Hulings also noted that there was no apparent decrease in number of specimens with increase in seasonal temperatures, and that the female populations tended to remain relatively constant.

C. pH

Little research has been done on the effects of pH on ostracods, although one paper (Klug, 1927) reported on tolerance ranges of two freshwater species. He found that those species could survive from 1 to 2 days at pH 4 and from 1 to 7 days at pH 10. None died at pH 6 to 8. Von Morkhoven (1962) pointed out that no work has been done to determine if any species can become adapted to pH values outside its normal range.

D. Depth

As brackish water environments are, as a rule, shallow, very little can be said about the depth distribution of species confined to those habitats. In the marine environment the large majority of species are found in the littoral and shallow marine zones, although ostracods are known from depths exceeding 3000 m. Increasing depth may affect ostracod ecology by reducing the amount of light penetration, which would, in turn, affect associated plant growth.

E. Substrates

Elofson (1941), in discussing the marine ostracods of Sweden, showed the importance of the role of the substrates as an environmental factor for many ostracod species, and included a detailed analysis of the influence exerted by the substrates on structural features of the valves.

Elofson's observations pointed out that: (1) Thin-shelled, hyaline, elongate, more or less laterally compressed, weak-hinged, and smooth carapace forms live mainly on plants. (2) The relatively few species that live exclusively in very coarse-grained bottom sediments often have the same type of carapace. (3) Burrowing species have predominantly smooth shells, which are more strongly calcified than those of nonburrowing benthonic forms. (4) Surface-crawling species often have a pronounced flattening of the ventral surface of the carapace. (5) Species that live in sandy sediments are usually shorter than plant-dwelling species, and the larger ostracods usually live on soft mud.

Von Morkhoven (1962) cautioned that Elofon's work was confined to a small group of marine benthonic forms from the Baltic Sea region, and that conclusions from that study might not be applicable to all other areas.

Kornicker (1964), in a seasonal study of the ostracods of Redfish Bay, Texas, noted that considerable differences in the sediment composition between stations had no readily apparent effect on those ostracod populations, but Kilenyi (1969) felt that the nature of the substratum had a decisive influence. He observed, in a study of the Thames Estuary, that the bottom sediments may be either prohibitive or selective on ostracod populations, and discussed in detail the interactions between grain size, current velocity, and interstitial existence of ostracod species.

Benson (1959) observed that the type of sediment in Todos Santos Bay and Estero de Punta Banda, Baja California, had a marked influence on the distributions of some assemblages and on the density of general populations—either by affecting the ostracods themselves or the algae with which they are associated.

F. Organic Production

Few data are available on organic production rates in relation to ostracod populations. However, McKenzie (1965) reported on ostracods from Scammon Lagoon, Baja California, and indicated that the organic production rate in the lagoon was about 50 mg of C per cubic meter per day. These figures, he said, were at least as high as others reported from the area, and higher than most. Phleger and Ewing (1962) suggested that this production rate may be related to the near-shore upwelling of water in the open ocean along this coast, and that as a consequence organisms thrive in those shallow protected lagoons close to a rich nutrient supply.

VI. Species versus Higher Categories

As pointed out by von Morkhoven (1962), the basic taxonomic unit for ecological as well as paleoecological studies is the species. He stressed that the species biotope is better defined and more restricted than that of the higher taxa, and pointed out that ostracod genera provide less detailed ecological information than their individual species. Benson (1959) had also wisely stressed that, as the genus is a subjective category and can change with the addition of new species, it is probably better to use species than genera in assessing ecological conditions.

Kaesler and Mulvany (1977), however, using Brillouin's equation from information theory to study diversity and community structure of recent and fossil ostracods, considered that diversity may be partitioned hierarchically among taxonomic categories and that species diversity may be viewed as the sum of diversities of super-families, families within super-families, genera within families, and species within genera. They determined that ostracods from 56 samples from Todos Santos Bay, Baja California, showed diversities at all taxonomic categories, followed the same general pattern of distribution within the bay, and they suggested that ecological studies of categories higher than the species level provide valid information—rather than being too generalized as one might expect. Malkin, in 1953, had also suggested that generic assemblages might be useful for indicating certain environments.

Kaesler (1966), using environmental and faunal data on ostracods from Todos Santos Bay, Baja California (collected by Benson, 1959), tested the applicability of quantitative methods of numerical taxonomy to biofaces analysis. He found that counts of specimens per species at each station were unreliable indicators of environmental similarity (particularly where the total population was considered) and that such factors as mixing, differential productivity, and differential removal of specimens influenced the counts. Biotopes were determined by clustering Q-matrices of simple matching coefficients, and biofaces were determined by clustering R-matrices of Jaccard coefficients. He found that after certain assumptions were satisfied, the use of numerical taxonomic methods of biofaces analysis

TABLE I

SELECTED REFERENCES TO ECOLOGICAL FACTORS RELATED TO FRESHWATER OSTRACODS AND/OR OSTRACOD ASSEMBLAGES

	Temperature	Salinity	pH	TDS	CaCO$_3$	SO$_4$	PO$_4$	Still water	Lakes	Running water (temporary)	Running water (permanent)	Light penetration	Depth	Substrate composition	Associated organisms	Seasonal data	Statistical analyses
Bhatia and Singh (1971)	X	X	X	X	X	X	X					X	X	X	X		
Bowen (1976)[a]	X													X		X	X
Hoff (1942)			X					X	X	X	X			X			
Mason (1939)	X																

[a] Particularly important freshwater pollution study.

TABLE II

SELECTED REFERENCES TO ECOLOGICAL FACTORS RELATED TO ESTUARINE OSTRACODS AND/OR
OSTRACOD ASSEMBLAGES

	Temperature	Salinity	pH	DO	Nutrients	Carbon production	Depth	Currents	Light penetration	Substrate composition	Associated organisms	Seasonal data	Species diversity	Abundance
Benson (1959)[a]	X	X		X	X		X			X	X	X		
Benson and Kaesler (1963)	X	X								X				
Benson and Maddocks (1964)	X	X					X			X				
Elliott et al. (1966)[a]	X	X		X					X	X				X
Keyser (1975)		X												
Keyser (1977)	X	X	X							X				
Kilenyi (1969)[a]		X		X						X				
Kilenyi (1971)		X												
King and Kornicker (1970)[a]	X	X								X	X	X	X	X
Kornicker (1964)[a]	X	X				X				X	X		X	X
Kornicker (1975)											X			
Kornicker (1977)	X	X					X			X	X		X	X
Kornicker and Wise (1962)	X	X					X							
Krutak (1975)											X			
McKenzie (1964)[a]		X	X		X	X	X	X		X	X			
McKenzie and Swain (1967)		X												
Pinto de Ornellas (1974)		X												
Tressler and Smith (1948)[a]	X	X	X	X					X					

[a]Papers containing particularly important data on methodology or data that might be especially relevant to pollution studies.

gave results closely similar to those based on qualitative interpretations. The quantitative methods are said to have the advantages that the results are objective and repeatable, computation is rapid, results may be expressed graphically, and choice of similarity level is clearly arbitrary and relative.

VII. Summary and Conclusions

Contrary to the thrust of most chapters in this volume and its companion "Pollution Ecology of Freshwater Invertebrates," emphasis has not been on individual ostracod species or genera; few are even mentioned here. The reason for this is simply that we feel little would be gained at this time by engaging in a

4. *Ostracods (Arthropoda: Crustacea: Ostracoda)*   141

detailed species-by-species discussion of ostracods and their environmental tolerances.

The purpose of this chapter is to point out that ostracods, at the species level, the generic level, or in various assemblages should be considered seriously as tools in contemporary ecological assessments of polluted aquatic environments. A large number of ecological studies have been made of "normal" ostracod communities for the purpose of extrapolating backward in geological time to estimate previous environmental conditions. The data contained in those studies, although collected for another purpose, form a large body of solid baseline data which could be analyzed in light of pollution work, augmented by more pertinent site-specific studies, and used to advantage in on-going pollution assessments.

Tables I–III summarize the kinds of ecological information contained in a number of references to freshwater, estuarine, and marine ostracods and/or assemblages—and note papers that appear to contain data of particular significance to pollution work.

TABLE III

<small>Selected References to Ecological Factors Related to Marine Ostracods and/or Ostracod Assemblages</small>

	Temperature	Salinity	pH	DO	PO₄	Alkalinity	Nutrients	Carbon production	Depth	Currents	Light penetration	Substrate composition	Associated organisms	Seasonal data	Species diversity	Abundance	Statistical analyses	Laboratory studies
Baker and Hulings (1966)	X							X	X				X					
Bate (1971)	X	X											X					
Benson (1959)ᵃ	X	X						X	X	X	X	X	X					
Benson and Coleman (1963)	X	X							X				X					
Benson and Kaesler (1963)	X												X					
Colman (1940)													X	X				
Ganning (1967)	X	X	X	X								X						X
Hagerman (1965)	X								X				X	X				
Kaesler (1966)																	X	
Kaesler and Mulvany (1977)ᵃ									X							X	X	X
McKenzie (1965)													X					
Puri *et al.* (1964)ᵃ	X	X	X	X	X	X			X	X	X	X						
Rosenfeld (1977)		X										X						
Skaumal (1977)	X	X							X				X					
Whatley and Wall (1969)ᵃ	X	X	X	X					X				X	X	X	X		
Williams (1969)									X					X	X			

ᵃPapers containing particularly important data on methodology or data that might be especially relevant to pollution studies.

References

Baker, J. H., and Hulings, N. C. (1966). Recent marine ostracod assemblages of Puerto Rico. *Publ. Inst. Mar. Sci., Univ. Tex.* **11**, 108–125.

Bate, R. H. (1971). The distribution of recent Ostracoda in the Abu Dhabi Lagoon, Persian Gulf. *Bull. Cent. Rech. Pau* **5**, Suppl., 239–256

Benson, R. H. (1959). Ecology of recent ostracodes of the Todos Santos Bay region, Baja California, Mexico. *Univ. Kans. Paleontol. Contrib., Arthropoda, Artic.* **1**, 1–80.

Benson, R. H. (1961). Ecology of ostracode assemblages. *In* "Treatise on Invertebrate Paleontology" (R. C. Moore and C. W. Pitrat, eds.), Vol. Q, pp. 56–63. Univ. of Kansas Press, Lawrence.

Benson, R. H., and Coleman, G. L., II (1963). Recent marine ostracodes from the eastern Gulf of Mexico. *Univ. Kans. Paleontol. Contrib., Arthropoda, Artic.* **2**, 1–52.

Benson, R. H., and Kaesler, R. L. (1963). Recent marine and lagoonal ostracodes from the Estero de Tastiota region, Sonora, Mexico (Northeastern Gulf of California). *Univ. Kans. Paleontol. Contrib., Arthropoda, Artic.* **3**, 1–34.

Benson, R. H., and Maddocks, R. F. (1964). Recent ostracodes of Knysna Estuary, Cape Province, Union of South Africa. *Univ. Kans. Paleontol. Contrib., Arthropoda, Artic.* **5**, 1–39.

Bhatia, S. B., and Singh, D. (1971). Ecology and distribution of some recent ostracodes of the Vale of Kashmir, India. *Micropaleontology* **17**, 214–220.

Bowen, M. (1976). Effects of a thermal effluent on the ostracods of Par Pond, South Carolina. *ERDA Symp. Ser.* **40**, 219–226.

Colman, J. (1940). On the faunas inhabiting intertidal seaweeds. *J. Mar. Biol. Assoc. U.K.* [N.S.] **24**, 129–183.

Elliott, H. A., Ellison, R., and Nichols, M. M. (1966). Distribution of recent ostracodes in the Rappahannock estuary, Virginia. *Chesapeake Sci.* **7**, 203–207.

Elofson, O. (1941). Zur Kenntnis der marinen Ostracoden Schwedens, mit besonderer Berücksichtiqung des Skagertaks. *Zool. Bidr. Uppsala* **19**, 215–534.

Ganning, B. (1967). Laboratory experiments in the ecological work on rockpool animals with special notes on the ostracod *Heterocypris salinus. Helgol. Wiss. Meeresunters.* **15**, 27–40.

Hagerman, L. (1965). The ostracods of the Øresund, with special reference to the bottom-living species. *Ophelia* **2**, 49–70.

Hoff, C. C. (1942). The ostracods of Illinois: Their biology and taxonomy. *Ill. Biol. Monogr.* **19**, 1–196.

Hulings, N. C. (1969). The ecology of the marine Ostracoda of Hadley Harbor, Massachusetts, with special reference to the life history of *Parasterope pollex* Kornicker, 1967. *In* "The Taxonomy, Morphology, and Ecology of Recent Ostracoda" (J. W. Neal, ed.), pp. 412–422. Oliver & Boyd, Edinburgh.

Kaesler, R. L. (1966). Quantitative re-evaluation of ecology and distribution of recent Foraminifera and Ostracoda of Todos Santos Bay, Baja California, Mexico. *Univ. Kans. Paleontol. Contrib., Pap.* **10**, 1–50.

Kaesler, R. L., and Mulvany, P. S. (1977). Approaches to the diversity of assemblages of Ostracoda. *Proc. Int. Symp. Ostracods, 6th, Saalfelden (Salzburg)* pp. 33–43.

Keyser, D. (1975). Ostracodes of the mangroves of south Florida, their ecology and biology. *Bull. Am. Paleontol.* **65**, 489–499.

Keyser, D. (1977). Ecology and zoogeography of recent brackish-water Ostracoda (Crustacea) from south-west Florida. *Proc. Int. Symp. Ostracods, 6th, Saalfelden (Salzburg)* pp. 207–222.

Kilenyi, T. I. (1969). The problems of ostracod ecology in the Thames Estuary. *In* "The Taxonomy, Morphology, and Ecology of Recent Ostracoda" (J. W. Neal ed.), pp. 251–267. Oliver & Boyd, Edinburgh.

Kilenyi, T. I. (1971). Some basic questions in the palaeoecology of ostracods. *Bull. Cent. Rech. Pau,* Suppl. **5**, 31–44.

King, C. E., and Kornicker, L. S. (1970). Ostracoda in Texas bays and lagoons: An ecologic study. *Smithson. Contrib. Zool.* **24**, 1-92.

Klug, A. B. (1927). The ecology, food relations and culture of freshwater Entomostraca. *Trans. R. Can. Inst.* **16**, 15-99.

Kornicker, L. S. (1964). A seasonal study of living Ostracoda in a Texas bay (Redfish Bay) adjoining the Gulf of Mexico. *Publ. Stn. Zool. Napoli* **33**, Suppl., 45-60.

Kornicker, L. S. (1975). Spread of ostracodes to exotic environs on transplanted oysters. *Bull. Am. Paleontol.* **65**, 129-139.

Kornicker, L. S. (1977). Myodocopid Ostracoda of the Indian River complex, Florida. *Proc. Biol. Soc. Wash.* **90**, 788-797.

Kornicker, L. S., and Wise, C. D. (1962). *Sarsiella* (Ostracoda) in Texas bays and lagoons. *Crustaceana (Leiden)* **4**, 57-74.

Krutak, P. R. (1975). Environmental variations in living and total populations of Holocene Foraminifera and Ostracoda of coastal Mississippi, U.S.A. *Am. Assoc. Pet. Geol. Bull.* **59**, 140-148.

McKenzie, K. G. (1964). Ostracod fauna of Oyster Harbour, a marginal marine environment near Albany, Western Australia. *Publ. Stn. Zool. Napoli* **33**, Suppl., 421-461.

McKenzie, K. G. (1965). Myodocopid Ostracoda (Cypridinacea) from Scammon Lagoon, Baja California, Mexico, and their ecologic associations. *Crustaceana (Leiden)* **9**, 57-70.

McKenzie, K. G., and Swain, F. M. (1967). Recent Ostracoda from Scammon Lagoon, Baja California. *J. Paleontol.* **41**, 281-305.

Malkin, D. S. (1953). Biostratigraphic study of Miocene Ostracoda of New Jersey, Maryland, and Virginia. *J. Paleontol.* **27**, 761-799.

Mason, I. L. (1939). Studies on the fauna of an Algerian hot spring. *J. Exp. Biol.* **16**, 487-498.

Phleger, F. B., and Ewing, G. C. (1962). Sedimentology and oceanography of coastal lagoons in Baja California, Mexico. *Geol. Soc. Am. Bull.* **73**, 145-182.

Pinto de Ornellas, L. (1974). *Minicythere heinii* Ornellas, gen. et sp. nov., from southern Brazil, and a characteristic ostracode association of brackish water environment. *An. Acad. Bras. Cienc.* **46**, 469-496.

Puri, H. S., Bonaduce, G., and Mallory, J. (1964). Ecology of the Gulf of Naples. *Publ. Stn. Zool. Napoli* **33**, Suppl., 87-199.

Remane, A. (1934). Die Brachwasser-Fauna (mit besonderer Berücksichtigung der Ostsee). *Verh. Dtsch. Zool. Ges.* **36**, 34-74.

Rosenfeld, A. (1977). Die rezenten Ostracoden-Arten in der Ostsee. *Meyniana* **29**, 11-49.

Skaumal, U. (1977). Preliminary account of the ecology of ostracods on the rocky shores of Helgoland. *Proc. Int. Symp. Ostracods, 6th, Saalfelden (Salzburg)* pp. 197-205.

Tressler, W. L., and Smith, E. M. (1948). "An Ecological Study of Seasonal Distribution of Ostracoda, Solomons Island, Maryland, Region," Publ. No. 71. Chesapeake Biol. Lab., Solomons Island, Maryland.

von Morkhoven, F. P. C. M. (1962). "Post-paleozoic Ostracoda: Their Morphology, Taxonomy, and Economic Use," 2 vols. Am. Elsevier, New York.

Wagner, C. W. (1957). Sur les Ostracodes du Quaternarie Récent des Pays-Bas et leur utilisation dans l'étude géologique des dépôts holcènes. Dissertation, Université de Paris, Mouton en Co., Den Haag.

Whatley, R. C., and Wall, D. R. (1969). A preliminary account of the ecology and distribution of recent Ostracoda in the southern Irish Sea. *In* "The Taxonomy, Morphology and Ecology of Recent Ostracoda" (J. W. Neal, ed.), pp. 268-298. Oliver & Boyd, Edinburgh.

Williams, R. (1969). Ecology of the Ostracoda from selected marine intertidal localities on the coast of Anglesey. *In* "The Taxonomy, Morphology, and Ecology of Recent Ostracoda" (J. W. Neal, ed.), pp. 299-329. Oliver & Boyd, Edinburgh.

Note Added in Proof

A recent paper (Athersuch, 1979) was brought to our attention after this chapter was in proof. Recognizing the possibility of using ostracods as indices of pollution, Athersuch undertook a study of the ecology and distribution of the littoral ostracods of Cyprus. Cyprus was chosen in order to evaluate the factors affecting ostracod distribution in unpolluted waters, so that later the effects of pollutants can be assessed against that baseline.

Athersuch noted that previous studies have shown that salinity, pH, dissolved oxygen, depth, temperature, and substrate are considered to be the most significant physical factors affecting ostracod distribution, and he discussed the use of scuba (self-contained underwater breathing apparatus) diving as being the most suitable means of collecting ostracod study material.

Considering only living ostracods, he recognized six main biotopes in Cypriot waters: (1) filamentous algae, (2) sea grasses, (3) calcareous algae, (4) encrusting algae, (5) sand, and (6) silt. Details of species composition for each biotope are given, and preliminary analysis suggests that the species fall into three groups: a cosmopolitan group found throughout the Mediterranean and the northeast Atlantic, a group endemic to the Mediterranean, and a small group so far known only from the Mediterranean. No evidence was found that minor variations in pH, dissolved oxygen, or salinity affect ostracod distributions in the study area.

Reference

Athersuch, J. (1979). The ecology and distribution of the littoral ostracods of Cyprus. *J. Nat. Hist.* **13,** 135–160.

CHAPTER 5

Copepods (Arthropoda: Crustacea: Copepoda)

JOHN K. DAWSON

I. Introduction

This chapter summarizes available data related to the effects of various pollutants on holoplanktonic estuarine copepods found to occur in the North American continent. Included also are data on the pollution of estuarine and nonestuarine species in neretic environments. Information on nonestuarine copepods is included, however, when little or no data were available on estuarine species. It is assumed that the effects of pollution on neretic copepods may give insight to the pollutants' effects on estuarine species.

145

II. Ecology of Estuarine Copepods

While the major emphasis of this chapter is on the effects of pollution on estuarine copepods, it is impossible to recognize these effects unless a fair amount is known about the "normal" ecology of estuarine copepods. As stated by Golubic (1971), "The basis for a biological analysis of the pollution should be a profound knowledge about the composition, distribution, and ecological inter-dependencies of the dominant communities of nonpolluted areas." While pristine areas are rare, if not impossible to find in the North American continent, a review of the knowledge of these interdependencies at the present level of pollution may aid in providing a basis for a biological analysis of future pollution levels.

Estuaries have long been known for their very high levels of production, and as Cronin *et al.* (1962) state, zooplankton standing crop is always greater in an estuary as compared to levels in the adjacent river and ocean. This "truism," however, has been questioned by Williams *et al.* (1968). Their study showed that the role of the intermediate step between primary and carnivore production traditionally played by zooplankton is doubtful in shallow water estuaries (1 m or less). They calculate that the zooplankton had a carbon demand of only a fraction of the net photosynthesis of phytoplankton. Qasim (1970) suggests an increase in the benthic herbivores in shallow estuaries probably compensates for the lack of zooplankton and most likely makes the system as efficient as others.

In estuaries that are not extremely shallow, zooplankton are the primary avenue in transferring energy of primary to carnivore production. Copepods and clado-cerans are the only two holoplanktonic groups which have adapted to the vicissitudes of estuarine life and are the mainstays in this energy transfer. Calanoid copepods are the dominant members of the estuarine zooplankton (Cro-nin *et al.*, 1962; Bakker and dePauw, 1975; Hulsizer, 1976). Cladocerans and cyclopoid copepods become abundant only for brief periods in confined areas according to Jeffries (1967), who has described estuarine holoplankton as "a monotonous assortment of calanoid copepods."

Various copepod groups have been characterized for their respective positions within estuaries. Wooldridge (1976) and Haertel and Osterberg (1967) described a three-position grouping where the most landward group is the low salinity or freshwater copepods. Members of this group penetrate brackish water by virtue of their being carried down from upstream populations. A second group, the one endemic to the estuary, is dominated by *Eurytemora,* which comprises up to 100% of the zooplankton. Few species are characteristic of this group, but concentrations in excess of 2500/m³ may be achieved. The third group is that of the saltwater intrusion, dominated by *Acartia* and *Pseudocalanus.* These genera are most abundant, for example, in the Columbia River estuary (Haertel and Osterberg, 1967) during the fall when low river flow and high tides combine for maximum saltwater intrusion. A similar nearly year-round situation occurs in Southern California, where the long dry season prevents a fixed salinity gradient. As a result, the dominant year-round copepods in Los Angeles Harbor are *Acartia* and *Paracalanus* (Dawson, 1976).

Another, perhaps more defined, classification of estuarine copepod groups is that of Jeffries (1967). He bases the four groups on their ability to reproduce and maintain populations over a range of salinity rather than as isolated occurrences at exceptional salinities. These groups are illustrated and described in Fig. 1.

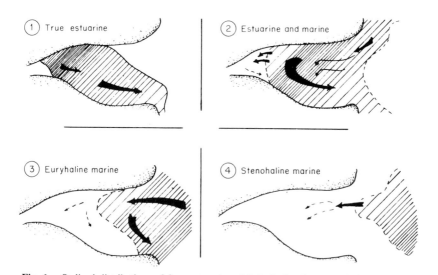

Fig. 1. Stylized distributions of four categories of holoplanktonic copepods in a hypothetical estuary. Darkened arrows indicate the drift of animals produced in areas shown by diagonal lines; the lines are closely spaced in the centers of propagation. Relative development of each component in an estuary is a function of salinity distribution and net circulation. 1, True estuarine—propagates only in brackish water; tolerance for reproduction under natural conditions of interplay between members of the community is approximately 5–30‰. Found in the open ocean as strays from less saline waters. 2, Estuarine and marine—propagates throughout a major portion of an estuary's length, usually spanning the gradient zone; reproduction is not limited exclusively to the marine zone and population development is usually limited by salinities less than 10‰; estuarine populations are maintained by indigenous recruitment, not dependent on influxes from offshore to maintain critical population densities. Ocean populations are generally most abundant near the coast. 3, Euryhaline marine—all stages in life history of the species usually found throughout the marine zone, but these are adventitious migrants from the ocean carried landward by tidal action; eggs, nauplii, and copepodites have an incomplete development, probably including molts; maintenance of the population is, however, dependent on continuous supply from the ocean. 4, Stenohaline marine—adults and other late copepodites occur infrequently near the mouth of the estuary, occasionally straying through the marine zone; nauplii and copepodite stages I–III are absent or very scarce; characterize open neritic waters. (From H. P. Jeffries, 1967, *Publ. Am. Assoc. Adv. Sci.* **83,** 500–508. Copyright 1967 by the American Association for the Advancement of Science.)

The dominant estuarine copepod genera are clearly *Acartia* and *Eurytemora* (Jeffries, 1967; Bakker and dePauw, 1975; Knatz, 1978) with *A. tonsa* Dana, 1849 and *A. clausi* Giesbrecht, 1889 the important former species and *E. affinis* (Poppe, 1880) and *E. americana* Williams, 1906 the significant latter species (Barlow, 1955; Cronin *et al.*, 1962; Heinle, 1972; Stickney and Knowles, 1975; Hulsizer, 1976). The two *Acartia* species alternate dominance in the mid-Atlantic state estuaries, competing twice a year probably between younger stages and eliminating the species less adapted to the prevailing conditions (Conover, 1956). *Acartia tonsa* is primarily abundant in the summer and fall and *A. clausi,* a more cold water form, is a winter–spring dominant (Jeffries, 1962c; Sage and Herman, 1972; Heinle, 1972; Hulsizer, 1976). Because of the temperature preference of each species, *A. clausi* is the most abundant of the two in more northern estuaries (Riley, 1967; Dawson, 1969; Hulsizer, 1976), while *A. tonsa* is predominant in estuaries further south (Deevey, 1960; Heinle, 1966; Dawson, 1976). Deevey (1960) discusses the geographic range and seasonal cycles of these two species. She believes that *A. tonsa* is limited in its geographic and seasonal distribution not as much by temperature directly as by the presence of dominant cold water species such as *A. clausi.* From this, Riley (1967) concludes that biological competition at suboptimal levels of temperature and salinity are more important upon survival than direct effects of either of these parameters.

Eurytemora affinis and *E. americana* are the chief "true-estuarine" (Jeffries, 1967) species. Deevey (1948) and Jeffries (1962a) identified *Eurytemora hirundoides* (Nordqvist, 1888) as important estuarine species; however, these identifications were apparently incorrect and should have been assigned to *E. affinis* (Deevey, 1960; Jeffries, 1967). Gurney (1931) questioned the validity of *E. hirundoides,* and regards this species to be within the range of variation of *E. affinis.* Wilson (1959) also believes that most American records of *E. hirundoides* (e.g., Cronin *et al.*, 1962) are probably *E. affinis.* Developmental stages of *E. affinis* have also been mistakenly identified as *E. hirundoides* (Davis, 1943). Katona (1971) gives a more complete description of these developmental stages.

The estuarine distribution and succession of *E. affinis* and *E. americana* are discussed at length by Jeffries (1967). These species typically dominate the upper estuary during the winter and spring where salinities are less than 15‰ (Cronin *et al.*, 1962; Jeffries, 1964).

The two preceding genera, i.e., *Eurytemora* and *Acartia,* are the holoplanktonic groups that make estuarine areas distinctly different from either the adjacent ocean or river and considerably richer than either (Cronin *et al.*, 1962). Other typically neritic euryhaline species also contribute to the estuarine copepod fauna—however, in usually reduced numbers or in only sporadic abundance.

The calanoid copepods include *Pseudocalanus minutus* (Krøyer, 1849), *Centropages hamatus* (Lilljeborg, 1853), *Temora longicornis* (Müller, 1792),

Paracalanus crassirostris Dahl, 1894, *Labidocera aestiva* Wheeler, 1899, and *Pseudodiaptomus coronatus* Williams, 1906 (Deevey, 1952b, 1960; Cronin *et al.*, 1962; Sage and Herman, 1972; Hulsizer, 1976). Adults of *P. coronatus* are not considered to be truly planktonic by Jacobs (1961) because of its close association with the bottom (Jeffries, 1959) and thus may be considered tycho-planktonic rather than holoplanktonic.

Of the cyclopoid copepods, *Oithona similis* Claus, 1866, *O. brevicornis* Giesbrecht, 1891, *O. spinirostris* Claus, 1863, and *O. plumifera* Baird, 1843 are all numerous (Deevey, 1952a,b, 1960; Heinle, 1972; Sage and Herman, 1972) and are all, especially *O. brevicornis,* too small to be quantitatively retained by standard mesh plankton nets (Deevey, 1960). For this reason their importance is probably underestimated by the numbers collected.

Harpacticoid copepods, which at times are important members of the estuarine plankton, are *Euterpina acutifrons* (Dana, 1852) and *Scottolana canadensis* (Willey, 1923) (Heinle, 1972; Stickney and Knowles, 1975). The latter species applies only to the pelagic nauplii since the adults are epibenthic in habitat.

One of the most pressing problems of indigenous planktonic existence within an estuary is that of maintenance of a population. The Delaware River estuary, for example, transports an average of 12,000 ft^3 of water per second to the ocean (Cronin *et al.*, 1962). This kind of transport down the estuary poses obvious problems for estuarine planktonic existence.

Factors that aid in the maintenance of species are both physical and biological in nature. During the winter and spring there is typically a layering effect of the freshwater flowing down river over intruded seawater lying on the bottom. This system is Pritchard's (1955) "B" type estuary. Copepods that exhibit diurnal vertical migration can take advantage of this type of system by being in the deep-water inflow during the day. *Acartia* is an excellent example of a species which undergoes diurnal vertical migration (Motoda *et al.*, 1971; Wooldridge, 1976; Furuhashi, 1976) and makes use of it in maintaining themselves within the estuary (Barlow, 1955; Haertel and Osterberg, 1967). During periods of low water flow, as in the summer, estuaries are often homogeneously mixed as described by Pritchard's "C" type system. Under such a system, vertical migration will not aid in estuarine copepod maintenance. However, with a fast enough reproduction rate, a species may maintain itself by use of the lateral flow system where there is a net downstream flow along the right side of the estuary (facing downstream) and a net upstream flow along the left (see Fig. 1). *Eurytemora* is a genus in which vertical migration was not shown to be a significant maintenance mechanism, according to Haertel and Osterberg (1967). The primary mechanism of maintenance in this genus is apparently reproductive increases since flooding produced a loss of plankton population, a result consistent with the shorter flushing time during flooding. Ketchum (1954) has shown that the net seaward movement determines the rate at which plankton must reproduce in order to maintain themselves within an estuary no matter what maintenance system is used. Other mechanisms may also come into play in retaining copepod populations in estuaries, but have not been thoroughly investigated as of yet. Cronin *et al.* (1962) suggested that the more slowly flushed shoal margins of an estuary may hold copepod populations in reservoirs and Barlow (1955) postulated that resting eggs on the bottom may help to alleviate the problem of estuarine maintenance in some species of copepod.

III. Thermal Pollution and Power Plant Operations

With increasing demands for electricity in this country, utility companies will be trying to locate more power plants where adequate amounts of cooling water are available. This results in an accelerated rate of infringement of heated effluent upon estuarine waters. The operation of these power plants, be it fossil fueled or nuclear, can have very definite effects on the copepod fauna of estuaries either by heating the receiving waters or by killing plankton in the entrained cooling water.

The effects of heated effluent discharged into an estuary will depend on the size, circulation, and flushing rate of the estuary. Addition of heated water will result in an extension of the temperature range to which estuarine copepods must adapt if they are to survive. Survival may be in doubt, however, when the normal estuarine temperature is near the thermal tolerance limit of the estuarine species. This appears to be the case in the Chesapeake Bay area for *Acartia tonsa* and *Eurytemora affinis,* which have similar upper thermal limits (30–35°C) and are not greatly influenced by acclimation (Heinle, 1969). Their upper thermal limits are near the normal summer temperature of this area (Cory, 1967). Survival of the two species is in obvious danger if the summer water temperatures are pushed further by more power plant operation. The upper thermal maxima for the species may be even more closely approached throughout a greater part of the year in lower latitude estuaries.

Heinle (1969) has shown that the effect of thermal tolerance can be seen in the growth of copepods. He indicates that near or below 30°C, the growth of *A. tonsa, Oithona brevicornis,* and *Canuella canadensis* (Willey, 1923), which is synonymous with *Scottolana canadensis* (Coull, 1972), was good, but temperatures above 30°C resulted in poor growth and survival with little or no reproduction. Copepod nauplii have also been shown to be less abundant in the area affected by a power plant water discharge (Tsuruta and Tawara, 1975).

In more northern latitude estuaries, a warm water discharge can augment the presence of lower latitude species and may create conditions which make possible the introduction of semitropical species. This has been shown to occur in Southampton waters, where semitropical *A. tonsa* has increased in abundance

(Conover, 1957; Raymont, 1964) and a related warm water species, *Acartia grani* (Sars, 1904), has appeared for the first time (Raymont, 1964). Other changes in the community structure can also result by thermal addition to estuaries. Heinle (1969) has noted that the two epibenthic copepod species, *Oithona brevicornis* and *Canuella canadensis* = *Scottolana canadensis,* appear to tolerate higher temperatures than the holoplanktonic species *A. tonsa* and *E. affinis.* Thus, there is a potential for a diversion of energy from the plankton to the benthos if temperature favors a more tolerant epibenthic species. Further ecological changes can result from the general tendency of a copepod species to be smaller when reared at higher temperatures. As a result, energy flow patterns can change as reported by Heinle (1969) studying the Patuxent estuary. He discusses the three classes of planktivores: (1) the passive feeders, such as medusae, ctenophores, and such fish as the menhaden; (2) the mechanically selective filter feeders, such as the alewives; and (3) the behaviorally selective filter feeders, such as silversides and anchovies. Among these three classes, he believes that a decrease in size of copepods would result in greater energy flow to the first class of predators.

Competition between species may also be altered by a temperature increase, creating suboptimal conditions for one species and thus allowing another to out-compete it. Temperature influence on competition has already been discussed between *Acartia tonsa* and *A. clausi,* where presumably a temperature increase in northern Atlantic estuaries would favor the former species at the expense of the latter. This potential for changing the species' competition and patterns of energy flow, as Heinle (1969) points out, is particularly great in estuaries where the number of individuals is sizable, but the diversity of species is typically low.

In addition to a general heating of receiving waters, power plant operations can have a significant effect upon the trophic food web of an estuary by the entrainment of zooplankton in the cooling water. Death of these organisms can result from a number of traumas, such as temperature rise shock, mechanical abrasion, rapid pressure change, and chemically induced biocidal action. This loss of biomass can be more than trivial, for Carpenter *et al.* (1974) have calculated a loss in production of 0.1 to 0.3% of the zooplankton in the eastern Long Island Sound by the operation of the Millstone plant. They have concluded that if the removal rate of the ten power plants in Long Island Sound is similar to that of the Millstone plant, then 0.05% of the total area of Long Island Sound would be affected. The entrainment and subsequent death of a large percentage of zooplankton can be approached from Enright's (1977) viewpoint that the power plant is a giant nonselective predator ingesting large amounts of zooplankton and producing an equivalent amount of "fecal" material of nearly equal nutritional value. He then points out that the replacement of biomass of predators which ordinarily feed on zooplankton by the power plant predation can be calculated along with the compensatory increase of decomposers and detritus feeders which live off the "fecal" material.

The extent to which copepods are killed by passing through the cooling system of power plants, as well as the cause of these mortalities, has been investigated in a number of studies (Carpenter *et al.,* 1974; Davies and Jensen, 1975; Heinle, 1976). Several studies indicate that death of entrained copepods is due to thermal shock. Reeve and Cosper (1970) showed a 50% mortality to *Acartia tonsa* from Biscayne Bay, Florida, when held at 32°C for 3 hours with ambient temperature at 21°C. In the summer, however, with a higher ambient temperature, there was more than 25% mortality at 36°C for 6 hours. Suchanek and Grossman (1971) found at the Northport plant on Long Island Sound a mortality of 100% when temperature was in excess of 34°C. Temperature shock, however, was not shown to be a significant factor in the death of estuarine-entrained copepods in other studies. Heinle (1976) studied the effects of copepod entrainment in three power plant operations. Although the Vienna power plant in the upper Nanticoke River estuary raised the temperature above the lethal limits of *A. tonsa, E. affinis,* and *S. canadensis* (26°–37.5°C), there was no significant reduction of live copepods in the absence of chlorination. At the Chalk Point plant on the Patuxent River estuary and the Morgantown plant on the Potomac River estuary, a temperature rise alone resulted in only slight to no mortalities, but with the addition of chlorination there were significant mortalities of the copepods. The conclusion is that chlorination was the probable cause of copepod mortalities, with an increasing order of sensitivity from *S. canadensis* to *E. affinis* to *A. tonsa.* Heat, however, was found to be lethal to copepod eggs and early nauplii at the Chalk Point plant. Heinle's (1976) study showed only the immediate effects of passage, while Reeve (1970) indicated that delayed effects were important in assessing the thermal effects on zooplankton. Death could not be determined with certainty for up to 24 hours following a temperature cycle simulating entrainment. Carpenter *et al.* (1974) did study the delayed effects of copepod passage through the Millstone plant. Vital staining indicated that less than 15% died immediately following passage, 50% died in 3½ days, and 70% died 5 days after passage, with a control showing 10% mortality. His study also indicated that chlorination and heat were not the main causes of copepod mortality and that mechanical or hydraulic stress were responsible for copepod loss at the Millstone plant. A further study (Davies and Jensen, 1975) at the power plant on the Indian River, Delaware, showed relatively little effect on the zooplankton entrained using mobility ratios of pre- and postentrainment as indicators. The low average temperature rise (6°C) and short travel time (2 minutes) may be responsible for the negligible immediate effects on entrained zooplankton. These mobility ratios do not indicate lack of subtle sublethal effects which Reeve (1970) suggests could include sterility. Such effects can only be predicted with long-term experiments.

Thus, it appears that mortality of entrained copepods through power plant cooling systems may depend to a large degree on acclimation temperature, temperature increase, transit time, biocide concentrations, and physical design of the cooling system relative to hydraulic shock and mechanical abrasion.

IV. Oil Pollution

The effects of an oil spill and subsequent oil pollution on estuarine copepods have received little attention to date. A number of studies, however, have been conducted on the more typically coastal marine copepod species. These studies will be reviewed with the assumption that much of what has been learned about oil pollution effects on coastal species may also be relevant to estuarine copepods.

There are three main areas of concern regarding the effects of oil spills on copepod fauna (Moore and Dwyer, 1974). These include direct lethal toxicity of the oil or oil fraction; sublethal disruption of physiological or behavioral responses, of which extremely little is known; and persistence and accumulation of oil within copepods, which may be passed up the food web trophic structure.

Direct lethal toxicity is to a large extent dependent on the type of oil pollution and its relative percentages of the most toxic fractions. The toxicity of an oil is a function of the di- and triaromatic hydrocarbon content, i.e., the naphthalene and other naphthalene-like compounds (Boylan and Tripp, 1971; Anderson *et al.*, 1974; Lee and Anderson, 1977). Moore *et al.* (1974) agreed with conclusions that the soluble aromatics of an oil produce the primary toxic effects in the marine environment. Other studies (Berdugo *et al.*, 1977) have indicated that naphthalene by itself is not a particularly toxic compound, but when mixed with other hydrocarbons, its toxicity is increased. Corner *et al.* (1976b) has pointed out, however, that during weathering of oil through oxidation there can be a formation of thiacyclones (sulfoxides) and alkylphenols. These changes can affect the toxicity to zooplankton.

Mironov (1969) tested the survival rates of *Acartia clausi, Paracalanus parvus* (Claus, 1863), and *Oithona nana* Giesbrecht, 1892 against various concentrations of crude oil. He determined that 0.001 ml/liter accelerated the death and 0.1 ml/liter killed all species within 1 day. These results, however, have been criticized by Corner *et al.* (1976b) on the basis that the oil composition relative to the naphthalene fractions was not stated and that only marginal differences existed between the controls and the lowest levels of oil concentration. Furthermore, the copepods appeared to be under stress since half of the control animals died in less than 1 week.

Although the majority of the hydrocarbons present in water are in solution, most hydrocarbons are accumulated through particulate material in the *Calanus* diet (Corner *et al.*, 1976a,b). This indicates the possibility that the particulate fraction may be the most important in terms of copepod accumulation. For this reason, bioassay procedures to determine the lethal effects of accumulated hydrocarbons should be done in terms of assessing the concentration of hydrocarbons present as both food for zooplankton as well as accumulations from the soluble form.

The rate and extent to which hydrocarbons are removed from copepods by either metabolism or depuration determine to a large extent the amount of hydrocarbons passed up the food web. Lee (1975) has found that, while there was a rapid loss of accumulated hydrocarbons from *Euchaeta japonica* Marukawa, 1921, *Calanus helgolandicus* (Claus, 1863), and *C. plumchrus* Marukawa, 1921 when placed into fresh seawater, trace levels were always present in the copepods. Similar results were found by Corner *et al.* (1976b) and Harris *et al.* (1977) for the estuarine copepod *Eurytemora affinis* as shown in Fig. 2. This figure indicates a rapid decline of ^{14}C-labeled naphthalene after the copepods were immersed for 24 hours at a concentration of 0.96 ppm and then placed into fresh seawater. The initial depuration was rapid (over half lost in 1 day), but some radioactivity persisted at the end of the 9-day experiment. Lee (1975) indicated that the rapid loss of hydrocarbons is not simply from desorption from sites on the surface, since *Calanus* fed ^{14}C-labeled naphthalene in *Biddulphia* cells also showed early rapid loss and a leveling off of radioactivity. However, the rate of ingested hydrocarbon depuration was found by Corner *et al.* (1976b) to be slower than that directly accumulated from seawater. Lee's (1975) general conclusion is that copepods are able to metabolize aromatic hydrocarbons, based on evidence that 90% of the radioactivity in the copepods was present as un-

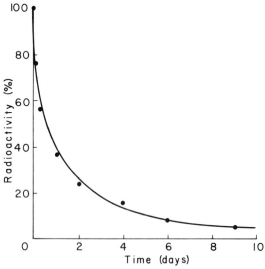

Fig. 2. Release of radioactivity by *Eurytemora affinis* previously treated with ^{14}C-labeled naphthalene. (From Corner *et al.*, 1976b, reproduced by permission.)

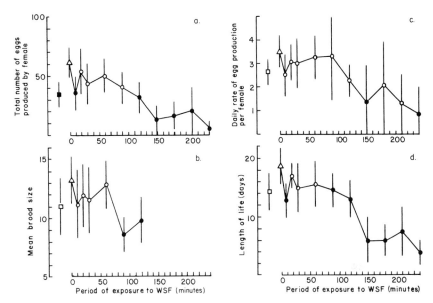

Fig. 3. Effect of exposure for different periods of time to water-soluble fraction of aromatic heating oil on subsequent length of life and egg production by female *Eurytemora affinis*. (a) Total number of eggs produced by each female, (b) mean brood size, (c) rate of egg production, (d) length of life. △, control animals; ○, mean values not significantly different from control; ●, mean values significantly different from the control (Mann–Whitney *U* test, 5% level); □, animals exposed for 24 hours to a solution of naphthalene at 1 mg/liter (mean not significantly different from control); ■, animals exposed for 24 hours to a solution of naphthalene at 1 mg/liter (mean significantly different from control at the 5% level). Means are calculated for 10 replicates, vertical bars show ± 95% confidence limits. (From Berdugo *et al.*, 1977, reproduced by permission.)

changed naphthalene, while over two-thirds of that released was in forms other than naphthalene. Similar results and conclusions were reached by Corner *et al.* (1976b) from feeding experiments of nauplii laced with ^{14}C-labeled naphthalene to *Calanus helgolandicus*.

Berdugo *et al.* (1977) have investigated the effects of these persistent low hydrocarbon levels after depuration in the estuarine copepod, *Eurytemora affinis*. They exposed the copepods to relatively high concentrations of the water-soluble fraction of aromatic heating oil to simulate the high levels which might be expected to occur in the immediate vicinity of an oil slick.

Total hydrocarbon levels of 0.52 mg/liter showed a 38% reduction of ingestion rate, while levels in excess of 1 mg/liter caused narcotization of the copepods. Although narcotization was reversible upon transfer to clean seawater, extended exposure to these levels of hydrocarbon resulted in mortality.

After relatively brief exposures to the water-soluble fraction, the subsequent lives of the copepods were observed relative to life span, total number of eggs produced by each female, mean brood size, and rate of egg production per female per day. The results of these statistics are illustrated in Fig. 3. A shorter life was shown for all exposed copepods with an approximately linear decrease in longevity with increased exposure time. The total number of eggs produced by each female showed, as expected, a similar pattern as the longevity study. The number of eggs contained in each egg sac was lower than the controls in all exposed copepods, but because of the large variance, only those females exposed for 90 and 120 minutes were significantly different from the controls. Increased exposure to the water-soluble fraction (WSF) resulted in a decreased egg production rate with differences significant at the 150- and 240-minute duration.

Similar studies were also conducted by Berdugo *et al.* (1977) after exposure of *E. affinis* to 24 hours of 1 mg/liter naphthalene. All studies indicated (Fig. 3) lowered values for exposed copepods, but only in the total number of eggs produced by each female was it significantly lower than the controls.

These studies indicate that exposure of *E. affinis* to concentrations of hydrocarbon for periods of less than 4 hours, as might be found under an oil slick, can have significant sublethal effects on reproduction during the subsequent life span of the female.

In long-term, low-level experiments of feeding naphthalene to *Eurytemora affinis*, Harris *et al.* (1977) has shown that, at concentrations of 1 μg/liter and 10 μg/liter for 9 and 15 days, respectively, there was no further uptake of hydrocarbon. Even at concentrations as high as 50 μg/liter, a leveling off of incorporated hydrocarbon was seen, which indicates a steady state within the copepod where hydrocarbon uptake equaled hydrocarbon loss through metabolism.

Harris *et al.* (1977) have indicated that a much higher concentration of naphthalene per body weight is accumulated by the estuarine copepod *Eurytemora affinis* than by the offshore species *Calanus helgolandicus* from exposure to equal concentrations. If this is true for small copepods in general, they concluded that it may have definite implications to fish feeding on these copepods. They have calculated that a herring requires about 2.4×10^4 *C. helgolandicus* or 4.2×10^5 *E. affinis* for weekly body weight maintenance. For an inshore area with a 1.0-μg naphthalene/liter concentration, the fish will ingest naphthalene to the extent of 3.24 μg when feeding on *C. helgolandicus* and 156.2 μg feeding on *E. affinis*. In this case, feeding on small copepods represents a 48-fold increased hydrocarbon intake.

It is interesting to note that while spilled oil may affect the life of copepods living in the area, copepods apparently also affect the oil spill by ingesting suspended particles of oil and immobilizing it within the fecal membrane upon defecation. Freegarde *et al.* (1971) found *Calanus finmarchicus* (Gunner, 1765) feeding on a fine suspension of Kuwait crude and releasing it in fecal pellets. Conover (1971) made similar observations of *C. finmarchicus* and *Temora longicornis* feeding on weathered bunker C oil particles down to a depth of 80 m.

There were no apparent toxic effects of this suspension feeding in either study and both authors concluded that the bacteria within the fecal pellet may hasten the decomposition of the hydrocarbon. It is important to note, however, that the crude oil concentration used by Freegarde *et al.* (1971) was within the range that Mironov (1969) found toxic to copepods over a longer test period. Parker *et al.* (1970) also confirmed the ingestion of oil by adding a fluorescent tracer to the oil and unambiguously confirmed the presence of considerable quantities of oil within the gut and fecal pellets of copepods.

The extent to which copepods can contribute to the immobilization of an oil spill will depend on the toxicity of the oil, degree of dispersion into suspended droplets, concentration of copepods present in the water, and feeding rate. Conover (1971) estimates that 0.21 metric tons of oil per day could be sedimented to the bottom by zooplankton feces for a total of 20% of the oil spill. Parker *et al.* (1970) estimate that a shoal of copepods numbering 2000/m^3 over an area 1 km^2 to 10 m deep could immobilize 3 tons of oil per day. Thus, it appears that copepods may be a very important natural agent leading to eventual degradation of oil spills.

V. Heavy Metal Pollution and Other Toxic Elements

Heavy metals such as copper, zinc, and cobalt are essential in trace amounts for normal development and growth among estuarine zooplankton. Such trace quantities are normally supplied to estuaries by the flow of rivers into the estuaries and ultimately to the sea. In populated areas, sufficient additional heavy metals may enter estuaries from a number of sources such as sewage, industrial effluents, street runoff, and atmospheric pollution to become toxic. It is important, therefore, to determine at what levels heavy metals and other substances as well become toxic to estuarine and marine organisms.

The fate of heavy metals within the estuary and marine environment has been reviewed by Bryan (1971). He discusses the three processes that remove heavy metals from water. If the concentration of metal is higher than the solubility of the least soluble metal-binding compound, then this metal will be removed by precipitation. Adsorption of metals occurs at the surfaces of particles such as suspended clays and phytoplanktonic organisms, and absorption of metals from water is a major avenue of heavy metal entry into estuarine organisms. Bryan (1976) has shown that heavy metal entry into benthic organisms is primarily by food and particle ingestion. If this is also true of estuarine copepods, then the adsorption of heavy metals into phytoplankton and subsequent ingestion by feeding copepods may be a secondary means of entry into zooplankton.

The detrimental effects of heavy metals depends on the type and concentration of metal pollution and the species of organism. The degree of metal toxicity cannot be rigidly stated since much will depend on species affected as well as other factors acting antagonistically or synergistically with the metal. Bryan (1971), however, states generally that in decreasing order of toxicity, mercury, silver, and copper are the most toxic; followed by cadmium, zinc, and lead; then by chromium, nickel, and cobalt. Factors that influence the toxicity of these metals include the form of the metal in water, e.g. soluble ion, complex, compound, chelate, or in the particulate form; presence of other metals or poisons acting synergistically; factors influencing the physiology of the organism, e.g. salinity, temperature, pH, dissolved oxygen; and condition of the organism, e.g., stage in life cycle, activity of animal, and acclimation to metals.

In addition to the lethal nature of many of these metals, sublethal effects may be at least as injurious to copepod populations since often these effects are overlooked or are too difficult to measure. Bryan (1971) lists three areas of sublethal effects: (1) morphological change; (2) inhibitory effects, such as change in growth rate and sexual development; and (3) behavioral change, which may result in lowered ability to escape predators or compete effectively.

The two heavy metals that have received the most attention relative to their effects on copepods are mercury and copper. While both metals are considered very toxic, mercury has been found to be between two and four times more toxic than copper in higher concentrations and about equal toxicity in lower concentrations of 1 to 10 μg/liter (Reeve *et al.*, 1976, 1977b). Corner and Sparrow (1956) have determined the relative toxicities of some heavy metal compounds for *Acartia clausi* (Fig. 4). With mercuric chloride as unity, ethylmercuric chloride = 2.0; mercuric iodide = 1.7; and copper sodium citrate = 0.01. As can be seen, the latter copper compound is two orders of magnitude less toxic than the former three mercury compounds.

Heavy metals are, as expected, most concentrated in estuarine and coastal waters with a declining concentration offshore. Windom *et al.* (1973) have shown this for mercury, where a high of 5.3 ppm was found in the New York Bight as opposed to 0.1–0.3 ppm in areas further offshore. They feel that a significant amount of mercury transfer was by the food web connecting nearshore plankton to offshore higher trophic levels rather than by direct transport by advection. There appears, however, to be little difference in sensitivities of inshore and offshore copepods. Reeve *et al.* (1976) showed that offshore copepod species of the Florida current were only slightly more sensitive to both mercury and copper than the inshore estuarine copepod, *Acartia tonsa*. These differences are minor in comparison to other factors affecting sensitivity. Reeve *et al.* (1976) also showed that short-term heavy metal toxicity is directly related to the biomass of the organism where larger sizes, either within or between species, are less affected by toxic chemicals than are smaller sizes. Windom *et al.* (1973), however, found no evidence of this in North Atlantic plankton. Within the species *Acartia tonsa*, Reeve *et al.* (1976) found a considerable

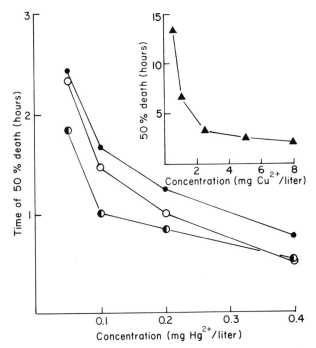

Fig. 4. Survival of *Acartia clausi* in seawater containing copper sodium citrate (▲——▲), mercuric chloride (●——●), ethylmercuric chloride (◑——◑), and mercuric iodide (○——○). (From Corner and Sparrow, 1956, reproduced by permission.)

variability in sensitivity to concentrations of copper. Over a 6-month period, the 24-hour LC_{50} measurements ranged from 100 to 300 ppb copper. This implies that there may be a wide genetic variability within a species' tolerance to heavy metals. If this is true, then a population of copepods which might be initially decimated by a concentration of heavy metal in a particular locality may subsequently return to their previous density over a long period of time.

Some investigators, such as Corner and Sparrow (1956), have determined the lethal concentrations of various heavy metal compounds as shown in Fig. 4. Such data are useful in comparing the toxic effects of various compounds on important estuarine species, but it is probably more realistic to study the sublethal effects of lesser concentrations of heavy metals which more closely approximate those concentrations found in nature. Factors most sensitive to heavy metal pollution appear to be feeding rate and egg production. While Reeve *et al.* (1976) found no mortality trend evident for *Acartia tonsa* from 5 to 50 ppb copper, there was a clear downward trend in egg production and feeding between 10 and 20 ppb. At 50 ppb copper, numbers of fecal pellets were half the control and egg production

John K. Dawson

was nearly eliminated (see Fig. 5). Reeve *et al.* (1976) determined that these two
parameters could be used to demonstrate sensitivity at copper concentrations
from 1 to 1.5 orders of magnitude lower than the 24-hour LC_{50} concentration
measured for the same population. Similar effects were shown for more oceanic
copepod species as well. Reeve *et al.* (1977b) report that sublethal effects on
nutrition and reproduction can be observed over periods of 5 to 10 days exposure
to heavy metal concentrations as low as 5 μg/liter. Reeve *et al.* (1977a) showed
that while these effects may be difficult to interpret on small copepods such as
Pseudocalanus, they point out that the effects on larger copepods such as
Calanus plumchrus is very striking. Egg production in the control was six times
that of the group exposed to 5 μg/liter copper and almost no egg production
occurred in the 10 μg/liter copper test. Whether the growth of adult females is
retarded or the adults did not produce as many eggs as a result of these copper
concentrations is not known. Reeve *et al.* (1977b) investigated the effects of
copper and mercury on respiration and excretion rates of zooplankton as indicators of sublethal stress. They concluded that these were very poor indicators
and that any consistent change in respiration and excretion rates usually indicated
imminent death.

Thus, according to Reeve *et al.* (1977b), biological effects can be demonstrated in a range which, in the case of copper, may be within an order of
magnitude of the environmental background levels of some inshore regions.
Biological effects such as egg production are limited to adult copepod popula-

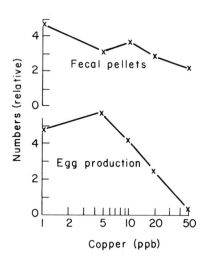

Fig. 5.   Fecal pellet and egg production rate of *Acartia tonsa* exposed to various concentrations
of copper over 4 days in the laboratory (numbers are relative). (From Reeve *et al.*, 1976, reproduced
by permission.)

tions and cannot be used for immature stages that may be more sensitive to heavy metal concentrations. Toxicity of copper to the prefeeding stages of *Euchaeta japonica* has been demonstrated by J. I. Lewis *et al.* (1972) and A. G. Lewis *et al.* (1973).

The above observations represent results of copepods exposed to mercury and copper ions and compounds separately in solution. In reality, however, many heavy metals may be present in a polluted estuary and have an effect on copepods—an effect greater than the sum of individual metals. Such synergistic effects have been shown by Corner and Sparrow (1956) for the estuarine copepod *Acartia clausi* exposed to mercury and copper. Their studies indicate that copper increases the permeability of the copepod to mercury poisons. Most of the present research has not considered the impact of copepod ingestion of phytoplankton polluted with heavy metals. Results of the study of Parrish and Carr (1976) with *Acartia clausi* would indicate that this was not an important factor in the introduction of mercury into the food web, but their results are sketchy and open to question.

Investigations by Knauer and Martin (1972), Williams and Weiss (1973), Corcoros *et al.* (1973), and Leatherland *et al.* (1973) concluded that in general there is little or no evidence that mercury concentrations are amplified in passing through trophic levels. Williams and Weiss (1973) believe that higher mercury levels found in fish and some benthic invertebrates result from their being long lived, and thus mercury concentrations within an organism may be a function of time.

Arsenic is another toxic element found in pelagic copepods. While the effects of this element have not been studied extensively, Kennedy (1976) found that copepods had the lowest concentrations of arsenic (1.4 ppm wet weight) as compared with the rest of the zooplankton. Leatherhead *et al.* (1973) and Bohn (1975) have found 6.0 and 14.5 ppm dry weight arsenic in their respective studies. Bohn (1975) concluded that arsenic was accumulated at higher trophic levels since copepods had about 10% of the arsenic concentrations found in fish and prawns which feed on the copepods.

With the introduction of acid–iron wastes into coastal environments, the effects on the abundance and distribution of copepods have been investigated. Grice *et al.* (1973), using three species of copepods, found the acid wastes caused substantial mortality in the laboratory. When these wastes were neutralized, however, no harmful effects were noted, implying that it was the acid and not the waste which was harmful. The concentrations used, however, did not reflect what is found in the field due to rapid mixing of the wastes. No mortality was found when acid wastes were diluted to what would be found in the field. Wiebe *et al.* (1973) showed no trends from zooplankton sampling that would suggest that acid wastes were important factors in zooplankton distribution. Significant trends were also lacking in the species diversity of the acid waste area compared to nonpolluted areas. These studies indicate that acid–iron wastes are not important factors affecting the distribution and abundance of zooplankton.

VI. Chlorinated Hydrocarbon Pollution

The presence of organochloride compounds has been indicated by Odum (1968) as the most serious pollution problem in estuaries. The extreme persistence in the ecosystem, sometimes remaining at harmful levels for many years, is one of the most hazardous aspects of the organochloride compounds. Addison (1976) categorizes chlorinated hydrocarbons into three groups. First are the PCB's, which were introduced in the late 1920s. These compounds have widespread uses which include heat transfer fluids, condenser dielectrics, and hydraulic fluids. Second is the DDT group of insecticides introduced in the mid-1940s; p,p'-DDT is certainly the most widely used member of this group. Other structurally related compounds such as p,p'-DDD is a metabolite of p,p'-DDT and still has insecticidal properties. The third group are the chlorinated naphthalenes, or the cyclodiene insecticides, which were introduced in the early 1950s. This group includes the insecticides aldrin, dieldrin, and endrin.

Almost no work has been done relating the concentrations of PCB's in estuarine copepods. Risebrough *et al.* (1973) studied the PCB residues in Atlantic shelf zooplankton and found a range of 2.4 to 260 ppm with a median value of 40 ppm. Takagi *et al.* (1975) showed that PCB's were not concentrated up the trophic structure from zooplankton to fish. While most PCB contamination is the result of runoff from the land, Jensen *et al.* (1972) has shown that serious PCB pollution may to some extent depend on boat and ship traffic. A considerable amount of PCB's can leach off the bottom of boats, as shown by their study of plankton collected from the bow and stern of the ship. They found a 4- to 14-fold increase in PCB from plankton collected in the wake. This phenomenon must be born in mind when collecting plankton to perform a PCB bioassay.

The uptake of DDT by marine copepods has been studied by several investigators. Harding (1977) measured the uptake and clearance of DDT with *Calanus finmarchicus* in order to predict DDT levels in marine copepods. This study did not account for the DDT accumulated through ingestion which Cox (1971) indicated to be significant in other zooplankton. Darrow and Harding (1975) showed that three times the number of *C. finmarchicus* and *C. hyperboreus* survived in the controls as compared to that exposed to DDT. A surprising result of this experiment was that over an 8-week period there was no significant metabolism of the DDT. This lack of metabolism may be due to the DDT being absorbed externally rather than incorporated into the animals, or to the very low temperature (6°C) of the experiment.

The concentrating effects of DDT up the trophic level food web have been shown to occur generally from plankton to birds (Woodwell *et al.*, 1967). But as Robinson *et al.* (1967) pointed out, while the concentration of pesticides in trophic levels shows an overall trend, it is not necessarily true from one particular trophic level to another. He found there was apparently no concentrating effect from planktonic crustacea to fish. Portmann (1975) and Harvey *et al.* (1973) also showed little evidence that DDT is concentrated from phytoplankton to zooplankton.

VII. Enrichment

The enrichment of estuarine waters with particulate organics may have beneficial or detrimental effects on the copepod population, depending on the quantity of organics introduced. Unregulated dumping of massive amounts of organics will, of course, have extremely detrimental effects by increasing the BOD so that dissolved oxygen drops to near zero. A closely monitored and regulated disposing of organics can, however, enhance the copepod population such that general benefits can be derived by this bioenhancement.

The nutritive role of detritus for zooplankton has been acknowledged by various authors (Darnell, 1967; Qasim and Sankaranarayanan, 1972). Heinle and Flemer (1975) have hypothesized that the estuarine copepod *Eurytemora affinis* must consume detritus to meet part of its nutrient requirements. The substantial production of *E. affinis* in March when primary production is not large has been attributed to the presence of detritus. Darnell (1967) also showed that areas of zooplankton abundance were correlated with centers of detritus rather than with phytoplankton abundance. This dietary role was shown when *E. affinis* reproduced if fed on detritus enriched with bacteria and protozoa (Heinle *et al.*, 1974). These diets were seldom equal to algae controls, which suggests that detritus provides only a part of the copepod's carbon requirement. The harpacticoid *Scottolana canadensis* was less able to utilize detritus and associated bacteria, but did obtain some energy from that source. It has been found by Odum and de la Cruz (1967) that the colonization of detritus by microorganisms substantially enhances its nutritive value to consumers. Thus, detritus may serve mainly as a substrate and food source for bacteria, and these microorganisms appear to be the component of detritus most often used by consumers.

Field studies have shown that the effects of treated sewage in Los Angeles Harbor (Brewer, 1976) resulted in a 20-fold increase of zooplankton largely made up of *Acartia tonsa* around the sewage outfall, as compared to the mean abundance in the rest of the harbor.

A pre- and postdomestic sewage abatement study by McNulty (1970) has indicated a zooplankton volume decrease to half of the preabatement volumes in some areas. My studies, as well as that of Anonymous (1972), have shown, however, that enormous amounts of paper fiber may be collected with plankton when sampled near sewage outfall areas. This result could strongly bias any pre- and postabatement zooplankton volume data.

McCrary (1970) reported on a study where the plankton, dominated by the copepods *Acartia tonsa, Pseudodiaptomus coronatus,* and *Oithona* sp., were grown in ponds with treated sewage as an intermediary system before entry into an estuary.

These studies indicate that sewage and other detrital wastes can be used to enhance the copepod population and consequently the entire food web if careful monitoring of such wastes are maintained.

Acknowledgments

Thanks are extended to Dr. Dorothy Soule and Geraldine Knatz of the University of Southern California for reviewing the manuscript. I am also appreciative of the cooperative help given by the staff of the Allan Hancock Foundation Library of Biology and Oceanography. My final thanks go to my wife, Janice, for the many hours she spent typing the manuscript.

References

Addison, R. F. (1976). Organochloride compounds in aquatic organisms: Their distribution, transport and physiological significance. *Soc. Exp. Biol. Semin. Ser.* **2**, 127–143.

Anderson, J. W., Neff, J. M., Cox, B. A., Tatem, H. E., and Hightower, G. H. (1974). Characteristics of dispersions and water soluble extracts of crude and refined oil and their toxicity to estuarine crustaceans and fish. *Mar. Biol.* **27**, 75–88.

Anonymous (1972). "The Effects of Waste Disposal in the New York Bight," Summary Final Report. U. S. Dept. of Commerce, Washington, D.C. (Prepared by National Marine Fisheries Service, Middle Atlantic Coastal Fisheries Center, Sandy Hook Lab., Highlands, New Jersey.)

Arnott, G. H., and Hussainy, S. U. (1972). Brackish water plankton and their environment in the Werribee River, Victoria. *Aust. J. Mar. Freshwater Res.* **23**, 85–97.

Bacon, P. R. (1971). Plankton studies of a Caribbean estuarine environment. *Caribb. J. Sci.* **11**, 81–89.

Baird, W. (1843). Notes on British Entomostraca. *The Zoologist* **1**.

Bakker, C., and dePauw, N. (1975). Comparison of plankton assemblages of identical salinity ranges in estuarine tidal and stagnant environments. II. Zooplankton. *Neth. J. Sea Res.* **9**, 145–165.

Barlow, J. P. (1955). Physical and biological processes determining the distribution of zooplankton in a tidal estuary. *Biol. Bull. (Woods Hole, Mass.)* **109**, 211–225.

Bayly, I. A. E. (1965). Ecological studies of the planktonic Copepoda of the Brisbane River estuary with special reference to *Gladioferens pectinatus* (Brady) (Calanoida). *Aust. J. Mar. Freshwater Res.* **16**, 315–350.

Berdugo, V., Harris, R. P., and O'Hara, S. C. M. (1977). The effect of petroleum hydrocarbons on reproduction of the estuarine planktonic copepod. *Eurytemora affinis,* in laboratory culture. *Mar. Pollut. Bull.* **8**, 138–143.

Bohn, A. (1975). Arsenic in marine organisms from W. Greenland. *Mar. Pollut. Bull.* **6**, 87–89.

Boylan, D. B., and Tripp, B. W. (1971). Determinations of hydrocarbons in seawater extracts of crude oil and crude oil fractions. *Nature (London)* **230**, 44–47.

Bradley, B. P. (1975). "Adaptation of Copepod Populations to Thermal Stress," Tech. Rep. 34. Water Resour. Res. Cent., University of Maryland, College Park.

Brewer, G. D. (1976). Enhancement studies: Zooplankton. *In* "Marine Studies of San Pedro Bay, California" (D. Soule and M. Oguri, eds.), Part 12, pp. 193–198. Harbors Environ. Proj., Inst. Mar. Coastal Stud., University of Southern California, Los Angeles.

Bryan, G. W. (1971). The effects of heavy metals (other than mercury) on marine and estuarine organisms. *Proc. R. Soc. London, Ser. B* **177**, 389–410.

Bryan, G. W. (1976). Some aspects of heavy metal tolerance in aquatic organisms. *Soc. Exp. Biol. Semin. Ser.* **2**, 7–34.

Cairns, J., Jr., Heath, A. G., and Parker, B. C. (1975). Temperature influence on chemical toxicity to aquatic organisms. *J. Water Pollut. Control Fed.* **47**, 267–280.

Carpenter, E. J., Peck, B. B., and Anderson, S. J. (1974). Survival of copepods passing through a nuclear power station on northeastern Long Island Sound, USA. *Mar. Biol.* **24**, 49–55.

Claus, C. (1863). "Die freilebenden Copepoden mit besonderer berucksichtigung der fauna Deutschlands, der Nordsee und des Mittlemeeres." Leipzig.

Claus, C. (1866). Die Copepoden-Fauna von Nizza. Ein Beitrag zur charakteristik der farmen und deren abänderungen "im Sinne Darwin's." Marburg und Leipzig.

Conover, R. J. (1956). Oceanography of Long Island Sound, 1952–1954. VI. Biology of *Acartia clausi* and *A. tonsa*. *Bull. Bingham Oceanogr. Collect.* **15**, 156–233.

Conover, R. J. (1957). Note on the seasonal distribution of zooplankton in Southampton Water with special reference to *Acartia*. *Ann. Mag. Nat. Hist.* [12] **10**, 63–67.

Conover, R. J. (1971). Some relationships between zooplankton and bunker C oil in Chedabucto Bay following the wreck of the ship ARROW. *J. Fish. Res. Board Can.* **28**, 1327–1330.

Corcoros, G., Cahn, P. H., and Siler, W. (1973). Mercury concentrations in fish, plankton and water from three western Atlantic estuaries. *J. Fish Biol.* **5**, 641–647.

Corner, E. D. S., and Sparrow, B. W. (1956). The modes of actions of toxic agents. I. Observations on the poisoning of certain crustaceans by copper and mercury. *J. Mar. Biol. Assoc. U. K.* **35**, 531–548.

Corner, E. D. S., Harris, R. P., Kilvington, C. C., and O'Hara, S. C. M. (1976a). Petroleum compounds in the marine food web: Short-term experiments on the fate of naphthalene in *Calanus*. *J. Mar. Biol. Assoc. U. K.* **56**, 121–133.

Corner, E. D. S., Harris, R. P., Whittle, K. J., and Mackie, P. R. (1976b). Hydrocarbons in marine zooplankton and fish. *Soc. Exp. Biol. Semin. Ser.* **2**, 71–105.

Cory, R. L. (1967). Epifauna of the Patuxent River estuary, Maryland, for 1963 and 1964. *Chesapeake Sci.* **8**, 71–89.

Coull, B. C. (1972). *Scottolana canadensis* (Willey, 1923) (Copepoda Harpactacoides) redescribed from the United States East coast. *Crustaceana (Leiden)* **22**, 209–214.

Cox, J. L. (1971). Uptake, assimilation and loss of DDT residues by *Euphausia pacifica*, a euphausid shrimp. *Fish. Bull.* **69**, 627–633.

Cronin, L. E., Daiber, J. C., and Hulbert, E. M. (1962). Quantitative seasonal aspects of zooplankton in the Delaware River estuary. *Chesapeake Sci.* **3**, 63–93.

Dahl, F. (1894). Die copepodenfauna des unteren Amazonas. *Ber. Naturforsch. Ges. Freiburg im Breisgau* **8**.

Dana, J. D. (1849). *Proc. Am. Acad. Arts Sci.* **2**, 26.

Dana, J. D. (1852). Crustacea. *In* "U. St. Exploring Expedition During the Years 1838–1842 Under the Command of Charles Wilkes," Vol. 13, pp. 1019–1262. Philadelphia.

Darnell, R. M. (1967). Organic detritus in relation to the estuarine ecosystem. *Publ. Am. Assoc. Adv. Sci.* **83**, 376–382.

Darrow, D. C., and Harding, G. C. H. (1975). Accumulation and apparent absence of DDT metabolism by marine copepods, *Calanus* spp., in culture. *J. Fish. Res. Board Can.* **32**, 1845–1849.

Davies, R. M., and Jensen, L. D. (1975). Zooplankton entrainment at three mid-Atlantic power stations. *J. Water Pollut. Control Fed.* **47**, 2130–2142.

Davis, C. C. (1943). The larval stages of the calanoid copepod, *Eurytemora hirundoides* Nordquist. *Nat. Resour. Inst., Chesapeake Biol. Lab., Publ.* **58**, 1–52.

Dawson, J. K. (1969). A study of the calanoid copepods of Stone Lagoon, California. Unpublished Thesis, California State University, Humboldt.

Dawson, J. K. (1976). Zooplankton distribution and seasonality. *In* "Environmental Investigations and Analysis for Los Angeles-Long Beach Harbors, Los Angeles, California 1973–1976," pp. 95–126. Harbors Environ. Proj., Inst. Mar. Coastal Stud., Allan Hancock Found., University of Southern California, Los Angeles.

Deevey, G. B. (1948). The zooplankton of Tisbury Great Pond. *Bull. Bingham Oceanogr. Collect.* **12**, 1–44.

Deevey, G. B. (1952a). A survey of the zooplankton of Block Island Sound, 1943–1946. *Bull. Bingham Oceanogr. Collect.* **13**, 65–119.

Deevey, G. B. (1952b). Quantity and composition of the zooplankton of Block Island Sound, 1949. *Bull. Bingham Oceanogr. Collect.* **13**, 120–164.

Deevey, G. B. (1960). The zooplankton of the surface waters of the Delaware Bay region. *Bull. Bingham Oceanogr. Collect.* **17**, 5–53.

Enright, J. T. (1977). Power plants and plankton. *Mar. Pollut. Bull.* **8**, 158–161.

Freegarde, M., Hatchard, C. G., and Parker, C. A. (1971). Oil spill at sea; its identification, determination and ultimate fate. *Lab. Pract.* **20**, 35–40.

Furuhashi, K. (1976). Diel vertical migrations suspected in some copepods and chaetognaths in the inlet waters, with a special reference to behavioral differences between male and female noted in the former. *Publ. Seto Mar. Biol. Lab.* **22**, 355–370.

Gabriel, P. C., Dias, N. S., and Nelson-Smith, A. (1975). Temporal changes in the plankton of an industrialized estuary. *Estuarine Coastal Mar. Sci.* **3**, 145–151.

Gaudy, R. (1977). Etude des modifications du métabolism respiratoire de populations d'*Acartia clausi* (Crustacea; Copepoda) après passage dans le circuit de refroidissement d'une centrale thermo-electrique. *Mar. Biol.* **39**, 179–190 (in French, English abstract).

Gibson, V. R., and Grice, G. D. (1977). Response of macro-zooplankton populations to copper: Controlled ecosystem pollution experiment. *Bull. Mar. Sci.* **27**, 85–91.

Giesbrecht, W. (1889). *Atti Accad. Naz Lincei, Mem., Cl. Sci. Fis., Mat. Nat.* [4] **5.**

Giesbrecht, W. (1891). *Atti Accad. Naz Lincei, Mem., Cl. Sci. Fis., Mat. Nat.* [4] **7.**

Giesbrecht, W. (1892). *Pelagische Copepoden Fauna Flora Golfes von Neapel* **19**, 1–831.

Golubić, S. (1971). Biological indication of the pollution in marine coastal environment. *Vie Milieu, Suppl.* **22**, 293–299.

González, J. G. (1974). Critical thermal maxima and upper lethal temperatures for the calanoid copepods *Acartia tonsa* and *A. clausi. Mar. Biol.* **27**, 219–223.

Grice, G. D., Wiebe, P. H., and Hoagland, E. (1973). Acid-iron waste as a factor affecting the distribution and abundance of zooplankton in the New York Bight. 1. Laboratory studies on the effects of acid wastes on copepods. *Estuarine Coastal Mar. Sci.* **1**, 45–50.

Gunner, J. E. (1765). Nogle smaa rare, mustendelen nye norske Søbdyr, beskrevene. *Skr. Kjobenhavn. Selsk.* **10**, 175.

Gurney, R. (1931). "British Fresh-water Copepoda," Vol. 1. Ray Society, London.

Haertel, L., and Osterberg, C. (1967). Ecology of zooplankton, benthos and fish in the Columbia River estuary. *Ecology* **48**, 459–472.

Haertel, L., Osterberg, C., Curl, H., and Park, P. K. (1970). Nutrient and plankton ecology of the Columbia River estuary. *Ecology* **50**, 962–978.

Harding, G. C. H. (1977). Uptake from sea water and clearance of ^{14}C-p,p'-DDT by the marine copepod *Calanus finmarchicus. J. Fish. Res. Board Can.* **34**, 177–182.

Harris, R. P., Berdugo, V., O'Hara, S. C. M., and Corner, E. D. S. (1977). Accumulation of ^{14}C-1-Naphthalene by an oceanic and an estuarine copepod during long-term exposure to low-level concentrations. *Mar. Biol.* **42**, 187–195.

Harvey, G. R., Brown, V. J., Backus, R. H., and Grice, G. D. (1973). Chlorinated hydrocarbons in open ocean Atlantic organisms. *In* "The Changing Chemistry of the Oceans" (D. Green and D. Jagner, eds.), pp. 177–186. Wiley, New York.

Heinle, D. R. (1966). Production of a calanoid copepod, *Acartia tonsa* in the Patuxent River estuary. *Chesapeake Sci.* **7**, 59–74.

Heinle, D. R. (1969). Temperature and zooplankton. *Chesapeake Sci.* **10**, 186–209.

Heinle, D. R. (1972). Free-living Copepoda of the Chesapeake Bay. *Chesapeake Sci.* **13**, Suppl., S117–S119.

Heinle, D. R. (1976). Effects of passage through power plant cooling systems on estuarine copepods. *Environ. Pollut.* **11**, 39–58.

Heinle, D. R., and Flemer, D. A. (1975). Carbon requirements of a population of the estuarine copepod *Eurytemora affinis. Mar. Biol.* **31**, 235–247.

Heinle, D. R., Flemer, D. A., Ustach, J. F., Murtagh, R. A., and Harris, R. P. (1974). The role of organic debris and associated micro-organism in pelagic estuarine food chains. *Tech. Rep., Water Resour. Res. Cent., Univ. Md.* **22**.

Heip, C. (1972). The reproductive potential of copepods in brackish water. *Mar. Biol.* **12**, 219–221.

Hodgkin, E. P., and Rippingale, R. J. (1971). Interspecies conflict in estuarine copepods. *Limnol. Oceanogr.* **16**, 573–576.

Hulsizer, E. E. (1976). Zooplankton of lower Narragansett Bay, 1972–1973. *Chesapeake Sci.* **17**, 260–270.

Hyland, J. L., and Schneider, E. D. (1976). Petroleum hydrocarbons and their effects on marine organisms, populations, communities, and ecosystems. *In* "Sources, Effects and Sinks of Hydrocarbon in the Aquatic Environment," pp. 463–506. Am. Inst. Biol. Sci., Washington, D.C.

Jacobs, J. (1961). Laboratory cultivation of the marine copepod *Pseudodiaptomus coronatus* Williams. *Limnol. Oceanogr.* **6**, 443–446.

Jeffries, H. P. (1959). Plankton biology of Raritan Bay. Ph.D. Thesis, Rutgers University, New Brunswick, New Jersey.

Jeffries, H. P. (1962a). Salinity-space distribution of the estuarine copepod genus *Eurytemora. Int. Rev. Gesamten Hydrobiol.* **47**, 291–300.

Jeffries, H. P. (1962b). Copepod indicator species in estuaries. *Ecology* **43**, 730–733.

Jeffries, H. P. (1962c). Succession of two *Acartia* species in estuaries. *Limnol. Oceanogr.* **7**, 354–364.

Jeffries, H. P. (1964). Comparative studies on estuarine zooplankton. *Limnol. Oceanogr.* **9**, 348–358.

Jeffries, H. P. (1967). Saturation of estuarine zooplankton by congeneric associates. *Publ. Am. Assoc. Adv. Sci.* **83**, 500–508.

Jensen, S., Renberg, L., and Olsson, M. (1972). PCB contamination from boat bottom paint and levels of PCB's in plankton outside a polluted area. *Nature (London)* **240**, 358–360.

Katona, S. K. (1971). The developmental stages of *Eurytemora affinis* (Poppe, 1880) (Copepoda, Calanoida) raised in laboratory cultures, including a comparison with larvae of *Eurytemora americana* Williams, 1906, and *Eurytemora herdmanni* Thompson and Scott, 1897. *Crustaceana (Leiden)* **21**, 5–20.

Kennedy, V. S. (1976). Arsenic concentrations in some marine organisms from Newfoundland and Labrador. *J. Fish. Res. Board Can.* **33**, 1388-1393.

Ketchum, B. H. (1954). Relationships between circulation and planktonic populations in estuaries. *Ecology* **35**, 191-200.

Knatz, G. (1978). Succession of copepod species in a middle Atlantic estuary. *Estuaries* **1**, 68-71.

Knauer, G. A., and Martin, J. H. (1972). Mercury in a marine pelagic food chain. *Limnol. Oceanogr.* **17**, 868-876.

Kontogiannis, J. E., and Barnett, C. J. (1973). The effect of oil pollution on survival of the tidal pool copepod, *Tigriopus californicus. Environ. Pollut.* **4**, 69-79.

Krøyer, H. (1849). Karcinologiske Bidrag. *Naturh. Tidsskr.* [2] **2**, 527-605.

Lance, J. (1963). The salinity tolerances of some estuarine planktonic copepods. *Limnol. Oceanogr.* **8**, 440-449.

Lance, J. (1964). The salinity tolerance of some estuarine planktonic crustaceans. *Biol. Bull. (Woods Hole, Mass.)* **127**, 108-118.

Leatherland, T. M., Burton, J. D., Culkin, F., McCartney, M. J., and Morris, R. J. (1973). Concentrations of some trace metals in pelagic organisms and of mercury in northeast Atlantic Ocean water. *Deep-Sea Res.* **20**, 679-685.

Lee, R. F. (1975). Fate of petroleum hydrocarbons in marine zooplankton. *Proc.—Conf. Prev. Control Oil Spills, 1975* pp. 549-553.

Lee, R. F., and Anderson, J. W. (1977). Fate and effect of naphthalenes: Controlled ecosystem pollution experiment. *Bull. Mar. Sci.* **27**, 127-134.

Lewis, A. G., Whitfield, P., and Ramnarine, A. (1973). The reduction of copper toxicity in a marine copepod by sediment extract. *Limnol. Oceanogr.* **18**, 324-326.

Lewis, J. I., Parrish, P. R., Patrick, J. M., and Forester, J. (1972). Some particulate and soluble agents affecting the relationships between metal toxicity and organism survival in the calanoid copepod *Euchaeta japonica. Mar. Biol.* **17**, 215-221.

Lilljeborg, W. (1853). "De crustaceis ex ordinibus tribus: Cladocera, Ostracoda et Copepoda, in Scania occurrentibus." Lund.

McConaugha, J. R. (1976). Bioassay investigations of the impact of wastes on the copepod *Acartia tonsa. In* "Marine Studies of San Pedro Bay, California" (D. Soule and M. Oguri, eds.), Part 12, pp. 215-225. Harbors Environ. Proj., Inst. Mar. Coastal Stud., University of Southern California, Los Angeles.

McCrary, A. B. (1970). Zooplankton and invertebrates. *In* "Studies of Marine Estuarine Ecosystems Developing with Treated Sewage Wastes" (H. T. Odum and A. F. Chestnut, principal investigators). Inst. Mar. Sci., University of North Carolina, Chapel Hill.

McNulty, J. K. (1970). Effects of abatement of domestic sewage pollution on the benthos, volumes of zooplankton, and the fouling organisms of Biscayne Bay, Florida. *Stud. Trop. Oceanogr.* **9**.

Marukawa, H. (1921). Plankton list and some new species of Copepoda from the northern waters of Japan. *Bull. Inst. Oceanogr.* No. 384.

Mironov, O. G. (1969). Effects of oil pollution upon some representatives of Black Sea zooplankton. *Zool. Zh.* **48**, 980-984 (in Russian, English summary).

Moore, S. F., and Dwyer, R. L. (1974). Effects of oil on marine organisms: A critical assessment of published data. *Water Res.* **8**, 819-827.

Moreira, G. S. (1975). Studies on the salinity resistance of the copepod *Euterpina acutifrons* (Dana). *In* "Physiological Ecology of Estuarine Organisms" (F. J. Vernberg, ed.) pp. 73-79. Univ. of South Carolina Press, Columbia.

Motoda, S., Kawamura, T., and Minoda, T. (1971). Diel variation of hydrography and abundance of a copepod, *Acartia clausi* in Oshoro Bay. *Bull. Plankton Soc. Jpn.* **18**, 86-88.

Müller, O. F. (1792). "Entomostraca seu Insecta testacea quae in aquis Daniae et Norwegiae reperit, descripsit et iconibus illustravit," Francofurti.

Naylor, E. (1965). Effects of heated effluents upon marine and estuarine organisms. *Adv. Mar. Biol.* **3**, 63–103.

Nordquist, O. (1888). Die Calaniden Finlands. *Bidr. Finl. Nat. Folk* **47**.

Odum, W. E. (1968). Pesticide pollution in estuaries. *Sea Front.* **14**, 234–245.

Odum, E. P., and de la Cruz, A. A. (1967). Particulate organic detritus in a Georgia salt march estuary ecosystem. *Publ. Am. Assoc. Adv. Sci.* **83**, 383–388.

Parker, C. A., Freegarde, M., and Hatchard, C. G. (1970). The effect of some chemical and biological factors on the degradation of crude oil at sea. *In* "Water Pollution by Oil" (Seminar sponsored by Institute of Water Pollution Control, Institute of Petroleum, London, England, and World Health Organization), Pap. No. 17.

Parrish, K. M., and Carr, R. A. (1976). Transport of mercury through a laboratory two-level marine food chain. *Mar. Pollut. Bull.* **7**, 90–91.

Perkins, E. J. (1974). "The Biology of Estuaries and Coastal Waters." Academic Press, New York.

Poppe, S. A. (1880). Ueber eine neue art der Calaniden-Gattung *Temora* Baird. *Abh. Naturwiss. Ver. Bremen* **7**, 55–60.

Portmann, J. E. (1975). The bioaccumulation and effects of organochloride pesticides in marine animals. *Proc. R. Soc. London, Ser. B* **189**, 291–304.

Pritchard, D. W. (1955). Estuarine circulation patterns. *Proc. Am. Soc. Civ. Eng.* **81**, 1–11.

Qasim, S. Z. (1970). Some problems related to the food chain in a tropical estuary. *In* "Marine Food Chains" (J. H. Steele, ed.), pp. 45–51. Univ. of California Press, Berkeley.

Qasim, S. Z., and Sankaranarayanan, V. N. (1972). Organic detritus of a tropical estuary. *Mar. Biol.* **15**, 193–199.

Raymont, J. E. G. (1964). The marine fauna of Southampton Water. *In* "A Survey of Southampton and its Region" (F. J. Monkhouse, ed.). Br. Assoc. Adv. Sci., Southampton.

Reeve, M. R. (1970). Seasonal changes in the zooplankton of south Biscayne Bay and some problems of assessing the effects on the zooplankton of natural and artificial thermal and other fluctuations. *Bull. Mar. Sci.* **20**, 894–921.

Reeve, M. R., and Cosper, E. (1970). The acute thermal effects of heated effluents on the copepod *Acartia tonsa* from a subtropical bay and some problems of assessment. *FAO Tech. Conf. Mar. Pollut. Effects Living Resour. Fishing* pp. 1–5.

Reeve, M. R., Grice, G. D., Gibson, V. R., Walter, M. A., Darcy, K., and Ikeda, T. (1976). A controlled environmental pollution experiment (CEPEX) and its usefulness in the study of larger marine zooplankton under toxic stress. *Soc. Exp. Biol. Semin. Ser.* **2**, 145–162.

Reeve, M. R., Gamble, J. C., and Walter, M. A. (1977a). Experimental observations on the effects of copper on copepods and other zooplankton: Controlled ecosystem pollution experiment. *Bull. Mar. Sci.* **27**, 92–104.

Reeve, M. R., Walter, M. A., Darcy, K., and Ikeda, T. (1977b). Evaluation of potential indicators of sublethal toxic stress on marine zooplankton (feeding, fecundity, respiration and excretion): Controlled ecosystem pollution experiment. *Bull. Mar. Sci.* **27**, 105–113.

Reish, D. J., Kauwling, T. J., and Mearns, A. J. (1976). Marine and estuarine pollution. *J. Water Pollut. Control Fed.* **48**, 1439–1459.

Reish, D. J., Kauwling, T. J., Mearns, A. J., Oshida, P. S., and Rosi, S. S. (1977). Marine and estuarine pollution. *J. Water Pollut. Control Fed.* **49**, 1316–1340.

Reish, D. J., Kauwling, T. J., Mearns, A. J., Oshida, P. S., Rosi, S. S., Wilkes, F. G., and Ray, M. J. (1978). Marine and estuarine pollution. *J. Water Pollut. Control Fed.* **50**, 1424–1469.

Riley, G. A. (1967). The plankton of estuaries. *Publ. Am. Assoc. Adv. Sci.* **83**, 316–326.

Risebrough, R. W., Vreeland, V., Harvey, G. R., Miklas, H. P., and Carmignani, G. M. (1973). PCB residues in Atlantic zooplankton. *Bull. Environ. Contam. Toxicol.* **8**, 345–355.

Robertson, E. D., and Bunting, D. L. (1976). The acute toxicity of four herbicides to 0–4 hour nauplii of *Cyclops vernalis* Fisher (Copepoda, Cyclopoida). *Bull. Environ. Contamin. Toxicol.* **16**, 682–688.

Robinson, J., Richardson, A., Crabtree, A. N., Coulson, J. C., and Potts, G. R. (1967). Organochloride residues in marine organisms. *Nature (London)* **214**, 1307–1311.

Sage, L. E., and Herman, S. S. (1972). Zooplankton of the Sandy Hook Bay area, N.J. *Chesapeake Sci.* **13**, 29–39.

Sars, G. O. (1904). Description of *Paracartia Grani, Bergens Mus. Arbok* **2**, 3–16.

Stickney, R. R., and Knowles, S. C. (1975). Summer zooplankton distribution in a Georgia estuary. *Mar. Biol.* **33**, 147–154.

Suchanek, T. H., Jr., and Grossman, C. (1971). Viability of zooplankton. *In* "Studies on the Effects of a Steam-electric Generating Plant in the Marine Environment at Northport" (G. C. Williams *et al.*, eds.), Tech. Rep. No. 9, pp. 61–74. Mar. Sci. Res. Cent., State University of New York, New York.

Takagi, M., Iida, A., and Oka, S. (1975). PCB contents in several species of zooplankton collected in the North Pacific Ocean. *Bull. Jpn. Soc. Sci. Fish.* **41**, 561–565 (in Japanese, English abstract).

Tsuruta, A., and Tawara, S. (1975). The influence of the warm cooling water from a fossil fueled power plant on oceanographic conditions and composition of plankton in Owase Bay. I. Water temperature in relation to distribution of micro-plankton. *J. Shimonoseki Univ. Fish.* **23**, 121–136 (in Japanese, English abstract).

Venezia, L. D., and Fossato, V. U. (1977). Characteristics of suspensions of Kuwait oil and Corexit 7664 and their short- and long-term effects on *Tisbe bulbisetosa* (Copepoda: Harpacticoida). *Mar. Biol.* **42**, 233–237.

Wheeler, W. M. (1899). The freeswimming copepods of the Woods Hole region. *Bull. U.S. Fish Comm.* **19**, 157–192.

Wiebe, P. H., Grice, G. D., and Hoogland, E. (1973). Acid-iron waste as a factor affecting the distribution and abundance of zooplankton in the New York Bight. II. Spatial variations in the field and implications for monitoring studies. *Estuarine Coastal Mar. Sci.* **1**, 51–64.

Willey, A. (1923). Notes on the distribution of free-living Copepoda in Canadian waters. *Contrib. Can. Biol.* **1**, 303–324.

Williams, L. W. (1906). Notes on marine Copepoda of Rhode Island. *Am. Nat.* **40**, 645.

Williams, P. M., and Weiss, H. V. (1973). Mercury in the marine environment: Concentrations in sea water and in a pelagic food chain. *J. Fish. Res. Board Can.* **30**, 293–295.

Williams, R. B., Murdoch, M. B., and Thomas, L. K. (1968). Standing crop and importance of zooplankton in a system of shallow estuaries. *Chesapeake Sci.* **9**, 42–51.

Wilson, M. S. (1959). The Calanoida. *In* "Ward and Whipple's Freshwater Biology" (W. T. Edmondson, ed.), pp. 738–794. Wiley, New York.

Windom, H., Taylor, F., and Stickney, R. (1973). Mercury in North Atlantic plankton. *J. Cons., Cons. Int. Explor. Mer* **35**, 18–21.

Woodwell, G. M., Wurster, C. F., Jr., and Isaacson, P. A. (1967). DDT residues in an east coast estuary: A case of biological concentration of a persistent insecticide. *Science* **156**, 821–823.

Wooldridge, T. (1976). The zooplankton of Msikaba estuary. *Zool. Afr.* **11**, 23–44.

Youngbluth, M. J. (1976). Zooplankton population in a polluted, tropical embayment. *Estuarine Coastal Mar. Sci.* **4**, 481–496.

CHAPTER 6

Crabs (Arthropoda: Crustacea: Decapoda: Brachyura)

AUSTIN B. WILLIAMS and THOMAS W. DUKE

I. Introduction

Biological research on estuarine brachyuran crabs has centered on a few species found literally at the doorstep of marine laboratories. The abundance and accessibility of these species for field study, coupled with their capability for successful manipulation in laboratory experiments, have helped to promote a good understanding of their systematic position, physiology, ecology, and reaction to environmental alteration. A few species having significant value in fisheries have been similarly studied with somewhat greater effort, but the combined list is short, and if further limited to the subject matter covered in this book, it becomes even shorter. Acknowledging this, we have chosen to limit discussion a step further by confining this chapter primarily to species occurring in North America, feeling that effects of pollution on crabs in estuaries can be amply demonstrated by those species, but including data for other species when it seems productive to do so.

Crabs can be viewed as having unaltered or "normal" life histories and ecology as well as altered life patterns possibly brought on by pollution in changing environments. Because we deal in detail with only a few species, we have chosen to discuss these two fields of interest differently, presenting what is viewed as "normal" as a set of specific vignettes which form a foundation for discussion of the observed effects of pollution. Included in the "normal" set are four species not specifically discussed in the section on observed effects of pollution but which are intimately associated with communities reported here and treated in results of pollution studies elsewhere.

II. Taxonomy

The classification of the infraorder Brachyura long accepted by American scientists is that elaborated by Rathbun (1918, 1930) in her set of monographs on crabs of the Western Hemisphere. Gradually that system is being modified to accommodate new findings from the fossil record, adult morphology of living species, and larval development, as well as physiological, biochemical, and ecological studies.

Among the Brachyura, the most primitive groups generally have not successfully invaded estuaries in North America. None of the dromiaceans or the oxystomian box crabs and their relatives (Oxystomata) are found in any but the most seaward reaches of high-salinity estuaries, if present at all. Moreover, the great spider crab section (Oxyrhyncha) is essentially a euhaline marine group, only a few of its members being regularly encountered in lower reaches of estuaries, and none have figured in pollution studies. But members of the section Cancridea, *Cancer,* are found to some extent in estuaries, while those of the more highly specialized section Brachyrhyncha, including certain swimming crabs, mud crabs, square-backed crabs, and fiddlers, make up the conspicuously successful estuarine group. Among these are species of concern in pollution studies (Table I).

TABLE I

Taxonomic Position and Habitat of Decapod Crustacean Species, Infraorder Brachyura, of Concern in Estuarine Pollution Studies

Taxon	Habitat
Infraorder Brachyura	
Section Cancridea	
Family Cancridae	
Cancer irroratus Say, Rock crab	Temperate–polyhaline
Cancer magister Dana, Dungeness crab	
Section Brachyrhyncha	
Superfamily Portunoidea	
Family Portunidae, "Swimming" crabs	
Subfamily Portuninae	
Callinectes sapidus Rathbun, Blue crab	Temperate–tropical–euryhaline
Carcinus maenas (Linnaeus), Green or shore crab	Temperate–polyhaline
Superfamily Xanthoidea	
Family Xanthidae	
Subfamily Xanthinae, "Mud" crabs	
Cataleptodius (=*Leptodius*) *floridanus* (Gibbes)	Tropical–polyhaline
Eurypanopeus depressus (S. I. Smith)	Temperate–mesohaline
Neopanope sayi (S. I. Smith)[a]	Temperate–mesohaline
Panopeus herbstii A. Milne Edwards	Temperate–tropical–mesohaline
Rhithropanopeus harrisii (Gould)	Temperate–oligo–mesohaline
Subfamily Menippinae	
Menippe mercenaria (Say), Stone crab	Warm temperate–subtropical–mesopolyhaline
Family Grapsidae	
Subfamily Varuninae	
Hemigrapsus nudus (Dana), Purple shore crab[a]	Temperate–polyhaline
Subfamily Sesarminae	
Sesarma cinereum (Bosc), Wharf crab[a]	Temperate–tropical–polyhaline–semiterrestrial
Sesarma reticulatum (Say), "Marsh crab"[a]	Temperate–polyhaline–semiterrestrial
Superfamily Ocypodoidea	
Family Ocypodidae	
Subfamily Ocypodinae	
Uca minax (Le Conte), Red jointed fiddler	Temperate–oligo–mesohaline–semiterrestrial
Uca pugilator (Bosc), Sand fiddler	Temperate–subtropical–mesopolyhaline–semiterrestrial
Uca pugnax (Smith), Mud fiddler	Temperate–mesopolyhaline–semiterrestrial

[a] Species intimately associated with communities reported here and pollution studies published elsewhere.

III. Normal Life History–Ecology

Brachyuran crabs are dioecious, mating as adults. All have a complex ontogeny, passing through embryonic stages in eggs spawned by the female and borne by attachment to her pleopods until hatching into free swimming larvae, or zoeae, quite unlike the adult. The zoeae, released in waters of the meso- to euhaline range of salinity, depending on species, undergo a series of molts before reaching an intermediate stage, the megalopa, sometimes called the postlarva. The megalopa metamorphoses into the first crab stage, which is followed by successive molts until an adult stage is reached. Laboratory culture methods elaborated in the past 25 years have advanced knowledge of larval development in decapod crustaceans to the point that specific larval sequences not only can be described but also can be studied experimentally. The indispensable element in these techniques was discovery of a suitable diet for the larvae (Broad, 1957), the one almost universally used being freshly hatched nauplii of the brine shrimp, *Artemia salina,* occasionally supplemented or replaced by fertilized sea urchin eggs, rotifers (*Brachionis*), and, rarely, wild plankton (Sulkin and Epifanio, 1975; Sulkin and Norman, 1976). Postlarval stages, always easier to manipulate, have not posed such a feeding problem. The life history–ecology of all species discussed below is known from field and laboratory studies.

A. FAMILY CANCRIDAE

Twelve species of *Cancer* occur in waters of the Continental Shelf of America north of Panama. Distribution of the group is limited to cold or cool temperate waters (Nations, 1975). Only two species are reviewed here. These are the Atlantic rock crab, *C. irroratus* Say, and the Pacific Dungeness crab, *C. magister* Dana. The first ranges from Labrador to off Miami, Florida, to a depth of 575 m, the second from Amchitka, Aleutian Islands, Alaska (Hoopes, 1973), to Magdalena Bay, Baja California (Waldron, 1958), and from the low tide mark to about 90 m. Both are primarily marine species, normally entering estuaries only in their lower or high salinity regions.

Cancer irroratus has limited value in commercial fisheries (1.4 million lb in 1974) (Bell and FitzGibbon, 1977), and *C. magister* provides a considerable fishery (over 23 million lb, 1971–1975 annual average) (Robinson, 1977).

The species have similar life histories. Sexually mature hard males mate with sexually mature females somewhat smaller than themselves (Scarratt and Lowe, 1972), each such mating male carrying a female in a courtship embrace before, during, and after she molts into a terminally mature phase, the pair copulating when she is in the postmolt soft condition (Hartnoll, 1969; Snow and Nielsen, 1966; Butler, 1960; Poole, 1967). The mating season for *C. magister* extends from late February to August in California, but ovigerous females are known nearly year round (Waldron, 1958); mating for *C. irroratus* extends through about the same period to October (Scarratt and Lowe, 1972), being seasonally timed at different latitudes, with shortened duration at the extremes of the range. Larval life of five zoeal stages and a megalopa last about 4 months (Poole, 1967; Sastry, 1977). The larvae of *C. irroratus* have been recorded in estuarine plankton by a number of workers (Sandifer, 1973), always in regions of high salinity. In the case of this latest survey, low concentrations occurred occasionally in lower Chesapeake Bay Stations, most abundantly off the mouth of the Bay in water of 23.3–32.3‰ salinity (mainly 25‰) from May to July and September to October, with one peak in May and a smaller one in October, mainly in the temperature range 13°–21°C. Sandifer (1975) felt that larvae of this species showed no distributional adaptations for retention in bays, as do larvae of some other species more or less restricted to estuarine habitats. Experimentally reared larvae survived best in 30‰ salinity at 15°C (Sastry, 1977), in 26–30‰ salinity at 10.6°C (Poole, 1966), and the most successful molt of a prezoeal stage in 30–32‰ salinity at 10.5° and 17.5°C (Buchanan and Millemann, 1969). Reed (1969) found that survival of *C. magister* larvae in culture was not significantly affected by temperatures and salinities approximating ocean ranges of these variables found within 40 miles of the Oregon coast during the zoeal period (optimally 10.0°–13.0°C, 25–30‰ salinity) but that reduced temperature and current transport might affect survival.

Both species may migrate into shallow high-salinity areas after the larval stages are past. Jeffries (1966) found *C. irroratus* in Narragansett Bay on sandy bottom; there was little mixing with populations of a companion species, *C. borealis*. Others have not found such clear separation of the latter two species on the Continental Shelf, but have noted some substrate selection (Musick and McEachran, 1972; Haefner, 1976); and Turner (1953, 1954) discussed seasonal aspects of distribution for Boston harbor as did Winget *et al.* (1974) for Lewes, Delaware, harbor. Jeffries noted that *C. borealis* has proportionately heavier chelae than *irroratus,* is less prone to burying in sand when disturbed, and generally shows a lower walking and activity rate, which he correlated with differences in serum phosphate concentrations possibly related to level of energy metabolism. Serum nitrogen, chloride, and glucose were also assayed. Comparison of walking ability in a treadmill showed *C. irroratus* to react optimally at 14°–18°C. *Cancer magister* prefers sand or sandy mud bottom, but may be found on a variety of substrates (Waldron, 1958).

Cancer is a predator (Squires, 1965; Mann, 1973; Caddy, 1973; Cornick and Steward, 1968) in turn preyed upon by fishes (Sikora *et al.,* 1972; Waldron, 1958) and the American lobster, *Homarus americanus* (Squires, 1965; Evans and Mann, 1977). There is considerable difference in size between the Pacific and Atlantic species, *C. magister* males reaching commercial size in 4 years (48–60 mm width in Oregon) and probably attaining a maximum age of 8 years (maximum width about 220 mm) (Waldron, 1958; Butler, 1961). Less definite evidence on age is available for the Atlantic crab, distribution of sizes in one set of samples (Haefner, 1976) for *C. irroratus* falling into two modal sizes for each sex correlated positively with depth (<50, $51-100$, and for males a third group up to 127 mm) but none approaching the size of of large *C. magister*.

At molting of *C. irroratus,* most water uptake occurs between peeler and soft crab stages when progressive growth in length, width, and weight results in increases of 18–23, 19–27, and 52–82%, respectively (Haefner and Van Engel, 1975). In tank experiments, crabs acclimated at 7°C approached isosmoticity near 28–30‰ salinity; crabs acclimated at 17.5°C were isosmotic near 24–26‰, but showed hyperregulation in less than 20‰ salinity. Beers (1958) concluded from structure of the green gland in *C. borealis* that its function is inefficient in salinities below 50% seawater, and hence limiting to estuarine tolerance of the species.

Experimentally, oxygen uptake in *C. magister* is independent of ambient O_2 tension down to about 50 mm Hg; further decrease caused a drop in O_2 uptake, but the crabs can endure exposure to tensions as low as 30 mm Hg for at least one-half hour (Johansen *et al.,* 1970). Such low conditions are probably never encountered naturally. Extraction of oxygen from water passed over the gills averages 16%.

In temperature tolerance experiments, Vernberg and Vernberg (1970) found that *C. irroratus* collected from the Cape Hatteras area in February died after 1 hour in water of 30°C, that many animals died after 2 hours at 32°C, but that animals were unaffected by 48-hour exposure to 10°C. Zoeae exposed to 4°C in 30‰ salinity survived for long periods with no harm, but zoeae at 20°C in water of 10‰ salinity were inactive after 1.5 hours.

Other aspects of the biology, such as functional organization of proprioceptors and hormonal control of chromatophores, are beyond the scope of this chapter.

B. Family Portunidae

Two species of the swimming crab family Portunidae are preeminent in the literature of biological research on estuarine species. *Callinectes sapidus* Rathbun, the blue crab, for which there is a major fishery in the United States (over 141 million lb, 1971–1975 annual average) (Robinson, 1977) and an unde-termined catch harvested in Central and South America, has the most extended distribution of any member in its genus. The native range, occasionally Nova Scotia, Maine, and northern Massachusetts to northern Argentina, including Bermuda and the Antilles, has been extended by introduction to northwestern Europe, the central and eastern Mediterranean Sea as well as the western Black Sea, and to Lake Hamana-ko, central Japan (Williams, 1974, review; T. Sakai, personal communication). The green or shore crab, *Carcinus maenas* (Linnaeus), having a minimal commercial value (Bell and FitzGibbon, 1977), is thought to have been introduced in North America, where it ranges from Northumberland Strait, Canada, to Virginia, and in Australia. Its native range extends from Norway, just beyond 70°N, southward through the southern Baltic Sea, the North Sea, and the British Isles to the Straits of Gibraltar. A close relative, *C. mediterraneus* Czerniavsky of the Mediterranean Sea, has often been considered a subspecies (Christiansen, 1969; Rice and Ingle, 1975). Though both the blue and green crab live in shallow water, their biology is distinct. Our ecological review represents only a facet of knowledge concerning them, which has otherwise been reviewed by Waterman (1960, 1961), Crothers (1967, 1968, 1969, 1970), Clay (1965), Tagatz and Hall (1971), and Williams (1974).

1. Callinectes sapidus

Callinectes sapidus lives from the water's edge to 35-m depths, occasionally 90 m, in fresh water to salinities as concentrated as 117‰; large males have been taken from salt springs over 180 miles from the sea in Marion County, Florida. The species is known to tolerate summertime mean daily O_2 tension as low as 0.08 mg/liter in 1-m-deep experimental estuarine tertiary sewage treatment ponds in North Carolina (Williams, 1974, review), but crabs confined in baited pots in Chesapeake Bay showed stress when O_2 concentrations were less than 0.6 mg/ liter at temperatures above 28°C (Carpenter and Cargo, 1957).

The species leads a migratory life, developmental phases being adapted to an estuarine–marine–estuarine cycle. Mating in North America may occur from May to October, usually in estuaries of reduced salinity, each sexually mature hard male carrying a sexually mature female in a courtship embrace before, during, and after she molts into a terminally mature phase. The pair copulates when she is in the postmolt soft condition (Churchill, 1919; Van Engel, 1958). Such females spawn in 2, or if winter intervenes, up to 9 months after migrating to higher salinity areas near the sea where incubation and hatching occur (Van Engel, 1958). Males remain behind. The eggs hatch in about 15 days at 26°C (Churchill, 1919). Larvae develop through seven zoeal stages, atypically an eighth, and a megalopa (Costlow *et al.,* 1959; Costlow and Bookhout, 1959), the staging and duration of development varying over a period of 31 to 69 days (Costlow, 1965, 1967). Development progresses at comparable rates in salinities between 20.1 and 31.1‰. Above 31.1‰ development is slowed, and below 20.1‰ larvae rarely complete the first molt. Several workers have shown that the zoeae move considerable distances to sea over the Continental Shelf (Nichols and Keney, 1963; Tagatz, 1968; Sandifer, 1973), but that return to estuaries may be largely as megalopae (Williams and Deubler, 1968; More, 1969; King, 1971; Williams, 1971; Naylor and Isaac, 1973) and young crabs (Van Engel, 1958). The megalopal stage lasts 6–20 days, and this stage is found in plankton more or less year round. Most of the off- and onshore movement appears to be a normal developmental feature, but longshore larval transport obviously leads to temporary dispersal extensions at fringes of the range. Back within estuaries the young crabs undergo 18 to 20 molts before reaching maturity about 14 months later (Van Engel, 1958), and maximum age attained is estimated to be 3 years. Most movement of adults seems to be confined within estuarine systems (Fischler and Walburg, 1962; Tagatz, 1968), but there are exceptions (Leahy, 1975).

Kalber (1970) showed that the early zoeae are good osmoregulators, that the late zoeae lose this ability, and that the megalopae, though poor regulators at first, become good ones by the fifth day, the osmoregulatory adaptations apparently being fitted to the sequence of salinity stresses normally experienced during development. Ability to invade fresh water is governed by adaptation to chlorinity, the crabs being able to adjust to lowered levels if changes are experienced slowly, as in the case of migrations (Odum, 1953). Osmoregulatory ability of adult males and females is essentially the same, hypersomotic in 5 and 50% seawater but hyposmotic in 100% seawater (Tagatz, 1969, 1971; Engel, 1977). Growth increment of females during the terminal molt shows no significant correlation with salinity (Haefner and Shuster, 1964).

The usual diet of postlarval *C. sapidus* includes fishes, gastropods, *Littorina,* oysters, clams, mussels, tunicates, miscellaneous benthic invertebrates, and plant material (Darnell, 1959, 1961; Dunnington, 1956; Menzel and Hopkins, 1956; Tagatz, 1968; Hamilton, 1976; Odum and Heald, 1972; Pearson and Olla, 1977). The species is not expressly a scavenger, as often claimed.

Fluctuations in population density are reflected in commercial fishery statistics. Pearson (1948), analyzing possible relationships of measurable environmental factors to catch, concluded that the greatest single influence on success of a year-class may be salinity. The species needs high-salinity estuarine water for optimum hatches. Ovigerous females in river–bay–lagoon systems seem to have more successful hatches in years when river runoff is low. Experimental evidence from the larval studies supports this view.

2. *Carcinus maenas*

Carcinus maenas lives in depths of usually less than 5 or 6 m, but rarely descends to 200 m, on a variety of substrates (Christiansen, 1969). There is some migration in the shallows (Naylor, 1960), crabs <35 mm wide remaining below low tide in all months of the year, and larger crabs tending to move upshore with high tides in summer to remain stranded beneath stones at intervening low tides, but with no such movements in winter. Passive water loss from the gills tends to cool crabs subaerially exposed to high temperatures, helping to promote their survival (Ahsanullah and Newell, 1977). The species lacks noticeably flattened fifth swimming legs so characteristic of most portunids. Crothers (1969, 1970) found variation in population density associated with habitat, with lower densities occuring on wave-beaten, exposed shores, and higher concentrations prevailing in protected bays with lowered salinity. The normal salinity range of the species is about 10–33‰.

Early molting of males may be correlated with the fact that hard males mate with females in the soft condition (Hartnoll, 1969), and hence would precede molt of the latter. Mating resembles that described for *Callinectes sapidus,* and may last for hours and be repeated if the pair is separated (Cheung, 1966, and others). Reaction of males to a displaying female may vary from pairing to rejection with injury to the female, probably depending on physiological state of the male. A single mating can fertilize more than one batch of eggs, but few soft females fail to mate. Females lack a terminal molt.

In Holland, the number of ovigerous females declines to near zero during the late summer–early fall molting period but abruptly thereafter rises to a level of 50–60% (Broekhuysen, 1936). This number persists through winter until June, when another brief drop occurs followed by some new spawning in July. These fluctuations correspond to phasing of maturity among females. Those that spawned in fall drop hatching eggs in the spring, molt, and then may cast a second clutch of eggs in summer. Overwintering juveniles may attain maturity in spring and spawn after that in early summer, but spawning varies from this pattern in other geographic areas. In the Baltic, berried females hide in deeper (8–10 m), more saline water until the larvae are hatched (Rasmussen, 1959). A large female (46 mm wide) is estimated to have 185,000–200,000 eggs in a clutch.

Experimentally, eggs will develop in a salinity range of 25.8–26‰ to 40.3–50.4‰ at 10°C, 41–53.1‰ at $\bar{x}$ 16.3°C, and 20‰ at 15.7–17°C. Cold of 1.4°C in low salinity areas kills eggs; the animals usually migrate to avoid such extremes. Upon hatching in darkness (Dries and Adelung, 1976), developing larvae pass through a sometimes abnormal prezoeal stage, four zoeal stages, and a megalopa (Williams, 1968; Rice and Ingle, 1975) in a little under 60 days at 15°C. In the western Atlantic, Deevey (1960) found larvae in Delaware Bay from June to August, and Hillman (1964) reported them from July to August and December in Narragansett Bay. Kurian (1956), finding larval occurrence in Europe correlated with temperature, reported larvae present the year round at Plymouth, but most abundant in spring and summer, while in The Sound off Norway they are present in summer, with megalopae persisting until fall. Rasmussen (1973) found that eggs spawned later than May–July may never develop to first crab off Denmark and that annual fluctuation in salinity–temperature governs success or failure of a year-class in those waters. Experimentally, *C. maenas* zoeae show a higher mortality in diluted seawater than megalopae, and metamorphosis of the latter is delayed or prevented by water of low salinity (Lance, 1964).

The green crab is an omnivore whose feeding is influenced by abundance, size, and kinds of food available (Ropes, 1968). Pelecypods formed a dominant part of the diet of crabs 30–59 mm wide in Plum Island Sound, Massachusetts, which opened and ate clams equal to their own width, only the largest clams resisting attack. Greater amounts of food in crabs caught during low tide and just after sunrise suggested that feeding is heaviest at night and at high tide. Activity and feeding seem reduced at water temperatures below 7°C; low salinity apparently did not influence feeding. Ovigerous females were relatively inactive, thus less destructive of clams than nonovigerous females, as were mating males and crabs either preparing to molt or assuming a new hardened integument. The green crab has had an economic impact on the fishery for the soft clam, *Mya arenaria*; decreases in the fishery, especially from clam farming, have been correlated with its predation (Glude, 1955). A marked increase in the soft clam fishery of New England has been linked to mortalities and declining abundance of this crab, which have been associated with cold winters and a general cooling trend (Welch, 1968).

In normal seawater there is active absorption of water by the crabs, but in dilute media this is suspended when osmoregulation begins, and possibly there is a fall in passive permeability of the gills to water (Webb, 1940). Ionic regulation is a resultant of the following processes: active absorption by the gills of Na, K, Ca, and chloride at a rate greater than that at which they are lost by diffusion; differential excretion by the antennary gland, which tends to conserve K and eliminate Mg and SO_4; and inward diffusion of Mg and SO_4 across the gills in accordance with concentration of the gradient.

C. FAMILY XANTHIDAE

The Xanthidae, richest in species of all crab families, are represented in North American estuaries by six species, about which some understanding of pollution effects is known. The common name "mud crabs" can be applied to part of these with some justification (Table I) but not to all. Differences between species groups are extensive—encompassing structure, reproduction and larval development, behavior, and physiology, and it is useful to recognize subfamilies.

1. Subfamily Xanthinae

Aside from distinctive niches and geographic ranges of the species in this group, enough biological features are shared that they can be generalized under the last species in the series.

Cataleptodius (=*Leptodius*) *floridanus* (Gibbes) (see Chapter 8, p. 264) ranges from Bermuda, the Bahamas, and the Florida Keys to São Paulo, Brazil, on a variety of reefs and other substrates in shallow water (Rathbun, 1930; Guinot, 1968; Hazlett *et al.,* 1977).

Eurypanopeus depressus (Smith) ranges from Massachusetts Bay through east and west Florida to southern Texas, as well as Bermuda, the West Indies, and Uruguay (Williams, 1965). It is abundant on oyster bars in a recorded salinity range of 4.5 to 20.4‰ in Chesapeake Bay (Ryan, 1956), and has been recorded in similar habitats elsewhere in depths to 48 m.

, *Neopanope sayi* (Smith) (=*N. texana sayi*) (Abele, 1972) ranges from Chaleur Bay, New Brunswick, and Cape Breton Island, Nova Scotia, Canada (Bousfield and Laubitz, 1972); to the Florida Keys (Abele, 1972), and has been introduced in Wales (Naylor, 1960). It frequents mud or oyster shell bottoms, as well as sea grass beds (Ryan, 1956; Marsh, 1973), from the low tide mark to 46 m.

Panopeus herbstii A. Milne Edwards ranges from Bermuda and Boston, Massachusetts, to Santa Catarina, Brazil (Williams, 1965). It has been introduced in Hawaii and has supposedly long been seen on western U. S. shores (Edmondson, 1962). The species inhabits a number of nearshore substrates—predominantly mud, shell, or stones, from the intertidal zone to 22 m and over a broad range of salinities.

Rhithropanopeus harrisii (Gould) presumably had a native range extending from the southwestern Gulf of St. Lawrence, Canada, to Veracruz, Mexico, and it has been introduced on the West Coast of the United States (Williams, 1965). An older and wider introduction in northwestern Europe, the Black Sea, Sea of Azov, and Caspian Sea is often regarded as a distinct subspecies, the Holland crab (Christiansen, 1969; Gadzhiev, 1963; Makarov, 1977). The hardy species thrives in a very broad range of salinities, an attribute that enhances the ease with which it has been transported and introduced; and with which it can be manipulated in laboratory experiments. More is known about at least the larval development of this species than any other in the subfamily.

Larval development of species in the subfamily Xanthinae includes four zoeal stages and a megalopa. Connolly (1925), using plankton, and Chamberlain (1962), using laboratory culture, demonstrated this for *R. harrisii*. Christiansen and Costlow (1975) reared the species from hatching through the first or second crab stages under experimental conditions in 11 combinations of salinity and cyclic temperatures (5, 20, 35‰ salinity at 20-25°, 25-30°, 30-35°C; and 25‰ S at 20-25° and 30-35°C). Larvae survived to megalopa plus first crab in all combinations except 5‰ salinity at 30-35°C. Best survival to megalopa (94%) and first crab (90%) was obtained in 20‰ salinity at 20-25°C. In all other combinations there was a reduction in survival to the first crab stage. Duration of larval stages was affected significantly by temperature, whereas the effect of salinity on mean days from hatching to the first crab was not consistent in different temperature cycles. The authors concluded that larval development of *R. harrisii* is strongly influenced by environmental factors and not solely related to genetic differences. Earlier, Costlow *et al.* (1966) found that time of development was not significantly affected by salinities of 5-35‰ in noncycled temperatures and suggested that the capacity to develop normally within this broad range of environments may have contributed to the wide distribution of the species. Aspects of osmoregulatory ability are comparable to those discussed for *Callinectes sapidus* (Jones, 1941; Kalber and Costlow, 1966, 1968; Capen, 1972), and Smith (1967) showed that inward permeability to water is decreased at lower salinities.

Ovigerous females occur during the warm months in Chesapeake Bay (Ryan, 1956). Zoeae are present from May to October in tributaries of the Bay, with peak occurrence from June to July (Herman *et al.*, 1968; Sandifer, 1973), the latter finding them common to abundant in 0–10‰ salinities but less abundant in the Bay proper. Others have shown similar occurrences elsewhere with some adjustment for latitudinal temperature differences (Bousfield, 1955; Tagatz, 1968; Williams, 1971). Sandifer (1975) contended that stratification of the larvae on bottom indicates that their retention in estuaries is the major means of recruitment for the species.

Ryan (1956) judged maturity to be reached in about 1 year, but molting and growth continue throughout life. Maximum carapace widths are about 21 mm for males and 16 mm for females. Maximum age has not been estimated.

Success of *R. harrisii* in the estuarine environment is emphasized by its abundance and role in the food web. Odum and Heald (1972) found more than 40 animals per square meter in holes, under leaves, logs, shells, and other debris in the bed of an estuarine stream draining a *Juncus* marsh in south Florida. The omnivorous diet was dominated by mangrove leaf detritus. Crustaceans such as small amphipods and harpacticoid copepods were eaten more often by small crabs. The crabs are preyed upon by hakes (Sikora *et al.*, 1972), the white catfish (Heard, 1975), and probably other predaceous fishes.

What is known of the life histories of other xanthines helps to round out knowledge of them collectively. Mating pairs of *Neopanope sayi* have been observed in laboratory tanks supplied with running seawater at Gloucester Point, Virginia, from late March until early September. Females mate in the hard condition, occasionally when ovigerous, and both sexes may copulate several times during a molt cycle (Swartz, 1972). The reproductive season is judged to be governed by temperature, mainly in the 24°–30°C range. Ovigerous females have been observed from April in South Carolina (Lunz, 1937) to October in Chesapeake Bay. Eggs hatch in about 10 days. Chamberlain (1957, 1961) discussed larval development, finding that it varied with temperature (14 days at 30°C, 27 days at 21°C) and food, as did Sulkin and Norman (1976). Sandifer (1973) found zoeae in lower Chesapeake Bay from June to October but concentrated in June, mainly in 20–25‰ salinities, and he summarized latitudinal variations in planktonic occurrence found by others from Narragansett Bay to Florida, emphasizing his (1975) view that retention in estuaries is reinforced by stratified densities and circulation. Swartz (1972) concluded that the crabs mature in 1 year and that some individuals survive to an age of 2 years.

Neopanope sayi is a predator on oysters, the barnacle *Balanus improvisus,* and young *Mercenaria mercenaria* clams (McDermott and Flower, 1952; Landers, 1954). In turn it is preyed upon by fishes.

Panopeus herbstii roughly parallels *N. sayi* in life history details. It is a species to which Costlow *et al.* (1962) early applied the idea of response surface analysis in an attempt to describe optimum conditions for larval development, hypothesizing that effect of temperature on successive larval stages limits the productive spawning period. Low temperatures favor the spring brood of larvae, prolonging larval development until warmer water produces favorable conditions for the megalopa, whereas larvae hatched in fall are not so favored and therefore suffer higher mortalities. Among various marsh crabs in Georgia, only *P. herbstii* was found active at temperatures below 12°C (Teal, 1959).

Menzel and Nichy (1958) found that, among xanthids only *P. herbstii* and *Menippe mercenaria* were large enough to kill significant numbers of adult oysters. McDermott (1960) concluded that *P. herbstii,* preying on oyster spat as well as barnacles *(Balanus improvisus),* is potentially the most destructive of five species of mud crabs on New Jersey oyster beds.

Walton and Williams (1971) studied populations of *Eurypanopeus depressus* on artificial reefs in three small experimental estuarine ponds in North Carolina, finding it to be the most abundant species of xanthid following heavy seeding with estuarine plankton a year before. The crabs matured in the ponds, 50% of females in one pond being gravid when sampling began in late June. This percentage remained high in samples during the summer, once being 100% for an estimated population of 276. There was gradual reduction in gravid females after the end of August. The size range in samples from Chesapeake Bay suggested to Ryan (1956) that maturity might be reached in the first summer after eggs have hatched and that molting continues after maturity is reached. Sandifer (1973) found larvae of this species only three times during his plankton survey of lower Chesapeake Bay, but others (Pinschmidt, 1964; Dudley and Judy, 1971), including authors cited above, have found them throughout the warm season. McDermott (1960) considered the species a predator on oyster spat in southern New Jersey but not so serious a pest as *P. herbstii.*

2. Subfamily Menippinae

The stone crab, *Menippe mercenaria,* stands alone among the xanthid crabs in size, fishery value, and reproductive specializations. It ranges from the Cape Lookout, North Carolina, area to Yucatan, Mexico, the Bahamas, Cuba, and Jamaica, the crabs shifting among various environments as they grow. The maximum recorded depth of occurrence is 51 m. The youngest crabs are found under shell fragments or in crevices among rocks in deeper saltier estuaries, or in fouling on buoys. Older crabs may be found more shallowly among oyster shells, in fouling about jetties, etc., or in turtle grass. There they live until nearly grown, when they may move to shoals to construct burrows, some apparently colonial, as much as 75 cm in length, just below the tide mark. But adults are also found on oyster reefs, among rocks along jetties, and on offshore reefs or wrecks (Williams, 1965, review; Powell and Gunter, 1968).

About 2.5 million lb of stone crabs were landed in Florida in 1974 (Bell and FitzGibbon, 1977), although unrecorded catches for home use are known elsewhere as well. The flesh is considered a delicacy.

Reproduction in *M. mercenaria* has been a subject for active investigation since early in this century. Savage (1971) observed and described mating of a male and female held in an outdoor aquarium. Both then and among mating pairs observed in the field, males were always with freshly molted or presumed freshly molted females. Females in laboratory experiments have produced more than 10 broods of viable eggs within the same intermolt over a 120-day period without copulation taking place in intervals between production of the different broods (Cheung, 1968; Yang, 1971), the interval between hatch of one brood and oviposition of the next being 2–3 days. Eggs of all these broods were successfully attached to hairs of pleopods and developed to hatch as normal zoeae. Moreover, females from the wild held in isolation through a molt were found to retain sperm through ecdysis, a special adaptation (late disposal of the old invaginating exoskeletal wall between sperm and new wall) enabling retention. Cheung (1969) also showed that ovarian development is closely correlated with water temperature, optimum development being at 28°C. Seasonally, spawning built to a peak from August to September in southern Florida, but dropped to a low in winter. Elsewhere, ovigerous females are known from May to August in North Carolina and August in Texas (Powell and Gunter, 1968).

Porter (1960) described larval development, including a prezoeal and six zoeal stages, but he considered the first and last of these to be atypical. Ong and Costlow (1970), with much more sophisticated equipment, reared larvae in 18 environments (six salinities, three temperatures). Both factors were found to affect rate of development and survival, optimal conditions being about 30°C in 30–35‰ salinity in which the megalopa was reached in 14 days and the first crab in 21 days with survival of 60–72%. Development was slower in both lower temperature and salinity. Yang (1971) successfully reared mass and individualized cultures, showing that there were five zoeal stages and a megalopa. Dudley and Judy (1971) found stone crab larvae in plankton off Beaufort Inlet, North Carolina, from June to August. Mootz and Epifanio (1974) determined an energy budget for larvae fed *Artemia salina* nauplii, and found that larval growth is exponential during zoeal stages but decreases during the megalopa even though consumption reaches a peak during this stage.

These crabs are powerful and active predators, feeding on large and small oysters (Menzel and Hopkins, 1956), barnacles (Powell and Gunter, 1968), gastropods, etc. The fingers of the chelae, if fastened upon a hard object, can hardly be pried apart; the strength of this grip by one male on a hard object was so great that it crushed its own chela and bled to death. Powell and Gunter judged stone crabs to have a strong influence on the life of sand and mud flats that are exposed by wind or lunar low tides because their water-filled burrows are refuges for small aquatic organisms that cannot withstand drying. Commensals found in burrows include sea anemones, tube worms, bivalve and gastropod mollusks, barnacles, amphipods, caridean and brachyuran decapod crustaceans, and fishes, to name a few.

D. FAMILY GRAPSIDAE

Three species of Grapsidae considered here all live in the intertidal zone, two in marshlands of the eastern and southern U. S. coast, and the third along beaches of western North America. Considerable divergence in ecology is hidden under systematic similarities; therefore the species are discussed separately.

1. Subfamily Varuninae

Hemigrapsus nudus. The purple shore crab ranges from Yakobi Island, Lisianska Strait, Alaska, to the Gulf of California (Hart, 1968). The species lives mainly in upper reaches of the intertidal zone on well-drained substrate, such as boulders situated over coarse sand or broken shell, in contrast to a smaller relative, *H. oregonensis,* which lives lower down on siltier substrate (Knudsen, 1964, review). The adults are herbivorous, gleaning microalgae from rocks, mainly at night, by the large chelae, although gut contents include some sand and incidental animal matter.

Mating pairs (both sexes in hard condition) occur in Washington from early December to early February (Knudsen, 1964), and ovigerous females are reported there and elsewhere from October to July (see also Boolootian *et al.,* 1959). There is almost no evidence of a second brood. Hart (1935) described five zoeal stages, the megalopa and first crab stage from laboratory rearing on a diet composed mainly of *Ostrea lurida* Carpenter larvae, and thus there is change in diet during maturation.

Aspects of the species' physiology have been studied at length in the last quarter century. Dehnel (1962) demonstrated the inability of *H. nudus* to hypoosmoregulate and demonstrated further a seasonal variation in regulatory ability. Hyperosmotic regulation in dilute salinities in summer is achieved by extra-renal mechanisms, for the blood and urine are isosmotic. Hyperosmotic regulation in winter, however, is accomplished by production of urine less concentrated than the blood, suggesting ion reabsorption by the kidney (Dehnel, 1966; Dehnel and Stone, 1964). Size had no effect on osmotic response and weight remained stable during experiments. Dehnel, further elaborating seasonal changes in ionic balance, showed that salinity affects metabolic response to temperature; oxygen consumption is highest in low temperature–low salinity combinations (Dehnel, 1960, 1966), the response being thought to result from work to maintain osmotic balance. In experiments, acclimation (1 week) to high temperatures generally increased resistance to lethal temperatures whereas acclimation to low salinity decreased it (5°C, 35% seawater, 50% survived at 28.81°C and 28.68°C for 12 and 24 hours, respectively). High temperature–high salinity (20°C, 75% seawater) was the most favorable combination for withstanding the high test temperatures (50% survival at 34.32°C and 33.45°C for 12 and 24 hours, respectively) (Todd and Dehnel, 1960). Blood glucose has been determined at relatively low levels (0.75–2.55 mg/100 ml), varying to some degree with molt cycle (McWhinnie and Scheer, 1958), but on the whole is relatively stable through changes in diet and season (Meenahshi and Scheer, 1960), although its precursors vary.

2. Subfamily Sesarminae

a. *Sesarma cinereum.*   The wharf crab, wood crab or square-backed fiddler crab, ranges from the Magothy River, Chesapeake Bay, Maryland, to Palm Beach, Florida, and from Collier County, Florida, to Veracruz, Mexico (Abele, 1973), and is absent from extreme southern Florida. The crabs inhabit supralittoral zones of marshes characterized by high salinity ($\bar{x} = 27.9‰$) and sandy substrates (Seiple, 1979), actively crawl about on wharves and jetties or rest in shallow burrows above tidemark along shores, and have often been found on vessels along the coast, hiding by day and coming forth at night to search for food.

There is little recorded information on food habits. Seiple (1979) showed that the species eats *Spartina* shoots in the same manner as that recorded by Crichton (1974) for *S. reticulatum.* Oler (1941) maintained captives for about 1 year in an aquarium on mud substrate, with tap water as moisture and vegetable matter as food, but the larger crabs cannibalized smaller crabs in the group, presumably at ecdysis.

Ovigerous females are known from January along the Potomac River (Williams, 1965), and mid-April (Seiple, 1979) to November in North Carolina. Females in North Carolina produce four to six egg batches in close synchrony with lunar phase, each female carrying her eggs approximately 1 lunar month, with peak reproductive activity occurring from April to June (Seiple, 1979). From summer hatchings, Costlow and Bookhout (1960) found that larvae fed *Artemia salina* nauplii passed through four zoeal and a megalopal stage before metamorphosis. Costlow *et al.* (1960) found that optimum salinities exist for each larval stage, but that development proceeds best in the 20–26.7‰ salinity range. Temperature was found to have more effect on length of larval development than on mortality, with higher temperature speeding development, but the authors concluded that salinity is the chief physical factor confining *S. cinereum* to estuaries. Records of larvae from plankton vary with site of sampling. Pinschmidt (1964), in Newport River near Beaufort, North Carolina, found early zoeae in low concentration from June to September in water of 19°–34°C and 7–36‰ salinity, but most numerous in August at 25°–31°C. Sandifer (1973) found the larvae rarely in York and Pamunkey rivers, Virginia, in June and August, the samples being from bottom water of 24.8°–26.6°C in 11.9–19.64‰ salinity. *Sesarma* larvae occur from June to September up to 6.5 km off Beaufort Inlet, North Carolina (Dudley and Judy, 1971), and Tagatz (1968) found larvae of this genus to be the second most abundant form in samples from St. John River, Florida, from April to October, peaking in August. Sandifer (1975) applied his theory of estuarine retention to this as well as other species with similarly distributed larvae.

Teal (1959) implied that *S. cinereum* acts more like a land animal than an aquatic one because the species was relatively inactive under water in experiments, thus holding its oxygen consumption down. There was some experimental evidence for thermal acclimation of metabolism, but more evidence for acclimation by selection of microclimate. Gray (1957) showed that the gill area is nearly double that of another semiterrestrial species, *Ocypode quadrata* (Fabricius).

b. Sesarma reticulatum. This marsh crab is distributed from Woods Hole, Massachusetts, to presumably somewhere on the east coast of Florida, and from Sarasota on the west coast of Florida to Calhoun County, Texas (Abele, 1973). The species borrows communally in muddy salt marshes, its holes as much as 0.75 m deep and occasionally intersecting those of *Uca* species (Crichton, 1960; Allen and Curran, 1974). Teal (1959) found *S. reticulatum* active on Georgia marshes when the tide was high or the sky cloudy. When the marsh was exposed the crabs were in burrows, usually near the top in air or water. The diet is primarily *Spartina* (Crichton, 1974). The species occurs prominently in marsh ecosystems (Day *et al.*, 1973; Subrahmanyam *et al.*, 1976), often along with *Uca pugnax* and *minax* (Teal, 1958; Allen and Curran, 1974), functioning not only as a herbivore but also as a burrower that tills the land, increases erosion, and turns over the cord grass more rapidly than the annual decay cycle could do it.

From summer spawning, Costlow and Bookhout (1962) described a larval development of three zoeal stages and a megalopa in laboratory rearings on a diet of *Artemia salina* nauplii. Larvae have been taken sparingly in plankton of lower Chesapeake Bay and its tributaries (Sandifer, 1973) from June to October, most abundantly from July to August, in a temperature range of 22.8°–27.9°C and salinities of, rarely 10‰, usually 15–20‰, or more. Eighty percent of the larvae were taken at the bottom. Sandifer (1975) regarded this depth distribution as an adaptation for retention in estuaries. The functional basis of the adaptation may be a demonstrated hyperosmoticity of larval stages through early megalopa over a wide range of salinities, increasing density of the larvae (Foskett, 1977).

Foskett also thought this hyperosmoticity might act to provide turgor pressure within the thin larval cuticle whereas hard-shelled adults that hyperregulate in salinities below 27.5‰ but hyporegulate above this, as most crabs, have no need for such a mechanism.

Teal (1959) found respiration rates higher in water than in air. Teal and Carey (1967) found that *S. reticulatum* can regulate its metabolism down to about 6% atmosphere (45 mm Hg); however, marsh crabs do not encounter low oxygen conditions in air, only in burrows in water. There, during periods of submergence, the respiration rate may decline with decreasing oxygen pressure, and below critical pressures the crabs must convert metabolism to fermentation. Lactic acid is probably formed from glycogen, to be oxidized at the next period of emergence. Gray (1957) found the gill area of *S. reticulatum* to be relatively low compared to that of other species living in similar habitats (*Uca pugnax* and *minax*). He found *S. recticulatum* more robust but less active than its congener, *S. cinereum*.

E. FAMILY OCYPODIDAE

Crane (1975), in her exhaustive monograph on fiddler crabs of the world, distilled years of study into a compendium of information on systematics, zoogeography, ecology, functional morphology, behavior, phylogeny, and evolution of the group. She recognized 79 species, many subdivided into subspecies, clustering them into a number of subgenera. Three of these species from the Western Hemisphere are so universally recognized in the biological literature as *Uca minax* (LeConte, 1855), *U. pugilator* (Bosc, 1802), and *U. pugnax* (Smith, 1870) that we feel constrained to retain this nomenclature, accepting taxonomic simplifications recommended by von Hagen (1976). His thesis is that any specialist will recognize these entities while appreciating the subtleties of variation, as did Crane, and that formal splitting seems premature and burdensome because attributes of the populations are still unclear and their importance has not yet been weighed.

Uca minax, the red-jointed fiddler, ranges along the eastern and southern coasts of the United States from the southern shore of Cape Cod, Massachusetts, to northeast Florida, and from the area of Yankeetown, northwest Florida, to Louisiana, and on to Matagorda Bay, Texas, if von Hagen's analysis treating *Uca rapax longisignalis* Salmon and Atsaides, 1968, as a synonym is accepted. *Uca pugilator,* the sand fiddler, ranges from Cape Cod, Massachusetts (rare on the north shore), southward around the tip of peninsular Florida and westward to Corpus Christi, Texas; single occurrences are on record from Old Providence Island, the Bahamas, and Santo Domingo. *Uca pugnax,* the mud fiddler, ranges from Provincetown, Massachusetts, to Daytona Beach, Florida. *Uca virens* Salmon and Atsaides, 1968, from the Gulf of Mexico coast, considered a subspecies of *U. pugnax* by Crane, is synonymized by von Hagen with *U. rapax* (Smith, 1870), a tropical species ranging southward to southern Brazil.

Uca species live in an environmental borderland along the edges of estuaries, and habitually stay away from the sea proper. The crabs are behaviorally terrestrial but physiologically aquatic (Herrnkind, 1968b), orienting themselves within their narrow environment on the strand between water and land by visual cues, and able to return to it even if displaced several meters into either open water or to land beyond the beach. They exhibit endogenous rhythms in activity and color changes that are associated with local tidal schedules as well as diel cycles (Barnwell, 1968). All are burrowers, some retreating into more or less impermanent burrows during periods of inundation and emerging during low water stands. Of species discussed here, *U. minax*, the largest, lives on muddy banks of marsh creeks and among *Spartina, Salicornia,* and freshwater herbs in marshes threaded by brackish to freshwater streams under tidal influence (Miller and Maurer, 1973). *Uca pugnax*, medium sized, frequents salt marshes and sheltered shores on mud to sandy mud substrate. *Uca pugilator,* the smallest of the three, lives on sheltered shores with sandy to sandy mud substrate more or less free from vegetation and often mixed with scattered shells and stones but with muddy surfaces for feeding nearby. The food of all consists largely of minute particulate matter, algae, bacteria, detritus, etc., from the surface of the substrate which is scraped up by the small chelipeds and passed to the mouth for separation of accepted organic and rejected inorganic components by various mouthparts, the rejectamenta being left behind as trails of rounded pellets (Schwartz and Safir, 1915; Miller, 1961). Feeding underwater, however, has been observed for *U. pugnax* and *minax,* and the latter will kill and consume either of the other two species (Teal, 1958). Males, having only one small chela, feed twice as long as females, animals of each sex otherwise having equal metabolic responses (Valiela *et al.,* 1974).

Life histories of the three species are similar. Crane (1975) discussed courtship and mating in great detail, showing that males attract females by waving display and sound production, i. e., drumming the major hand against the substrate either on the surface or in burrows (Salmon and Horch, 1972), and that copulation in the hard condition (Hartnoll, 1969) takes place in the burrow of the male or, in *U. pugilator* and *pugnax,* on the surface at night, and even under water in captivity (Herrnkind, 1968a). Apparently the huge claw of the males is reserved for display (threat, attraction, etc.), not for handling females. Both sexes show a periodicity in gametogenesis (Young, 1974). Courtship and mating are influenced by temperature, with some evidence of latitudinal adaptation (Crane, 1975). Near New York, for example, *U. minax* fed and moved above ground during low tide in mid-June but did not display when air temperature during the preceding night fell below 21°C, even though daytime temperature then regularly reached above 29°C. On Cape Cod in early June several *U. pugilator* waved at low intensities on a sunny day when air temperature was 19°C and the previous nighttime low had been 10°C. Maximum breeding activity for that population was in mid-June followed by notably less activity in July even though temperature was higher. Never was the activity on Cape Cod equivalent to that seen in more southern populations. Spawning following mating, therefore, is most extended at low latitudes. Ovigerous female *U. pugilator* have been reported from early July to mid-August on Long Island (Schwartz and Safir, 1915), March to midsummer in Virginia (Williams, 1965), April to October near Miami, Florida (Herrnkind, 1968a), and Hedgpeth (1950) reported zoeae taken in a plankton net on May 20 at Long Lake, Texas. Females enter the water to let eggs hatch.

Uca pugilator, representative of all, has five zoeal stages and a megalopa, each of these subdivisions of larval life lasting 3–4 weeks for a total development time of 6–8 weeks (Hyman, 1920; Herrnkind, 1968a). The larvae are predatory, capturing prey (*Artemia* nauplii, etc.) on or between the telson spine by violent flapping of the abdomen, whereupon flexure of the body brings food to the mouth. The megalopa swims at first but gradually adopts a benthic existence. Zoeae of the three species have not been distinguished in plankton, but Sandifer (1973) found zoeae of *Uca* sp. to be the most commonly collected and abundant decapod larvae in lower Chesapeake Bay, with concentrations of $>100/m^3$ not unusual. The zoeae were taken over a salinity range of 0.06–32.34‰ at temperatures of 19.4° to 30.8°C, but most numerously in the York and Pamunkey rivers tributary to the Bay, mainly at salinities <5‰ at 26°–30°C. Larvae first appeared in June, peaked in July, and disappeared in October. All of the zoeal stages were collected but stage I was most numerous, predominantly at the surface. Stage V was more numerous in bottom than in surface samples. Sandifer found that the work of others (reviewed) reflected his findings with allowance for latitudinal variation, although workers in the Beaufort, North Carolina, area found larvae at higher salinities (Pinschmidt, 1964; Dudley and Judy, 1971).

Recently Christy (1978) proposed that mating, incubation, hatching, and larval development in *U. pugilator* are synchronized with tides such that male display reaches peaks at spring tides, and mating occurs in male burrows where females remain during the entire incubation period of about 2 weeks, emerging to release hatching larvae at neap tides. Ensuing larval development is so timed that the megalopae will be ready to metamorphose when they are swept into the estuary on spring tides. There are thus two breeding populations allied with new and full moon tidal phases. Nonbreeding males and females, or displaying males, constitute the feeding droves that emerge at each low tide.

Further topics summarized below under the species in which they have been studied most include salinity and substrate associations, metabolic response to temperature, and other environmentally influenced physiological changes.

1. Uca pugilator

Most attention has been focused on the sand fiddler. The first two crab stages following metamorphosis are relatively weak and adapted to clinging, but following them the crab starts to assume the familiar structural and behavioral characteristics of the species, burrowing during high tides first in the wet intertidal area and later along the strand near or above the high tide line (Herrnkind, 1972). From that refuge the population emerges between high tides to carry on its activities in droves on the exposed intertidal surfaces. Rhythmic activities in such an environment follow naturally, but their behavioral complexities lie mainly beyond the scope of this chapter (see Crane, 1975). An example of internal rhythms is that of blood glucose level tied to the diurnal cycle, high in late afternoon, low in early morning. Example levels measured in percent milligrams are 5:30 AM, 7.86; 1:30 PM, 8.06; 5:30 PM, 15.41 (Dean and Vernberg, 1965).

Testing salinity tolerance, Teal (1958) found that 50% of *U. pugilator* died after 3.5 days in fresh water, but over 50% survived more than 10 days in seawater of 7‰ salinity. Given a choice of fresh water or seawater of 30‰ salinity, the crabs chose seawater but preferences of females were less strong than males, as was true of *U. pugnax*. He concluded that this species lies between *U. minax* and *U. pugnax* in its tolerance of fresh water, and can survive soakings of the *Salicornia* marsh with rain between spring tides.

Teal (1958) also concluded that *U. pugilator* cannot feed properly where sand is absent, because in choice of substrate experiments (sand or mud above or under water) it burrowed in sand above water and when restricted to an unfavorable low muddy marsh did not survive. In company of either of the other species, *U. pugilator* reduced its burrows in sand above water by 50%.

Respiration rates are higher underwater than in air (Teal, 1959) probably because of increased activity necessary to aerate the gills. Like *Sesarma,* this species and *U. pugnax* can regulate their metabolism in air under experimental conditions down to 1 or 2% atmosphere (8–15 mm Hg), but low oxygen pressures are encountered in nature only in burrows in water and there the crabs, like *Sesarma,* can go into oxygen debt until emergence at the next low tide.

Latitudinal effects of temperature are manifest in metabolic response. Rate of metabolism in a Massachusetts population was higher at 1.4°C than in a Florida population, but at 15°C the difference was not significant. The northern population was less sensitive to temperature change and more resistent to low temperature than the Florida population (Démeusy, 1957). Edwards (1950) earlier had shown metabolic differences in the populations at 20°C. Teal (1959) found that crabs from Georgia showed no adjustment of respiration for temperature acclimation above 25°C, but that below 20°C there was some evidence of acclimation though not so well developed as in *U. pugnax*. Summer temperatures of 45°C on open *Distichlis–Salicornia* flats in Georgia prevent *U. pugilator* from permanently occupying these areas (Teal, 1958). Metabolic activity in *Uca* of the temperate zone exhibits a seasonal cycle and this cyclic change must be taken into account in comparing physiologic activity of relatives at different latitudes. Metabolic response of fiddler crabs has real significance in their distribution (Vernberg, 1959), manifesting itself to some degree even in the larval stages (Vernberg and Costlow, 1966).

Beyond respiratory responses, there is a significantly larger number of cells per unit volume and a higher titer of blood protein in *U. pugilator* at 30°C than at 10°C. Clotting time is significantly lengthened in the 10°C crabs (Dean and Vernberg, 1966). These physiological effects are associated with activity levels mediated by temperature.

2. Uca minax

The red-jointed fiddler, often found in *Spartina* marshes in the area immediately preceding the *Salicornia–Distichlis* zone (Vernberg, 1959) and occasionally at the edge of low woodlands, has lived in fresh water for more than 3 weeks (Teal, 1958), and when offered choice of fresh or seawater, chose fresh water. In choice of substrate experiments (mud or sand) it chose mud either above or under water, but dug few burrows when competitive *Uca* species were present.

Teal (1959) found *U. minax* to have the lowest rate of oxygen consumption among a number of marsh crab species (*U. pugnax* and *pugilator, Sesarma cinereum* and *reticulatum, Eurytium limosum,* and *Panopeus herbstii*). He (also Vernberg, 1959) showed that this species in all probability does not acclimate respiratory rate to changes in temperature. Moreover, its gill area per gram of body weight is lowest among U.S. East Coast *Uca* (Gray, 1957).

3. Uca pugnax

The mud fiddler's name derives from its preference for a muddy marsh environment, often well shaded (Schwartz and Safir, 1915), but it is perhaps somewhat excluded from areas where there is an abundance of halophyte roots or where substrate is too fluid to support burrows (Kraeuter and Wolf, 1974). Burrows can extend to depths of 60 cm (Pearse, 1914). Results of salinity tolerance experiments are consistent with this type of distribution (Teal, 1958); 50% of *U. pugnax* placed in fresh water died within 1.5 days, whereas 50% mortality occurred after 3 days in seawater of 7‰ salinity. Given a choice of fresh or seawater of 30‰ salinity, the crabs chose seawater.

As implied above, *U. pugnax* exhibits one of the most highly developed thermal adaptabilities among marsh crabs tested (Teal, 1959), and its abundance may be explained in part by its ability to regulate its metabolism over a wide range of temperature (Vernberg, 1959; Vernberg and Tashian, 1959).

IV. Pollution Ecology

Estuarine crabs occupy an ecological niche in the coastal environment that unfortunately is especially susceptible to man-made and natural pollutants. Sources of pollutants include industrial and municipal effluents entering rivers that transport the pollutants to the coastal environment—pesticides transferred by agricultural runoff, pesticides directly applied to control noxious insects such as mosquitoes, atmospheric exchange, and transported ocean-outfall discharge (Fig. 1). Whatever the source, pollutants can be concentrated in the crab's habitat, becoming part of the bottom sediment by precipitation, adsorption onto suspended or bottom sediments, and bioaccumulation. Conversely, pollutants can be diluted and dispersed by current and tidal action, biological transport, and vaporization. Recent surveys of pollutant residues in coastal zone organisms indicate that at least a portion of the pollutants entering this area is accumulated and often concentrated by estuarine brachyurans.

Crabs may be adversely affected by direct toxicity or indirect damages. Indeed, such influences can adversely affect a crab fishery if the amount of concentration in the species sought exceeds guidelines established by the Food and Drug Administration for seafood products.

For this discussion, we selected a few groups of pollutants of current interest to the scientific community: pesticides, heavy metals, oil, thermal discharges, and radionuclides. The list neither encompasses all groups of pollutants nor all the pollutants within a particular group; rather, it represents some of the major contaminants entering the crabs' environment.

A. Pesticides

Pesticides and other exogenous organics continually enter the marine environment through the atmosphere, river discharge, terrestrial runoff, and direct application. In many instances these materials, or their "breakdown" products, enter the biogeochemical cycles operating in the marine environment and are accumulated by organisms such as crabs. Often residues of these chemicals can be detected in the organism in minute (parts per trillion) concentrations by sophisticated analytical techniques, but their effect and their mechanisms of effect are essentially unknown. We do know, however, that pesticides are used to eliminate noxious or harmful arthropods (insects), and, consequently, these chemicals are usually harmful to nontarget arthropods (crustaceans). Pesticide use is changing in the United States in that less persistent, more specific chemicals are replacing more persistent organochlorines, such as DDT and its family. For purposes of this chapter, pesticides will be separated and discussed according to chemical structure, i.e., organochlorines, organophosphates and carbamates,

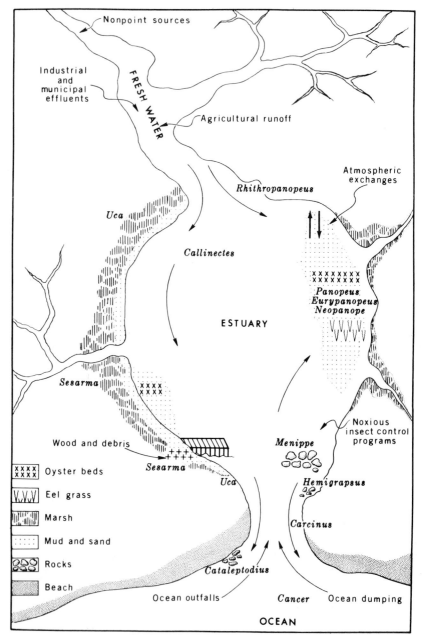

Fig. 1. Diagrammatic representation of an estuary showing generalized major habitats, sources of circulated materials, and relative position occupied by adults of brachyuran genera.

juvenile hormones, and insect growth regulators. The application of mirex to the coastal environment will be used as a case history because it illustrates the impact of a rather controversial pesticide on crabs and their environment.

1. Chlorinated Hydrocarbons

Chlorinated hydrocarbons, persistent pesticides that have been used in this country especially since the 1950s, include such compounds as DDT and its metabolites, dieldrin, mirex, and Kepone. These compounds generally are insoluble in water and much more soluble in lipids. Because of ubiquitous use in this country and persistence, these chemicals often appear as residues in marine organisms, including brachyuran crabs. For example, Albright *et al.* (1975) reported nondetectable to 22.0 ppb heptachlor epoxide and 23.0–295.8 ppb *p,p**-DDE in the tissue of *Cancer magister* from selected areas in the Georgia Straits near British Columbia. The sources of these chlorinated hydrocarbons were believed to be surface runoffs from aerial sprays, ground water, direct application for insect control, and domestic and industrial effluents. DDT and dieldrin residues from crabs and other organisms from San Antonio Bay, Texas, were reported by Petrocelli *et al.* (1974). Blue crabs, *Callinectes sapidus,* analyzed by a gas–liquid chromatograph with an electron capture detector, were found to contain rather high incidence of DDT, possibly caused by their ommnivorous or opportunistic feeding habits. Of 81 blue crabs analyzed, 94% contained DDT, 32% dieldrin, and 28% both chemicals. The amount of DDE ranged from nondetectable to 68.20 μg/kg or ppb DDE, and dieldrin ranged from nondetectable to 44.60 ppb. Modin (1969) examined residues of DDE, DDD, and DDT in Dungeness crab, *Cancer magister,* from the San Francisco region and found values of DDE from 40 to 62 ppb in stripped females and 54 to 430 ppb in ova. Commercial fishermen speculated that pesticide pollution might be a factor contributing to decline of the Dungeness fishery at San Francisco.

Biomagnification of dieldrin residues in blue crabs was studied by Petrocelli *et al.* (1975). Blue crabs, fed contaminated clam meat, concentrated dieldrin residues from 4.7 to 6.8 times the daily dose after 10 days of feeding. The authors warned that these animals were fed small amounts of food contaminated by dieldrin at relatively low levels over a short period of time. Under natural conditions in the·field, animals consuming contaminated foods for several months or years may have a much higher concentration factor. In any event, it was clearly shown that dieldrin could move through a two-step food chain and be bioaccumulated by blue crabs.

Sheridan (1975) detected DDT and its metabolites in five out of six major organs in blue crabs collected from the York River in Virginia in 1972. Concentrations were highest in the hepatopancreas (55–217 ppb) but below the limit of detection in heart and serum. Variable residues also occurred in gonad samples (9–67 ppb), gill (6–31 ppb), and edible claw and back fin muscles (1–13 ppb).

DDT itself was not detected in any of the organs. Laboratory studies were conducted to determine the accumulation of DDT in the crab's organs and the transfer between organs, as well as the rate of metabolism and loss of DDT and its metabolites. Response to concentrations of 0.01, 0.1, and 1.0 ppm DDT varied among organs and suggested that DDT was transported from gills to the hepatopancreas via the blood stream. Since DDD concentrations were observed in gill and the hepatopancreas immediately after exposure to DDT, it was suggested that DDT is rapidly dechlorinated to DDD. The apparent site of DDT metabolism was the hepatopancreas. DDE concentrations in edible claw and back fin muscles increased with time after exposure to DDT.

Epifanio (1973) determined the uptake of another chlorinated hydrocarbon, dieldrin, by larvae of *Cataleptodius* (=*Leptodius*) *floridanus*. One group of larvae was exposed to 0.5 ppb in seawater and fed *Artemia salina* nauplii uncontaminated with dieldrin while another group reared in seawater and contaminated with dieldrin was fed nauplii contaminated with 213 ppb dieldrin. In the study, no distinction was made between dieldrin absorbed on the cuticle and dieldrin stored in other tissues. The larvae accumulated dieldrin 19.1 times as fast from 0.5 ppb in seawater as from 213 ppb in food; Epifanio thought that this rate of uptake can differentially affect the toxicity of dieldrin to crab larvae.

Other information developed in the laboratory has indicated that chlorinated hydrocarbons are extremely toxic to estuarine crabs. For example, Epifanio (1971) showed some effects of dieldrin on the development of two species of crabs, *C. floridanus* and *Panopeus herbstii*; 10 ppb in the water prohibited both species from completing their normal development, but the toxicity of dieldrin to *C. floridanus* larvae was more dependent on stage of development at exposure than length of exposure. Also, the effect of dieldrin on the molting process of that species was not as pronounced as the effect on survival. The author surmised that concentrations of 1 ppb dieldrin in seawater in the natural environment would change the larval population structure of these species of crabs. Juvenile *Callinectes sapidus* fed, molted, and grew for 9 months in seawater containing 0.25 ppb DDT but could survive only a few days in water containing in excess of 0.5 ppb (Lowe, 1965). Weis and Mantel (1976) determined the effects of chronic exposure to DDT on limb regeneration and molting in *Uca pugilator* and *Uca pugnax* after single limb and multiple autotomy. Autotomy of either one or seven limbs was induced by pinching the merus with scissors. The *R* value [(length of limb bud × 100)/carapace width] was used as an index for comparing crabs of different sizes. Results showed that DDT levels of approximately 10 ppb accelerated limb regeneration in fiddler crabs, particularly crabs stimulated to regenerate by multiple autotomy. Molting time was also shortened in these crabs by the presence of DDT. The stimulation of regeneration phenonoma, causing heightened excitation of the central nervous system, was suggested as an effect of DDT.

The recent tragedy at Hopewell, Virginia, where the chlorinated hydrocarbon, Kepone, was improperly handled and disposed, not only caused adverse human health effects but also had an impact upon aquatic organisms in the James River/Chesapeake Bay. Estuarine crabs were among the organisms most adversely affected. The Governor of Virginia recently closed fishing for male blue crabs, *Callinectes sapidus,* because of high residues of Kepone in their tissue and the prospect of Kepone reaching man through consumption of the organisms. Laboratory studies were conducted to determine the toxicity of Kepone to blue crabs and bioaccumulation of this chemical by the crabs under controlled laboratory conditions. Schimmel *et al.* (1979) conducted two long-term studies to determine toxicity, uptake, and distribution of Kepone in blue crabs. Kepone was administered to crabs in seawater (0.03 or 0.3 μg Kepone/liter of water) or fed in contaminated oyster meat containing 0.25 μg Kepone/gm meat. Uptake of Kepone in this instance was primarily through contaminated oysters. When crabs were held in Kepone-free seawater and fed Kepone-free oysters for 28 days, no loss of the insecticide was evident. Adverse effects were evident in crabs fed oysters that contained 0.25 μg Kepone/gm oyster. In separate experiments, crabs, fed Kepone-contaminated James River oysters that contained approximately 1.9 μg Kepone/gm oyster and then Kepone-free oysters for 90 days, still contained detectable concentrations of pesticide in tissues analyzed at the experiment's conclusion. Also, blue crabs that ate oysters containing Kepone in concentrations similar to those found in oysters from the James River died in greater numbers or molted less frequently than crabs fed Kepone-free oyster meat. Schimmel *et al.* suggest, from this data base, that Kepone may be a factor in the present decline in the James River crab fishery.

Two field studies demonstrating the effect of chlorinated hydrocarbons on fiddler crabs, *Uca pugnax,* are of particular interest (Krebs *et al.,* 1974). Test plots in a marsh near West Falmouth, Massachusetts, were prepared by adding weekly doses of a commercial fertilizer manufactured from sewage sludge. To one plot was added sewage containing 25.2 gm/m^2 (HF) and to the other 8.4 gm/m^2 (LF). Two untreated plots served as control. The first indication of an effect was observed in the fall, when a lower density of crabs began to overwinter in plots treated with the higher dose of sewage contamination. The number of crabs dead or inactive on the marsh surface increased in both treated plots as compared to controls. Residue analysis of *U. pugnax* samples showed low but fairly uniform amounts of DDE; these concentrations were about three times the level present in the sediment. Chemical analysis of the sediment also revealed aldrin and dieldrin. The residues of dieldrin in live crabs correlated well with the degree of locomotor impairment observed in the crabs, but aldrin was independent of impairment. Krebs *et al.* concluded that insecticide contaminants were responsible for drastic reduction in the test fiddler crab population.

The interaction of temperature and DDT residues on *C. sapidus* mortality was studied by Koenig *et al.* (1976) in DDT-contaminated and uncontaminated salt marshes in the northern Gulf of Mexico. Residue samples in the crabs and sediment were analyzed on a Varian 2100 gas chromatograph with a tritium electron-capture detector. DDT in the contaminated marsh originated from a dog fly (*Stomoxys calcitrans*) control program in which certain areas were sprayed once or twice per year. The uncontaminated area was some distance away and residue analysis indicated that the inhabitants of the controlled area were virtually pesticide-free. It was observed that blue crab mortalities occurred in the DDT-contaminated population, whereas no mortalities were observed in uncontaminated blue crab populations when there was a simultaneous decrease in water temperature. The total amount of DDT in blue crabs from the contaminated marsh was variable but each amount was higher than residues from the uncontaminated control areas. Analysis of the tissue of the dead and dying crabs revealed DDT residue concentrations as high as 39.0 ppm in the hepatopancreas and 1.43 ppm in muscle tissue. Although this study did not indicate how reduced temperatures interact with body burden to produce the toxic symptoms, three mechanisms were suggested: (1) DDT residues are released from stored lipids during periods of reduced temperature; (2) low temperature favors the DDT-nerve membrane complex; and (3) acute effects could be amplified through impairment of behavioral characteristics of the crabs in low temperatures.

2. Mirex and Crabs: A Case History

The impact of mirex (a chlorinated organic pesticide used to control fire ants in the southeastern United States) on estuarine crabs is illustrative of the capacity of the crabs to accumulate chlorinated hydrocarbons and their sensitivity to these chemicals. Mirex was developed by Allied Chemical Company to replace dieldrin and heptachlor, two persistent chlorinated hydrocarbons that were formerly used to eradicate fire ants but were especially toxic to nontarget aquatic organisms. Mirex bait was prepared by adding mirex to soybean oil and sorbing the resultant mixture onto corn-cob grits. The bait was usually applied by aerial spraying. Unfortunately, mirex was not adequately tested prior to being dispersed throughout the coastal environment. For example, routine 48-hour bioassays did not reveal toxic properties of mirex to crabs and other estuarine crustaceans. After the mirex spraying program began, subsequent studies showed that crabs could accumulate mirex from the environment and that this accumulation could be toxic.

Several field studies showed bioaccumulation of mirex after application to coastal and other aquatic habitats. McKenzie (1970) detected mirex in 22 out of 50 crab samples examined from a coastal area in South Carolina. Residue levels varied from a low of 0.005 to 0.209 ppm. In a cooperative effort among the states of North Carolina, South Carolina, Georgia, and Florida, Mahood *et al.* (1970) examined 1950 blue crabs and found that 35% contained detectable levels of mirex. Borthwick *et al.* (1974) conducted a field evaluation of the accumulation and movement of mirex in selected estuaries in South Carolina in conjunction with aerial application of the pesticide. The data revealed that some mirex was translocated from treated lands and high marsh to estuarine biota. Biological concentrations of mirex occurred especially in predators such as raccoons and birds; blue crabs collected in this study contained from 0 to 0.60 ppm of mirex. A follow-up study conducted 2 years after aerial application showed that the mirex level in animals had dropped considerably. Blue crabs collected at two freshwater stations in June, 1972, contained an average of 0.026 ppm. None of these field studies was designed to show the effect of mirex on crabs in the natural environment other than through visual observations of mortalities in the field during sampling.

Parallel studies conducted in the laboratory aided in determining the impact of mirex residues in nature. McKenzie (1970) and Lowe *et al.* (1971) studied the effects of mirex on estuarine organisms, including blue crabs and fiddler crabs. Independently, these investigators found that mirex exhibited a delayed toxicity to crabs. McKenzie found that adult and subadult (76–127-mm carapace width) blue crabs were not affected by doses 10 times the standard application rate of mirex. Juvenile blue crabs of less than 76 mm width were extremely sensitive to mirex bait and the toxicity was temperature dependent. Lowe *et al.* allowed juvenile blue crabs to eat one particle of mirex bait. Four weeks after the bait was fed to the crabs, whole-body residue analysis was conducted by electron-capture gas chromatography. Nineteen of 25 crabs exposed to the bait were paralyzed or dead at the end of the experiment; 92% contained residues of mirex. The residues in surviving crabs averaged 0.99 ppm. Fiddler crabs, *U. pugilator,* were exposed in a similar manner. The crabs readily picked up and ingested particles of the mirex bait; paralysis or death occurred within one to several weeks, depending on the amount of bait eaten and the temperature of the experimental environment. Residues of mirex in fiddler crabs averaged about 0.30 ppm.

The accumulation of mirex from water and its distribution in adult blue crabs were determined by Schoor (1974). Adult crabs were exposed from 15 minutes to 16 hours to mirex labeled with carbon-14. Mirex residues were found in the hepatopancreas, brain, thoracic ganglion, muscle, and hemolymph serum. Schoor suggested that mirex was absorbed through the gills because of the manner in which it appeared in the hemolymph and hepatopancreas. In experiments with juvenile blue crabs, Leffler (1975) studied the effect of ingested mirex and DDT. Mirex proved to be a much more potent stomach poison than DDT. Subacute levels of mirex (0.02 to 0.2 ppm) caused pronounced metabolic rate elevation. This elevation was observed at all test concentrations above the minimum level eliciting the response. The possibility that mirex and other chlorinated hydrocarbons could act synergistically to affect the capacity of juvenile blue crabs to obtain food and oxygen to accommodate both normal growth and increased metabolism was discussed, suggesting that ingested mirex resulted in outright poisoning, metabolic rate elevation, decreased muscular coordination, and reduced carapace thickness/width ratio. It was concluded that this chemical is potentially disastrous with respect to blue crab populations.

The effects of mirex on development and survival of two other crabs in laboratory cultures, *Rhithropanopeus harrisii* and *Menippe mercenaria,* were differential in that an increase in duration of developmental stages with increasing concentrations appeared in *R. harrisii,* but no such effect was noted in *M. mercenaria* (Bookhout *et al., * 1972). Conversely, survival was reduced in relation to increased concentrations in both species, but *M. mercenaria* was much more sensitive than *R. harrisii.* Any extension of the duration of crab developmental stages could have ecological implications if crabs in various stages of development are subjected to adverse temperature, salinity, or other environmental conditions.

In experiments conducted at the community or systems level, simulated marsh systems were constructed with populations of various estuarine organisms, including crabs. Four 28-day seasonal experiments were conducted by Tagatz *et al.* (1975). To simulate movement of mirex bait from terrestrial to saltwater environments, mirex was leached from fire ant bait by fresh water mixed with salt water and added to communities of blue crabs, pink shrimp (*Penaeus duorarum*), grass shrimp (*Palaemonetes pugio*), and sheepshead minnows (*Cyprinodon variegatus*) at concentrations of 0.01–0.50 ppb. Leaching was greatest in summer and least in spring. Greatest mortality occurred in the summer experiments at the highest water temperature and concentrations of mirex. The least mortality occurred in spring at next to the lowest temperature and at the lowest concentration. The earliest deaths of organisms occurred in blue crabs after 6 days of exposure; small juvenile crabs were more sensitive to leached mirex than were large juveniles. Mirex did not appear to affect growth or frequency of molting in crabs. All exposed animals concentrated mirex; blue crabs concentrated the chemical 2300 times the amount in the water. Sand substrate contained mirex up to 770 times that in the water. In a similar experiment, mirex bait was added to simulated marsh systems containing marsh plants, oysters, blue crabs, fiddler crabs, and two species of cyprinodontids (Cripe and Livingston, 1977). Samples of water, bait, and animals were analyzed periodically; all animals concentrated mirex. Three photoproducts of mirex accumulated on the bait particles. Oysters and one species of fish, but not crabs, accumulated one of the photoproducts. Samples were analyzed with a Varian 2100 gas chromatograph with two tritium detectors. The rates of uptake of mirex from the water appeared very similar in blue crabs and fiddler crabs, but blue crabs took up a much greater quantity than fiddler crabs. This result was attributed to differences in lipid concentrations of the viscera and muscle tissues of the two species. The results of these system studies indicated again that application of mirex bait to a marsh area could result in rapid accumulation in resident organisms.

As a result of information gained in laboratory and field studies, the aerial application of mirex was limited to within 12 miles of the coastline, and efforts were made to avoid applying mirex to areas near rivers, streams, and other aquatic environments. It is possible that if full-scale aerial application of mirex bait had been permitted and the material spread indiscriminately over marsh and estuarine areas, severe effects on crab populations could have occurred. It is important that monitoring continue to determine that mirex applied under more restricted conditions still does not reach unacceptable levels in coastal regions.

3. Organophosphates and Carbamates

The trend in pesticide production and usage in this country is moving away from organochlorine chemicals to "second generation" chemicals, such as organophosphates and carbamates. The latter are often preferred over organochlorines because they are generally less persistent in the environment and often are more specific for target organisms. Because of the chemical characteristics of organophosphates and carbamates, their residues are difficult to measure and interpret, and require different analytical approaches (enzyme activity versus chemical analysis) (Coppage and Duke, 1971). These chemicals, however, can be extremely toxic to nontarget organisms such as crabs.

Bookhout and Costlow (1976) studied the effect of malathion on the complete larval development of *Rhithropanopeus harrisii* and *Callinectes sapidus* from the time of hatching until the first crab stage. *Rhithropanopeus harrisii* suffered a reduction in survival of larvae in each increase of concentration of malathion from 0.011 to 0.02 ppm. These concentrations also lengthened duration of zoeal and megalopa development. Survival of blue crab larvae decreased in concentrations from 0.02 to 0.08 ppm, but duration of development lengthened.

Caldwell (1977) found the zoeal stages of the Dungeness crab, *Cancer magister,* to be most sensitive to malathion and carbofuran. The lowest concentration of each chemical affecting survival was malathion, 0.02 μg/liter and carbofuran, 0.05 μg/liter. Sublethal effects of these pesticides either were not observed or were only slightly below the reported values for survival.

Studies of the effects of sevin on this species showed similar results. For example, Buchanan *et al.* (1970) reported that early larval stages were more sensitive to sevin than juveniles and adults. At 1.0-mg/liter concentration, sevin did not affect egg hatching in seawater but prevented molting of prezoeae to zoeae. Molting was delayed at concentrations as low as 0.0001 mg/liter. Data also were accumulated on the movement of sevin through a simple food chain and the resulting effects of the bioaccumulation of the chemical.

Several pertinent experiments have been completed on the impact of these chemicals on crabs maintained under field conditions. Tagatz *et al.* (1974) found no effect of malathion on caged blue crabs at an application of 57 gm/ha through a thermal fog and ultra-low-volume aerosal spray. Dursban, when applied manually at 0.05 lb/acre, had no effect on fiddler crabs, but the crabs accumulated a residue of 4.62 ppm (Marganian and Wall, 1972). However, when the same amount of Dursban was applied by duster, numerous fiddler crabs were killed. In experimental sites treated with Abate in New Jersey, populations of fiddler crabs decreased about 20% (Ward and Howes, 1974). The decrease was attributed either to direct toxicity of Abate leached from a 2% granular formation or to a behavioral effect—disappearance of the escape reaction to predators. The possibility of a behavioral effect was studied further (Ward *et al.*, 1976) by treating open marsh areas with Abate, then screening some affected fiddler crabs from predators while leaving others exposed. There was a significant reduction in the unprotected fiddler population, giving credence to the sublethal behavioral effect theory.

4. Juvenile Hormone Mimics or Analogs and Insect Growth Regulators

If organochlorine chemicals are considered first-generation pesticides, and the less persistent organophosphates and carbamates as the second generation, then juvenile hormone mimics or analogs and growth regulators can be characterized as third-generation pesticides. The mimics or analogs are used to slow or prevent metamorphosis, or to alter specific life stages of insects. Because of the phylogenetic relationship between insects and crustaceans, these products would be expected to affect crabs as well as insects adversely, and the growth regulators can interfere with cuticle formation. This was borne out in a study of the effect of two compounds, MON-0585 and methoprene, on the development of larvae of *Rhithropanopeus harrisii* and *Callinectes sapidus* (Costlow and Bookhout, 1977). MON-0585 concentrations of 1.0 and 0.1 ppm did not affect survival of the zoeae of *R. harrisii* at salinities of 20 and 35‰. However, when the salinity was reduced to 5‰ a concentration of 1.0 ppm resulted in total mortality within the zoeal stages; 10 ppm was lethal and no zoeae developed beyond the first stage in all three experimental salinities. The time required for development from hatching to the megalopa, megalopa to crab, and hatching to final metamorphosis of the crab was not affected by concentrations of either 1.0 or 0.1 ppm MON-0585 in experimental salinities in which development occurred. Megalopa of *C. sapidus* were killed in salinities of 20‰ and 35‰ containing 10 ppm MON-0585, and 1 ppm killed 60% regardless of salinity. Methoprene, at a concentration of 0.1 ppm and a temperature cycle of 20–25°C, reduced survival from 20 to 25% in all salinities tested. However, juvenile crab stages I, II, III, and IV did not show any effect when maintained in the same combinations (Costlow and Bookhout, 1977). An additional study by Costlow (1977) showed that 1.0 ppm methoprene resulted in total mortality of larval stages of *R. harrisii,* usually within the first 2 days of hatching. Christiansen *et al.* (1977) showed a significant reduction in survival of zoeae of *R. harrisii* with increasing concentrations of hydroprene, and an increase in duration of zoeal stages as the concentration of hydroprene increased.

Other laboratory studies have used behavior as an indicator of sublethal effect of juvenile hormone mimics on larval crabs (Forward and Costlow, 1976). The effect of MON-0585 on linear swimming speeds and phototaxis of *R. harrisii* was observed under controlled conditions. Swimming speeds were varied under microscopic illumination light as larvae were subjected to various concentrations of the chemical. The larvae decreased in sensitivity throughout developmental stages I through IV; in stage I, 0.1 ppm caused a significant increase in swimming speed over control; the same effect was created by 0.5 ppm in stages II and III and by 1.0 ppm in stage IV. The magnitude for both positive and negative phototaxis responses was depressed when the larvae were exposed to the higher test concentrations of MON-0585. The measured behaviors of these larvae were affected by 0.1 ppm MON-0585; early zoeal stages were the most sensitive since both swimming and phototaxis were affected at this concentration in both acute and chronic tests. Since the behavior patterns used as criteria in these studies are necessary for survival of larvae, Forward and Costlow suggested that caution must be exercised when using MON-0585 in areas where the larval crabs are present.

The growth regulator Dimlin has been shown to have a significant effect on survival of *Rhithropanopeus harrisii* larvae at concentrations of 1 ppb in water (Christiansen *et al.*, 1979), the authors suggesting that Dimlin may interfere with cuticle formation in crab larvae. Costlow (1977) discussed the effect of Dimlin on the development of the stone crab, *Menippe mercenaria,* and the blue crab, *Callinectes sapidus. Menippe mercenaria* could not survive 0.5 ppb Dimlin but *C. sapidus* was slightly more tolerant. Data from these and other experiments indicate that levels of this chemical which are toxic to insects are also toxic to zoeal stages of crabs.

Most studies on toxic organics report on parent compounds. The ability of breakdown products emanating from many of these compounds to persist in the marine environment is not known. Therefore, effects of these degradation products are not known, particularly those related to juvenile hormones. We are especially lacking in this information (see Table II).

B. THERMAL EFFECTS

Thermal discharges or thermal pollution from power-producing plants located in coastal and near-coastal environments can have an adverse effect on benthic populations, such as estuarine crabs. Much "baseline information" is available

TABLE II

SUMMARY OF PESTICIDE EFFECTS ON SEVEN SPECIES OF BRACHYURAN CRABS

Reference	Compound	Organisms	Exposure	Bioaccumulation	Observed effect
McKenzie (1970)	Mirex	Blue crabs		Residue levels varied 0.0005–0.209 ppm	Detected mirex in 22 of 50 sampled
Mahood et al. (1970)	Mirex	Blue crabs			Examined 1950; 35% contained detectable levels of mirex
Borthwick et al. (1974)	Mirex	Blue crabs	Aerial spraying	0–0.60 ppm	Two-year follow-up study at two freshwater stations contained average of 0.26 ppm
McKenzie (1970)	Mirex	Blue/fiddler crabs			Adults and subadults were not affected by doses 10 times the standard application. Juveniles of less than 76 mm were extremely sensitive
Lowe et al. (1971)	Mirex	Juvenile blue crabs	Hot soybean oil with mirex on corncob grits. Each crab exposed to one particle of bait for 4 weeks. Whole body residue analysis made on each crab.		Nineteen of 25 exposed paralyzed or dead at experiment end; 92% contained residues of mirex
		Fiddler crabs	Same as for juvenile blue crabs		Paralysis or death occurred within 1 to several weeks, depending on amount eaten and temperature; mirex in

204

Reference	Chemical	Concentration	Organism	Effects
				pancreas, brain, thorax ganglion, muscle and hemolymph serum
Leffler (1975)	Mirex/DDT	0.02 to 2 ppm mirex	Juvenile blue crabs	Caused metabolic rate elevation. Elevation observed at all test concentrations. Suggest possibility of ingested mirex resulting in outright poisoning, decreased muscular coordination, and reduced carapace thickness-width ratio. Mirex more potent stomach poison than DDT
Bookhout et al. (1972)	Mirex	0.1 to 10 ppb zoeal stage; 0.01 to 0.1 ppb megalopal stage	*R. harrisii* and *Menippe mercenaria*	Survival reduced in relation to increased concentration in both but *M. mercenaria* more sensitive that *R. harrisii*. Greater the concentration from 0.01 to 1.0 ppb, the larger percentage of extra 6th zoeal stages in *mercenaria* but not in *harrisii*.
Tagatz et al. (1975)	Mirex	Less than 0.3 μg/liter mirex leached from fire ant bait by fresh water and mixed with salt water	Blue crabs, pink shrimp, grass shrimp, and sheepshead minnows	Blue crabs concentrated mirex 2300 times amount in water. Sand substrate contained mirex up to 770 times that in water

(Continued)

TABLE II (*Continued*)

Reference	Compound	Organisms	Exposure	Bioaccumulation	Observed effect
Cripe and Livingston (1977)	Mirex bait (0.3% mirex)	Blue crabs, oysters, and fiddler crabs	Analyzed on Varian 2100 gas chromatograph with 2 tritium detectors		Application of mirex to marsh could result in rapid accumulation in resident organisms
Albright *et al.* (1975)	Heptachlor epoxide	Dungeness crabs, *Cancer magister*	Sources believed to be surface runoffs from aerial sprays, ground water, direct application for insect control	22.0 ppb	
	Paraprime DDE	*C. magister*		23.0–295.8 ppb	
Petrocelli *et al.* (1974)	DDT; dieldrin	Blue crabs		81 analyzed: 94% contained DDT, 32% dieldrin, and 28% contained both	
Modin (1969)	DDE, DDD, DDT	Dungeness crabs		40–62 ppb in stripped females and 54–430 ppb in ova	
Petrocelli *et al.* (1975)	Dieldrin	Blue crabs	Fed small amounts of contaminated food at low levels over short period of time	Fed contaminated clam meat with dieldrin residues from 4.7–6.8 times daily doses after 10 days of feeding	Dieldrin could move through two-step food chain and be bioaccumulated by crabs
Sheridan (1975)	DDT	Blue crabs		Highest concentration in hepatopancreas (55–217 ppb) and below level of detection in heart and serum. Residues in gonad samples variable	DDT itself not detected in any of organisms

206

Reference	Chemical	Species	Exposure	Effects
Epifanio (1973)	Dieldrin	*Cataleptodius floridanus*	0.5 ppb in seawater and fed *Artemia salina* nauplii not contaminated with dieldrin; another group in seawater contaminated with dieldrin but fed nauplii contaminated with 213 ppb	levels low in gill (6–31 ppb) and edible claw and back fin muscles (1–13 ppb) Accumulated dieldrin 19.1 times as fast from 0.5 ppb in seawater as from 213 ppb in food
Epifanio (1971)	Dieldrin	*Cataleptodius floridanus* and *Panopeus herbstii*	10 ppb	Prevented both species from completing normal development but toxicity to *C. floridanus* larvae more dependent on stage of development at exposure than length of exposure
Lowe (1965)	DDT	Juvenile blue crabs		0.25 ppb (molted and grew for 9 months); but survived only a few days with excess of 0.5 ppb
Weis and Mantel (1976)	DDT	*Uca pugilator; U. pugnax*	10 ppb accelerated limb regeneration in fiddler crabs	Autotomy of either one or seven limbs induced by pinching merus with scissors. Molting time shortened; heightened excitation of central nervous system

(Continued)

TABLE II (*Continued*)

Reference	Compound	Organisms	Exposure	Bioaccumulation	Observed effect
Schimmel *et al.* (1979)	Kepone	Blue crabs	0.03 or 0.3 μg/liter in seawater or in contaminated oyster meat containing 0.25 μg Kepone/gm meat	Uptake primarily through contaminated oysters	Crabs held in Kepone-free water and fed Kepone-free oysters for 28 days; no loss of insecticide evident. Adverse effects evident in crabs fed oyster with 0.25 μg/gm Kepone.
Krebs *et al.* (1974)	Chlorinated hydrocarbons	*Uca pugnax*	25.2 and 8.4 gm/m^2		Lower density of crabs over winter in plots with higher dose of sewage contamination. Death increased in both plots. Low but uniform amounts of DDE; concentration three times level in sediment. Aldrin and dieldrin also present in sediment
Koenig *et al.* (1976)	DDT and temperature	Blue crabs	DDT in contaminated marsh originated from dog fly control program (area sprayed 1 or 2 years). Dead and dying revealed 39.0 ppm in hepatopancreas and 1.43 ppm in swimmeret muscle		Mortalities occurred at same time as decrease in water temperature

on the thermal limits of marine organisms (Kinne, 1964), and of crabs in particular (Tagatz, 1969), including field observations on the impact of thermal effluents on crab populations.

Detailed biochemical studies of the effect of thermal stress on serum glucose and protein levels in blue crabs have shown that thermal stress altered levels of serum glucose and that this alteration could be used as a physiological indicator of thermal stress (Lynch, 1974). Other serum constituents, such as total serum protein levels, were not affected by thermal stress. Lynch proposed an index of conditions for blue crabs based on serum glucose concentrations that could vary, depending on a particular situation, even when the cause of the stress is not known. Leffler (1972) studied some effects of temperature on the growth and metabolic rate of juvenile blue crabs in the laboratory. Growth, of course, was shown to be dependent upon temperature. Crabs maintained at temperatures of 34° and 27°C grew faster than those in temperatures of 13°, 15°, and 20°C. Mortality was directly proportional to temperature between 13° and 34°C and was very high during ecdysis at higher temperatures. Leffler suggested that crabs living in regions of heated discharge (because of their mobility) could extend their growing season without decreasing size by actively avoiding areas of higher temperatures. Additional studies in the laboratory conducted by Burton *et al.* (1976) exposed blue crabs and mud crabs to rapid 5°C increases above ambient. The experimental animals were maintained at that temperature for a specific time; then the temperature was allowed to return to ambient over another specified time. Burton *et al.* pointed out the loose use of the term "thermal stress" in the literature, and defined "thermal stress" for their experiments as a stimulus that produces a significant deviation from normal physiological whole-animal consumption(Q_{0_2} or weight specific oxygen consumption responses)—not as a stimulus that necessarily causes a sublethal or prelethal effect. No significant changes in the Q_{0_2} patterns for blue crabs and mud crabs were noted at any acclimation temperature tested as a result of thermal exposure. Mihursky *et al.* (1974) completed extensive work on temperature–salinity effects on osmoregulation in crabs. Among their results are indications that size, but not sex differences, affected osmoconcentration, and that smaller crabs showed greater tolerance to thermal stress at high salinities than did larger crabs.

Copeland and Davis (1972) studied the effect of temperature on blue crabs in ambient pool experiments and found a large increase in the volume of blue crabs per pool during a spring experiment, particularly in heating pools. In summer and winter experiments, the change in volume of blue crabs, or biomass, was less than that observed during the spring experiments. Sewage added to the pools did not affect mortality; crabs in pools receiving sewage were significantly larger than those receiving heat alone—an effect related to the additional food available to the crabs in pools receiving the sewage. Studies by Gallaway and Strawn (1975) on seasonal abundance and distribution of blue crabs in a Galveston Bay area receiving hot water discharge from a generating station provided evidence that crabs avoided the hottest water except for feeding during hot months of the year. Abundance and distributional patterns in their samples indicated that the discharge in this area had no detrimental effect on the blue crab population.

C. Oil

Crabs as well as other estuarine organisms can be adversely affected by oil and and oil products. Oil can enter the estuarine environment through accidental spills, normal operations of oil tankers, offshore production, and disposal of oil waste material. In addition, some oil originates from atmospheric transport which includes hydrocarbons from combustion engines. Oil can affect crabs through direct toxicity, destruction of habitat, and damage to a fishery resource through tainted crab meat. Dolan *et al.* (1970) summarized these potential effects:

1. Low-boiling, saturated hydrocarbons can cause cell damage and death, especially in larval and juvenile stages of marine organisms.
2. Higher-boiling, saturated hydrocarbons occur naturally in some organisms but may interfere with chemical communication used by animals as they feed and reproduce.
3. Olefinic hydrocarbons, found in gasoline and other refined products, will combine with chlorine and other elements to produce toxic mixtures.
4. Aromatic hydrocarbons are probably the most acutely toxic to marine organisms other than petroleum fractions, including benzene, toluene, and phenols.
5. Higher boiling aromatics include benzpyrene and 1,2, benzanthracene which have been isolated from crude oil.
6. Nonhydrocarbons include nitrogen, oxygen, sulfur, and metal compounds.

Several field observations have confirmed that crabs can accumulate different fractions of oil and that the oil and oil products are toxic. For example, a relatively minor oil spill in West Falmouth, Massachusetts, released approximately 700 tons of No. 2 fuel oil into coastal waters. Blumer *et al.* (1970) reported the toxic effects of this spill on crabs and other marine organisms, showing that the coastal area has been slow to recover and that the extent of some damage has intensified with time.

Laboratory studies have shown the capacity of crabs to accumulate oil. *Callinectes spadius* accumulated paraffinic and aromatic hydrocarbons including benzopyrene, fluorene, napthalene, and other fractions (Lee *et al.*, 1976). When given radiolabeled hydrocarbons in food, the crabs assimilated 2–10% and excreted the remainder. Hydrocarbons, when taken up through water, were eliminated chiefly through fecal material. The hepatopancreas accumulated most of the hydrocarbons; 20 days after exposure to labeled hydrocarbons, only the hepatopancreas contained radioactivity. The studies suggest that the hepatopancreas was the site of hydrocarbon metabolism; no evidence was presented to show the storage of hydrocarbon by any other tissues. Burns (1976) examined the fiddler crab, *Uca pugnax,* for its capacity to metabolize foreign hydrocarbons. It was determined, by using aldrin epoxidation rates in microsomes from various body tissues, that *Uca pugnax* has the microsomal mixed function oxidase system necessary to oxidize foreign hydrocarbons. Rates of oxidation were slow, compared to those for insecticides in freshwater invertebrates. Based on *in vivo* naphthalene oxidation rates, metabolism alone could not clear body tissues of foreign hydrocarbons within the life span of the crab. Burns suggested that even under ideal conditions this lack of biochemical ability to metabolize foreign hydrocarbons partially could account for sensitivity of the organism to this type of chemical pollution.

Other laboratory studies have been directed toward determining sublethal effects of oil and other petroleum products on crabs. Takahashi and Kittredge (1973) studied sublethal effects of crude oil and petroleum products containing water-soluble components and the effect of these components on chemoreceptors in marine organisms. Many species (including crabs) depend upon chemoreception to locate food and sexual partners. Obviously, adverse influences on these important life functions could cause effects that would not be revealed in normal toxicity studies. These investigators found that exposure of an intertidal crab, *Pachygrapus crassipes* (a grapsine whose general ecology was not outlined above), to the water-soluble extracts of two crude oils inhibited the feeding response and the mating stance of males when exposed to the female's sex pheromone. The inhibition was found to be persistent when the crude oil concentration of the water-soluble fraction was below 10^{-8} in concentration (assuming that 1% of the crude oil application was extracted into seawater). Further studies showed that monoaromatic hydrocarbons were effective as inhibitors of chemoreception for relatively short periods of time—30 minutes to 1 hour; however, polynuclear hydrocarbons (such as naphthalene and binaphthyl) inhibited the crabs for 8 to 11 days, anthracene for 13 days. Evidently, crabs are sensitive through chemoreception to a minute concentration of aromatic hydrocarbons in seawater.

Blue crabs have also been exposed under laboratory conditions to No. 2 fuel oil near Narragansett, Rhode Island. Melzian (1977) exposed adult male blue crabs to four concentrations of seawater containing 1, 5, 10, and 20 ppm of No. 2 fuel oil in a continuous flow-through bioassay system. The experiments were conducted from August to October. The LC_{50} for 96 hours for the No. 2 fuel oil was 14.1 ppm and the LC_{50} 7-day value, 10.5 ppm. Preliminary experiments showed that the acute toxicity of No. 2 fuel oil may vary seasonally in that animals were more sensitive to 10 ppm No. 2 oil in seawater with a temperature of 20.0°C, than those exposed to the same concentration in seawater of 5.6°C. Preliminary electron microscopic studies of the antennules of a single crab that survived for 60 days in a 1-ppm exposure showed no differences from control animals. However, the aesthetascs (chemosensory hairs) of the dosed animal were infected with filamentous bacteria or fungi. The author suggested that this "infection" might have occurred because of a lowering of immunological resistance of exposed animals. These results are especially interesting with respect to the previously reported experiments of Takahashi and Kittredge.

Oil can taint or lower the palatability of crabs and other marine organisms. A panel of experts experienced in organoleptic tests for determining taste in marine organisms followed guidelines proposed by the American Society for Testing Materials to determine the effect of two crude oils that had previously been tested for LC_{50}'s with penaeid shrimp (Knieper and Culley, 1975). The panel's results showed a significant decrease in tainting taste with decreasing concentrations of the crude oil. Threshold values for each oil were 620 and 1250 ppm. The same lighter fractions of the crude oil that are most toxic to shrimp also may be the part of the crude oil that imparts an oily taste to crab meat. There were some indications in preliminary test trials that proximal portions of the appendages and the anterior muscles were more affected by the oil than distal appendages and posterior muscles.

D. METALS

Metals, among elements that normally occur in marine crabs, are required for physiological processes, but can become toxic in excess amounts. Unfortunately, crabs can accumulate excess or toxic quantities of metal from seawater containing metal contamination from industrial wastes and other sources. Much information is available on metal residues (particularly mercury and cadmium) found in crabs and the effects of these metals on their welfare.

Studies by Fowler *et al.* (1975) related the many tons of trace metals annually released in southern California coastal basins to tissue burdens of trace metals in bottom-dwelling marine species, including the crabs *Cancer anthonyi* Rathbun (parallels northern half of *C. magister* range) and *Mursia gaudichaudii* H. Milne Edwards (marine crab ranging from near San Francisco to central Chile). These analyses were performed by proton-induced X-ray emission analysis and by atomic absorption spectroscopy. X-Ray examination of muscle samples from *C. anthonyi* showed high levels of potassium and calcium followed in abundance by bromine, zinc, iron, arsenic, strontium, selenium, and other trace metals. Surprisingly, arsenic and selenium were in higher concentration in those samples collected from noncontaminated regions than those collected from contaminated regions. Atomic absorption data from *M. gaudichaudii* showed muscle concentration similar to that found in the *C. anthonyi*. Considerable variation was found among concentrations of the various metals in tissues; not all samples contained all the metals. A wide variation in the concentration of the metals in tissues of the animals is not unexpected because of the migratory habits of these crabs and because of environmental factors such as transport of metals by the California currents. These findings indicate that muscle and viscera of bottom-dwelling marine organisms living in an area polluted by heavy metals reflect this pollution, although they did not show conclusively that the levels in the tissue were a reflection of levels in the bottom sediments.

A preliminary survey of mercury and other metals in tissues of two species of crabs from Fraser River mudflats, British Columbia, was conducted by Parsons *et al.* (1973), who separated the Fraser River mudflat community into two areas, Sturgeon Bank and Roberts Bank, and analyzed the mercury content in muscle of *Cancer magister* by a flameless cell on an atomic absorption spectrophotometer. On a dry-weight basis, crabs of similar carapace width from Sturgeon Bank contained more mercury than those from Roberts Bank. For example, for a carapace width of 147 mm, crabs from Sturgeon Bank contained 3.7 ppm mercury and those from Roberts Bank contained only 0.32 ppm. The Vancouver City sewer outfall empties into Sturgeon Bank and the differences in mercury content might be due to this source of metal pollution. The mercury content of younger crabs on Sturgeon Bank was much lower than the older ones, and this was attributed to feeding habits. Young crabs are much more specific in their food requirements while the older crabs have greater variety in their diet, including detritus.

Several species of crabs and other decapod crustaceans from Plymouth, England, were analyzed for zinc and copper content by Bryan (1968). The majority of values, obtained through dithizone method and atomic absorption spectrophotometry, ranged from 20 to 35 μg metal/gm tissue. Bryan attributed this rather narrow range to the fact that the concentrations of these metals are regulated in the body fluids. Zinc and copper distribution within the body of the crabs, as well as methods for excreting these metals, are discussed.

Many data are available describing the toxicity of various metals, particularly mercury and cadmium, to crabs. For example, Calabrese *et al.* (1977) reviewed the effects of cadmium, mercury, and silver on marine organisms, including crabs. These and other workers in the field stress the need for knowledge of sublethal effects in order to determine "true" toxicity. These effects include such criteria as changes in oxygen consumption, osmoregulation, regeneration, ecdysis, and behavior. Also, effects from changes in temperature and salinity regimes on a crab's response to metal stress are most important. The following publications were selected to indicate the "state of the art" in determining lethal and sublethal effects of metals on marine crabs.

Knowledge of the accumulation and loss of mercury by crabs is most useful in determining the toxic effects of this metal. Vernberg and O'Hara (1972) investigated the effects of various temperature and salinity regimes on the accumulation of mercury by the fiddler crab, *Uca pugilator*. Although the total mercury content of gill and hepatopancreas remained essentially the same, the amount of the metal in each tissue varied under the different environmental conditions. Some insight into the mechanisms of mercury toxicity in crabs was noted, as the efficiency in the transport of mercury from gills to hepatopancreas increased with higher temperature. This might explain the toxicity of mercury to crabs at low temperature, as determined in another experiment by Vernberg and Vernberg (1972).

The biological half-life of inorganic mercury in *Cancer magister* was studied by Sloan *et al.* (1974). Inorganic mercury was added to seawater in the form of mercuric nitrate in a dosing aquarium. Crabs were removed after a specified time, placed in unpolluted seawater, and then analyzed at various time intervals. Mercury content was determined with an atomic absorption spectrophotometer containing a cold vapor apparatus. Analyses of data indicated a 25-day half-life of mercury under these experimental conditions. Three theoretical models were tested by computer program and the results caused the authors to question the validity of the use of the negative exponential "biological half-life." It would be interesting to compare the biological half-life of mercury taken up through food, as opposed to water only.

In study of the effect of mercury on larval *Uca pugilator* (DeCoursey and Vernberg, 1972), larvae were exposed to mercuric chloride in seawater at test concentrations of 180, 1.8, and 0.018 ppb. Analyses of mercury were accomplished with an atomic absorption spectrophotometer. Larval stages were significantly affected at 180 ppb of mercuric chloride within 24 hours, and the other test concentrations had deleterious effects but did not kill the larvae. Although these larvae were able to survive, attrition rate was greatly accelerated when compared to controls. It was also noted that sensitivity was related to developmental stage. Adult crabs can survive 180 ppb at ideal conditions of temperature and salinity for several hours, but stage V larvae survived only 6 hours, while stage I survived 24 hours. DeCoursey and Vernberg suggested that larvae would be even more susceptible to mercury in the environment where they would be subjected to additional temperature and salinity stresses.

Adult *Uca pugilator* exposed under laboratory conditions similar to those used with the larvae showed much less sensitivity to mercury than larvae (Vernberg and Vernberg, 1972). Crabs were exposed to 0.18 ppm of mercuric chloride, under optimum conditions of 25°C and 30‰ salinity, to determine direct toxicity. These conditions were varied to determine possible synergistic effects among mercury, temperature, and salinity. Mercury was analyzed by a Mercury

Analyzer System-50. Adult crabs accumulated mercury from the water rapidly in gill tissue in lesser amounts than in hepatopancreas and green gland tissues. Preliminary studies showed that the adult crabs could survive in 0.18 ppm mercury for at least 2 months under optimum conditions but temperature and salinity stress shortened this time considerably. The crabs withstood low temperature and high temperature better than low salinity coupled with low temperature. It appeared that mercury adversely affected the metabolic rate of males more than females.

The sublethal effect of mercury and cadmium on *Uca pugilator* under stress from temperature and salinity (Vernberg and DeCoursey, 1977) has been analyzed in terms of survival, tissue uptake, metabolism, behavior, microscopic anatomy, and enzymatic activity. The crabs were exposed to mercuric chloride at concentrations of 180, 1.8, and 0.018 ppb, and cadmium chloride at various concentrations. Mercury content in water and animals was determined with an Atomic Absorption Spectrophotometer Model 303, Mercury Analyzer System 50, or Coleman Mass Spec 30. Radioactive cadmium was detected by scintillation. Mercury was more toxic to males than females and, as reported previously, both sexes were found to withstand mercury in combination with high temperature and low salinity better than low temperature and low salinity. However, cadmium was most toxic at high temperature and low salinities. Fifty percent of the males exposed to 180 ppb mercury died within 5 days, and those exposed to 4000 ppb cadmium died within 6 days. The mode of action of both cadmium and mercury was probably related to bioaccumulation on the gills with subsequent breakdown of osmoregulatory or respiratory functions. The greater sensitivity of males to the effects of mercury was not explained in this study, but it was suggested that differences in metabolic rates of the sexes could be a contributing factor.

In crabs exposed to a mixture of mercury (180 ppb) and cadmium (1000 ppb) for 72 hours, the uptake of mercury was more influenced by the presence of cadmium than cadmium by mercury, in gill tissue, and a lesser amount in hepatopancreas tissue. Cadmium appeared to eliminate or severely slow the transport of mercury from gills to the hepatopancreas.

Metabolic rates established for adult male and female crabs were essentially the same. After sublethal levels of mercury were added, the metabolic rate of crabs was decreased, and the rate for males was significantly lower than for females after 2 days of exposure. Cadmium quickly altered the compensatory metabolic temperature response. This mechanism is, of course, necessary for crabs to adjust to environmental factors in the natural environment.

The effects of mercury on the response of crab larvae to light showed that stage I larvae were strongly photopositive regardless of temperature or salinity, and their phototactic response was not altered at mercury levels of 1.8 ppb. The

TABLE III

SUMMARY OF METAL EFFECTS ON ESTUARINE BRACHYURAN CRABS

Reference	Metal	Organisms	Exposure	Bioaccumulation	Observed effect
Fowler et al. (1975)	Trace metals	Cancer anthonyi	Environmental	High levels of K, Zn, Fe, As, Sr, Se, and others	
Parsons et al. (1973)	Hg	Cancer magister	Environmental	0.32–3.7 ppm	
Bryan (1968)	Zn and Cu	Crabs and other decapod crustaceans	Environmental	20–35 μg/gm	
Calabrese et al. (1977)	Cd, Hg, and Ag	Marine, including crabs			
Vernberg and O'Hara (1972)	Hg	Uca pugilator	0.18 ppm Hg^{2+} 24, 48, and 72 hours, various temperature–salinity regimes		Improved efficiency in transport of Hg from gills to hepatopancreas with higher temp
Sloan et al. (1974)	Hg	Cancer magister	$Hg(NO_3)_2 \cdot H_2O$ added to dosing aquaria		25-day half-life of Hg
DeCoursey and Vernberg (1972)	Hg	Uca pugilator	$HgCl_2$ in seawater; test concentrations of 180, 1.8, and 0.018 ppb		Within 24 hours larvae significantly affected by 180 ppb; sensitivity related to development stage
Vernberg and Vernberg (1972)	Hg	Uca pugilator	0.18 ppm of $HgCl_2$ under 25°C and 30‰ salinity	Rapid in gill tissue, less in hepatopancreas and green gland	Adult crab can survive in 0.18 ppm Hg for 2 months with optimum conditions

216

Reference	Metal	Species	Conditions	Tissue/Site	Effects
Vernberg and DeCoursey (1977)	Hg, Cd	*Uca pugilator*	$HgCl_2$: 180, 1.8, 0.018 ppb. $CdCl_2$, various concentrations Mixture of 180 ppb Hg and 1000 ppb Cd for 72 hours	Gills, with breakdown of osmoregulatory or regulatory functions Greater amount of Hg in gills, much less in hepatopancreas	50% males exposed to 180 ppb of Hg died within 5 days; those exposed to 4000 ppb Cd died within 6 days
Eisler (1971)	Cd	*Carcinus maenas*	$CdCl_2$: 20°C, 20‰, 96 hours		$LC_{50} = 4.1$ ppm
Collier et al. (1973)	Cd	*Eurypanopeus depressus*	$CdCl_2$: 25‰, 21°C, 1–12 ppm		$LC_{50} = 4.9$ ppm O_2 consumption varied greatly among individuals. No variance in O_2 consumption in different test concentrations
Thurberg et al. (1973)	Cd, Cu	*Cancer irroratus* *Cancer maenas*	$CdCl_2$ $CuCl_2 \cdot H_2O$ Various increasing concentrations, 48 hours		Mortality >1 ppm Cd Mortality >5 ppm Cu Survived 8 ppm Cd Survived 40 ppm Cu
Hutcheson (1974)	Cd	*Callinectes sapidus*	Cd: 11.14 ppm, 0.11 ppm, 0.0001 ppm	Gill, hepatopancreas, carapace	
O'Hara (1973)	Cd	*Uca pugilator*	96 hours at 10‰ and 30°C		$LC_{50} = 6.8$ ppm exposure lethal

response in stage III larvae, however, was reduced by this concentration of mercury.

This study showed that under optimum temperature and salinity regimes, adult fiddler crabs can tolerate higher levels of mercury and cadmium than are tolerable when these factors are at stress levels. Also, larvae were more sensitive to these heavy metals than adults and, therefore, provided a much more sensitive indicator of effects.

Other information is available on lethality and physiological responses of crabs to cadmium. Eisler (1971) exposed various marine species, including *Carcinus maenas* (the green crab), to cadmium chloride in acute static tests. Crabs were found to be less sensitive than the sand shrimp, *Crangon septemspinosa,* and more sensitive than the blue mussel, *Mytilus edulis.* The 96-hour lethal concentration for 50% of the test crabs was 4.1 ppm. A similar lethal concentration value for 72-hour, 4.9 ppm, was determined by Collier *et al.* (1973) for the mud crab, *Eurypanopeus depressus.* These investigators also studied oxygen metabolism of whole animals and gill tissues alone. The oxygen consumption in whole animals varied greatly among individuals, and there were no differences in overall oxygen consumption among various test concentrations. Consumption in gill tissue alone decreased as cadmium concentrations increased. In green crabs, *Carcinus maenas,* and rock crabs, *Cancer irroratus,* exposed to cadmium chloride in seawater (Thurberg *et al.,* 1973), cadmium depressed gill-tissue consumption in both species and elevated the blood-serum osmolality of green crabs. Interestingly, green crabs survived 8 ppm cadmium, whereas mortality of rock crabs occurred above 1.0 ppm. Tests with copper showed a graded loss of osmoregulation in both species exposed to increasing concentration of copper (most pronounced at lower salinities).

The influence of temperature and salinity on the uptake of cadmium in *Callinectes sapidus* and *Uca pugilator* has been studied in addition to toxicity to fiddler crabs. Hutcheson (1974) found gills and hepatopancreas to be the major sites of accumulation of cadmium in blue crabs, independent of the concentration in water. Accumulation of cadmium in the carapace reached an apparent equilibrium with amounts in the water, probably because of saturation of available sites. At 10 ppm cadmium in the water, crabs accumulate cadmium more rapidly at lower salinities. At these low salinities, rate of uptake is proportional to temperature. At lower concentrations of cadmium in water (0.11 and 0.20 ppm), the metal was not taken up over background levels in the claw. This is important in that cadmium would not be transported to man from the hardshell crab, because the claw makes up a great part of crab meat eaten by man. Similar results were obtained by O'Hara (1973) for the fiddler crab. The same temperature and salinity regime resulted in maximum toxicity of cadmium to fiddler crabs. A concentration of 6.8 ppm cadmium was lethal to 50% of the test population in 96 hours at 10‰ salinity and 30°C (see Table III).

E. RADIONUCLIDES

Radionuclides can enter the marine environment through effluents from nuclear reactors, fallout from nuclear testing, and numerous minor sources. At present, relatively little radioactivity enters the marine environment, but it is a potential source of pollution as the nation's energy requirements and the number of nuclear reactors increase. Radionuclides reaching the marine environment can remain in the water, precipitate, and become part of the sediments, or can be accumulated and transported by plants and animals.

Various monitoring studies have shown that crabs can accumulate and retain radioactivity from their environment. Parchevskii and Skolova (1971) measured the strontium-90 content of some Black Sea organisms, including the shore crab, *Carcinus maenas*. Although the amount of strontium-90 increased from 1965 through 1968 and could be represented by the mathematical expression $C = 171.8 + 3.9t$, the rate of increase during this time was less than the rate of increase of other Black Sea organisms recorded from 1961 to 1965. Among the various radionuclides in Dungeness crabs from the Columbia River reported by Toombs (1965), zinc-65 was readily detected and averaged about 15 pCi/gm. This average would require a person to eat in excess of 1000 lb of crabs per year to exceed the recommended permissible intake rate of 8×10^6 pCi/year. A later survey showing seasonal distribution of zinc-65, manganese-54, and chromium-51 in Dungeness crabs was conducted by Tennant and Forster (1969). It is difficult to compare their results with those of Toombs because of differences in methods used to determine radioactivity in tissues (ashed versus wet tissue analysis). However, the 1969 survey showed a positive correlation between specific activity (amount of zinc-65) and the amount of stable zinc and river discharge. Most of the radioactivity in the soft tissue was from zinc-65, chromium-51, and manganese-54. Observations of specific activity of the three nuclides gave some insight into rates of turnover and environmental "reservoirs" for the metals.

The capacity of mud crabs, *Panopeus herbstii,* as part of an estuarine community to accumulate zinc-65 under various salinities, pH's, temperatures, and concentrations of zinc in the water was studied by Duke *et al.* (1967) in several 15-day experiments. All factors, except the total concentration of zinc, significantly affected accumulation of zinc-65, clarifying, to some extent, differences in accumulation of this nuclide by crabs in the field. These laboratory and field data, however, do not directly address the question of the effect of radioactivity on the crabs.

Little attention has been devoted to the effects of accumulated radiation on crabs, although the effects of gamma radiation from external sources have been studied in some detail. The fiddler crab, *Uca pugnax,* was subjected to graded doses of cobalt-60 radiation by Rees (1962). The lethal dose for 50% of the exposed population over 30 days was 8209 rads. This value is two to three times higher than values for the grass shrimp, *Palaemonetes pugio,* irradiated in a companion experiment (Rees, 1962). Although the study was not specifically designed to determine sublethal effects, it was noted that molting did not occur in crabs that received dosages of 9750 rads and above, but occurred in crabs receiving 4785 rads and below.

The radiation sensitivity of three species of fiddler crabs, *Uca pugilator, U. pugnax,* and *U. minax,* was determined by Engel (1973). Each of the three species of crabs was irradiated with cobalt-60 and maintained under the same experimental conditions. When mean survival times were calculated at 60 days for the three species of crabs, *U. pugilator* and *U. minax* showed a dose-dependent pattern of survival time, i.e., decreasing survival with increasing dose. The relation was not as clear with *U. pugnax* because mean survival time was independent of dose above 4000 rads. Several explanations were discussed for different survival time for each of the species over the 60-day test period. The size of the organisms, salinity tolerance, and oxygen consumption rate all were rejected as possible explanations. Observed interspecific differences in the radiation sensitivities of crustaceans are not due to simple physiological or phylogenic relationships, but may be the result of complex interactions between the organism and its environment. Also, it is difficult to compare data on the effect of ionizing radiations on the survival of marine species because of the potential for environmental factors (such as temperature and salinity) to interact with radiation to alter expression of a particular effect.

Experiments concerned with interactions of radiation, salinity, and temperature on blue crab were conducted by Engel *et al.* (1973). This animal must osmoregulate to survive in the estuarine environment; it is possible that interactions of salinity, temperature, and radiation may have deleterious effects on the osmoregulatory control. Ionic composition and free amino acid level of hemolymph in the crabs were significantly affected by salinity and temperature, and radiation also interacted significantly with these factors to further alter the regulatory pattern. Radiation interacted with salinity and temperature, causing alteration in ionic regulation after irradiation.

In an investigation of the effects of radiation of juvenile blue crabs, *Callinectes sapidus,* Engel (1967) exposed the crabs to single and continuous doses of γ irradiation. For those given single doses the LC_{50} (30) was 51,000 rads. Definite behavioral effects were noted after the single exposures. For example, the crabs quit eating and became less antagonistic. Later, the crabs became sluggish and uncoordinated and eventually died. A neurological disorder was noted that was dose dependent. The crabs that were exposed continuously to irradiation had reduced survival and growth only at the dose rate of 29.0 rad/hour. Continuous exposure apparently caused death of the crabs through unsuccessful molting. At the lowest dose rate of 3.2 rads/hour, the growth rate of the crabs was greater than the other irradiated groups or the controls. This enhancement of growth may have been actually an acceleration of the aging process.

The understanding of the radiation biology of marine decapod crustaceans, and in particular the crabs, is incomplete. Recent data indicate that some crustaceans may be as sensitive to acute doses of radiation as mammals (Engel, in press), and that the primary reason for the extremely high LD_{50}'s is that sufficient time was not allowed in past investigations for damage to fully manifest itself. Thus, possibly, it would be inappropriate to assume that all crabs have sensitivities (LD_{50}) in the kilorad range.

V. Conclusions

Research suggests that estuarine crabs are sensitive to many forms of man-made pollution, either being acutely affected or serving as indicators through sublethal effects. We have discussed effects of only a few major pollutants on crabs, and have not cited other important hazards such as alterations of the habitat resulting from dredging, channelization, the addition of large amounts of fresh water, and other industrial activities. The animals and their habitats will continue to receive pollution stress in the future, but efforts are being made to counteract these stresses. Recent information on the number of chemical compounds or entities indicates that about 63,000 chemicals are in common use (Maugh, 1978). Undoubtedly, some of these will find or have found their way into the natural environment and the habitat of estuarine brachyurans. As our quest for energy sources heightens, the probability increases for spills and exposure of marine crabs to oil and its components. Also, as the world requirement for food increases, the use of pesticides will probably increase, at least in the immediate future.

Conversely, recent evidence indicates that programs to improve water quality have gained momentum. New federal legislation to control the testing and release of toxic substances (Toxic Substance Control Act of 1977) is now in effect. The data base concerned with the effect of pollutants on crabs has increased in volume and in quality during the past few years. There is no reason to believe that this trend will not continue in the future.

Because of the importance of crabs in the coastal zone and the sensitivity of these animals to many pollutants, crabs will probably be used more in future water quality studies as "indicator species." Because of their sensitivity to many pollutants, crabs can often be used to predict adverse influences on the environment and to highlight the impact of these substances on themselves.

In summary, many forces acting to impact the coastal environment adversely are offset by other factors that offer some hope for maintenance and improvement. The authors view this improvement with cautious optimism.

Acknowledgments

We thank Jackie Holley for assistance in preparing the pollution ecology section, B. B. Collette and M. E. Tagatz for critical reading of the manuscript, María Diéguez for drafting the illustration, and Virginia Tucker for final typing. A number of colleagues have also aided with critical suggestions and references.

References

Abele, L. G. (1972). A reevaluation of the *Neopanope texana-sayi* complex with notes on *N. packardii* (Crustacea: Decapoda: Xanthidae) in the northwestern Atlantic. *Chesapeake Sci.* **13**, 263–271.

Abele, L. G. (1973). Taxonomy, distribution and ecology of the genus *Sesarma* (Crustacea, Decapoda, Grapsidae) in eastern North America, with special reference to Florida. *Am. Midl. Nat.* **90**, 375–386.

Ahsanullah, M., and Newell, R. C. (1977). The effects of humidity and temperature on water loss in *Carcinus maenas* (L.) and *Portunus marmoreus* (Leach). *Comp. Biochem. Physiol. A* **56**, 593–601.

Albright, L. J., Northcote, T. G., Oloffs, P. C., and Szeto, S. Y. (1975). Chlorinated hydrocarbon residues in fish, crabs, and shellfish of the lower Fraser River, its estuary, and selected locations in Georgia Strait, British Columbia—1972-73. *Pestic. Monit. J.* **9**, 134–140.

Allen, E. A., and Curran, H. A. (1974). Biogenic sedimentary structures produced by crabs in lagoon margin and salt marsh environments near Beaufort, North Carolina. *J. Sediment. Petrol.* **44**, 538–548.

Barnwell, F. H. (1968). The role of rhythmic systems in the adaptation of fiddler crabs to the intertidal zone. *Am. Zool.* **8**, 569–583.

Beers, J. R. (1958). An histological and histochemical study of the green gland of *Cancer borealis.* *Anat. Rec.* **131**, 531–532.

Bell, T. I., and FitzGibbon, D. S., eds. (1977). "Fishery Statistics of the United States, 1974," Stat. Dig. No. 68. Natl. Oceanic Atmo. Adm., Natl. Mar. Fish. Serv., U. S. Dept. of Commerce, Washington, D.C.

Blumer, M., Souza, G., and Sass, J. (1970). Hydrocarbon pollution of edible shellfish by an oil spill. *Mar. Biol.* **5**, 195–202.

Bookhout, C. G., and Costlow, J. D., Jr. (1976). Effects of mirex, methoxychlor, and malathion on development of crabs. *U.S. Environ. Prot. Agency, Off. Res. Dev. [Rep.] EPA* **EPA-600/3-76-007**, 1–85.

Bookhout, C. G., Wilson, A. J., Jr., Duke, T. W., and Lowe, J. I. (1972). Effects of mirex on the larval development of two crabs. *Water, Air, Soil Pollut.* **1**, 165–180.

Boolootian, R. A., Giese, A. C., Farmanfarmaian, A., and Tucker, J. (1959). Reproductive cycles of five west coast crabs. *Physiol. Zool.* **32**, 213–220.

Borthwick, P. W., Cook, G. H., and Patrick, J. M., Jr. (1974). Mirex residues in selected estuaries of South Carolina, USA, June 1972. *Pestic. Monit. J.* **7**, 144–145.

Bousfield, E. L. (1955). Ecological control of the occurrence of barnacles in the Miramichi estuary. *Natl. Mus. Can., Bull.* **137**, 1–69.

Bousfield, E. L., and Laubitz, D. R. (1972). Station lists and new distributional records of littoral marine invertebrates of the Canadian Atlantic and New England regions. *Natl. Mus. Nat. Sci. Publ. Biol. Oceanogr.* No. 5, pp. 1–51.

Broad, A. C. (1957). The relationship between diet and larval development of *Palaemonetes. Biol. Bull. (Woods Hole, Mass.)* **112**, 162–170.

Broekhuysen, G. J., Jr. (1936). On development, growth and distribution of *Carcinides maenas* (L.). *Arch. Neerl. Zool.* **2,** 257-399.

Bryan, G. W. (1968). Concentration of zinc and copper in the decapod crustaceans. *J. Mar. Biol. Assoc. U. K.* **48,** 303-321.

Buchanan, D. V., and Millemann, R. E. (1969). The prezoeal stage of the Dungeness crab, *Cancer magister,* Dana. *Biol. Bull. (Woods Hole, Mass.)* **137,** 250-255.

Buchanan, D. V., Millemann, R. E., and Stewart, N. E. (1970). Effects of the insecticide sevin on various stages of the Dungeness crab, *Cancer magister. J. Fish. Res. Board Can.* **27,** 93-104.

Burns, K. A. (1976). Hydrocarbon metabolism in the intertidal fiddler crab *Uca pugnax. Mar. Biol.* **36,** 5-11.

Burton, D. T., Richardson, L. B., Margrey, S. L., and Abell, P. R. (1976). Effects of low ΔT powerplant temperatures on estuarine invertebrates. *J. Water Pollut. Control Fed.* **48,** 2259-2272.

Butler, T. H. (1960). Maturity and breeding of the Pacific edible crab, *Cancer magister* Dana. *J. Fish. Res. Board Can.* **17,** 641-646.

Butler, T. H. (1961). Growth and age determination of the Pacific edible crab *Cancer magister* Dana. *J. Fish. Res. Board Can.* **18,** 873-891.

Caddy, J. F. (1973). Underwater observations on tracks of dredges and trawls and some effects of dredging on a scallop ground. *J. Fish. Res. Board Can.* **3,** 173-180.

Calabrese, A. F., Thurberg, P., and Gould, E. (1977). Effects of cadmium, mercury, and silver on marine animals. *Mar. Fish. Rev.* **39,** 5-11.

Caldwell, R. S. (1977). Biological effects of pesticides on the Dungeness crab. *U.S. Environ. Prot. Agency, Tech. Rep.* **68-01-0188,** 1-180.

Capen, R. L. (1972). Studies of water uptake in the euryhaline crab, *Rhithropanopeus harrisi. J. Exp. Zool.* **182,** 307-319.

Carpenter, J. H., and Cargo, D. G. (1957). Oxygen requirement and mortality of the blue crab in the Chesapeake Bay. *Chesapeake Bay Inst. Johns Hopkins Univ., Tech. Rep.* No. **13,** pp. 1-22.

Chamberlain, N. A. (1957). Larval development of *Neopanope texana sayi. Biol. Bull. (Woods Hole, Mass.)* **113,** 338.

Chamberlain, N. A. (1961). Studies on the larval development of *Neopanope texana sayi* (Smith) and other crabs of the family Xanthidae (Brachyura). *Chesapeake Bay Inst., Johns Hopkins Univ., Tech. Rep.* No. 22, pp. 1-35.

Chamberlain, N. A. (1962). Ecological studies of the larval development of *Rhithropanopeus harrisii* (Xanthidae, Brachyura). *Chesapeake Bay Inst., Johns Hopkins Univ., Tech. Rep.* No. 28, pp. 1-47.

Cheung, T. S. (1966). An observed act of copulation in the shore crab, *Carcinus maenas* (L.). *Crustaceana (Leiden)* **11,** 107-108.

Cheung, T. S. (1968). Trans-molt retention of sperm in the female stone crab, *Menippe mercenaria* (Say). *Crustaceana (Leiden)* **15,** 117-120.

Cheung, T. S. (1969). The environmental and hormonal control of growth and reproduction in the adult female stone crab, *Menippe mercenaria* (Say). *Biol. Bull. (Woods Hole, Mass.)* **136,** 327-346.

Christiansen, M. E. (1969). Crustacea Decapoda Brachyura. "Marine Invertebrates of Scandinavia," No. 2. Universitetsforlaget, Oslo.

Christiansen, M. E., and Costlow, J. D., Jr. (1975). The effect of salinity and cyclic temperature on larval development of the mud-crab *Rhithropanopeus harrisii* (Brachyura: Xanthidae) reared in the laboratory. *Mar. Biol.* **32,** 215-221.

Christiansen, M. E., Costlow, J. D., Jr., and Monroe, R. J. (1977). Effects of the juvenile hormone mimic ZR-512 (Altozar ℞) on larval development of the mud-crab *Rhithropanopeus harrisii* at various cyclic temperatures. *Mar. Biol.* **39,** 281-288.

Christiansen, M. E., Costlow, J. D., Jr., and Monroe, R. J. (1978). Effects of the insect growth regulator Dimlin® (TH 6040) on larval development of two estuarine crabs. *Mar. Biol.* **50,** 29–36.

Christy, J. H. (1978). Adaptive significance of reproductive cycles in the fiddler crab *Uca pugilator:* A hypothesis. *Science* **199,** 453–455.

Churchill, E. P., Jr. (1919). Life history of the blue crab. *Bull. U.S. Bur. Fish.* **36,** 95–128.

Clay, E. (1965). "Literature Survey of the Common Fauna of Estuaries. 16. *Carcinus maenas* L." *Imperial Chemical Industries, Ltd. Brixham Laboratory,* PVM45/A/916. Brixham Lab., Imp. Chem. Ind. Ltd.

Collier, R. S., Miller, J. E., Dawson, M. A., and Thurberg, F. P. (1973). Physiological response of the mud crab *Eurypanopeus depressus* to cadmium. *Bull. Environ. Contam. Toxicol.* **10,** 378–382.

Connolly, C. J. (1925). The larval stages and megalops of *Rhithropanopeus harrisii* (Gould). *Contrib. Can. Biol.* [N. S.] **2,** 327–333.

Copeland, B. J., and Davis, H. L. (1972). "Estuarine Ecosystems and High Temperatures." Water Resour. Res., Inc., Raleigh, North Carolina.

Coppage, D. L., and Duke, T. W. (1971). Effects of pesticides in estuaries along the Gulf and southeast Atlantic coasts. *In* "Proceedings of the 2nd Gulf Coast Conference on Mosquito Suppression and Wildlife Management, held in conjunction with annual meetings of Louisiana Mosquito Control Association and Gulf States Council on Wildlife Fisheries and Mosquito Control, New Orleans, La., October 20–22, 1971," pp. 24–31.

Cornick, J. W., and Stewart, J. E. (1968). Pathogenicity of *Gaffkya homari* for the crab *Cancer irroratus. J. Fish. Res. Board Can.* **25,** 795–799.

Costlow, J. D., Jr. (1965). Variability in larval stages of the blue crab, *Callinectes sapidus. Biol. Bull. (Woods Hole, Mass.)* **128,** 58–66.

Costlow, J. D., Jr. (1967). The effect of salinity and temperature on survival and metamorphosis of megalops of the blue crab *Callinectes sapidus. Helgol. Wiss. Meeresunter.* **15,** 84–97.

Costlow, J. D., Jr. (1977). The effect of juvenile hormone mimics on development of the mud crab, *Rhithropanopeus harrisii* (Gould). *In* "Physiological Responses of Marine Biota to Pollutants" (F. J. Vernberg *et al.,* eds.), pp. 439–457. Academic Press, New York.

Costlow, J. D., Jr., and Bookhout, C. G. (1959). The larval development of *Callinectes sapidus* Rathbun reared in the laboratory. *Biol. Bull. (Woods Hole, Mass.)* **116,** 373–396.

Costlow, J. D., Jr., and Bookhout, C. G. (1960). The complete larval development of *Sesarma cinereum* (Bosc) reared in the laboratory. *Biol. Bull. (Woods Hole, Mass.)* **118,** 203–214.

Costlow, J. D., Jr., and Bookhout, C. G. (1962). The larval development of *Sesarma reticulatum* Say reared in the laboratory. *Crustaceana (Leiden)* **4,** 281–294.

Costlow, J. D., Jr., and Bookhout, C. G. (in press). Second generation pesticides and crab development. *Environ. Prot. Agency, Proc. Symp. Environ. Res.*

Costlow, J. D., Jr., Rees, G. H., and Bookhout, C. G. (1959). Preliminary note on the complete larval development of *Callinectes sapidus* Rathbun under laboratory conditions. *Limnol. Oceanogr.* **4,** 222–223.

Costlow, J. D., Jr., Bookhout, C. G., and Monroe, R. (1960). The effect of salinity and temperature on larval development of *Sesarma cinereum* (Bosc) reared in the laboratory. *Biol. Bull. (Woods Hole, Mass.)* **118,** 183–202.

Costlow, J. D., Jr., Bookhout, C. G., and Monroe, R. (1962). Salinity–temperature effects on the larval development of the crab *Panopeus herbstii* Milne-Edwards, reared in the laboratory. *Physiol. Zool.* **35,** 79–93.

Costlow, J. D., Jr., Bookhout, C. G., and Monroe, R. J. (1966). Studies on the larval development of the crab, *Rhithropanopeus harrisii* (Gould). 1. The effect of salinity and temperature on larval development. *Physiol. Zool.* **39,** 81–100.

Crane, J. (1975). "Fiddler Crabs of the World. Ocypodidae: Genus *Uca.*" Princeton Univ. Press, Princeton, New Jersey.

Marsh, G. A. (1973). The *Zostera* epifaunal community in the York River, Virginia. *Chesapeake Sci.* **14,** 87–97.

Maugh, T. H. (1978). Chemicals: How many are there? *Science* **199,** 162.

Meenahshi, V. R., and Scheer, B. T. (1960). Metabolism of glucose in the crabs *Cancer magister* and *Hemigrapsus nudus. Anat. Rec.* **137,** 381.

Melzian, B. D. (1977). "The Exposure of Blue Crabs, *Callinectes sapidus,* to Water Soluble Hydrocarbons Derived from #2 Fuel Oil," Annu. Rep., U.S. Environ. Prot. Agency, Environ. Res. Lab., Narragansett, Rhode Island.

Menzel, R. W., and Hopkins, S. H. (1956). Crabs as predators of oysters in Louisiana. *Proc. Natl. Shellfish. Assoc.* **46,** 177–184.

Menzel, R. W., and Nichy, F. W. (1958). Studies of the distribution and feeding habits of some oyster predators in Alligator Harbor, Florida. *Bull. Mar. Sci. Gulf Caribb.* **8,** 125–145.

Mihursky, J. A., Kennedy, V. S., McErlean, A. J., Roosenburg, W. H., and Gatz, A. J. (1974). The thermal requirements and tolerances of key estuarine organisms. *Tech. Rep. Water Resour. Res. Cent., Univ. Md.,* **26,** 1–153.

Miller, D. C. (1961). The feeding mechanism of fiddler crabs, with ecological considerations of feeding adaptations. *Zoologica (N.Y.)* **46,** 89–100.

Miller, K. G., and Maurer, D. (1973). Distribution of the fiddler crabs, *Uca pugnax* and *Uca minax,* in relation to salinity in Delaware rivers. *Chesapeake Sci.* **14,** 219–221.

Modin, J. C. (1969). Residues in fish, wildlife and estuaries. *Pestic. Monit. J.* **3,** 1–7.

Mootz, C. A., and Epifanio, C. E. (1974). An energy budget for *Menippe mercenaria* larvae fed *Artemia* nauplii. *Biol. Bull. (Woods Hole, Mass.)* **146,** 44–55.

More, W. R. (1969). A contribution to the biology of the blue crab (*Callinectes sapidus* Rathbun) in Texas, with a description of the fishery. *Tex. Parks Wildl. Dep., Tech. Ser.* No. 1, pp. 1–31.

Musick, J. A., and McEachran, J. D. (1972). Autumn and winter occurrence of decapod crustaceans in Chesapeake Bight, U.S.A. *Crustaceana (Leiden)* **22,** 190–200.

Nations, J. D. (1975). The genus *Cancer* (Crustacea: Brachyura): Systematics, biogeography and fossil record. *Nat. Hist. Mus., Los Angeles City Sci. Bull.* No. 23, 1–104.

Naylor, E. (1960). A North American xanthid crab new to Britain. *Nature (London)* **187,** 256–257.

Naylor, E., and Isaac, M. J. (1973). Behavioural significance of pressure responses in megalopa larvae of *Callinectes sapidus* and *Macropipus* sp. *Mar. Behav. Physiol.* **1,** 341–350.

Nichols, P. R., and Keney, P. M. (1963). Crab larvae (*Callinectes*), in plankton collections from cruises of M/V *Theodore N. Gill,* South Atlantic coast of the United States, 1953–54. *U. S., Fish Wildl. Serv., Spec. Sci. Rep.—Fish.* No. 448, pp. 1–14.

Odum, H. T. (1953). Factors controlling marine invasion into Florida fresh waters. *Bull. Mar. Sci. Gulf Caribb.* **3,** 134–156.

Odum, W. E., and Heald, E. J. (1972). Trophic analyses of an estuarine mangrove community. *Bull. Mar. Sci.* **22,** 671–738.

O'Hara, J. (1973). Cadmium uptake by fiddler crabs exposed to temperature and salinity stress. *J. Fish. Res. Board Can.* **30,** 846–848.

Ong, Kah-sin, and Costlow, J. D., Jr. (1970). The effect of salinity and temperature on the larval development of the stone crab, *Menippe mercenaria* (Say) reared in the laboratory. *Chesapeake Sci.* **11,** 16–29.

Parchevskii, V. P., and Sokolova, I. A. (1971). Strontium-90 content of some Black Sea organisms in 1965–1968. *Dokl. Akad. Nauk SSSR* **199,** 69–71.

Parsons, T. R., Bawden, C. A., and Heath, W. A. (1973). Preliminary survey of mercury and other metals contained in animals from the Fraser River mudflats. *J. Fish. Res. Board Can.* **30,** 1014–1016.

Pearse, A. S. (1914). On the habits of the *Uca pugnax* (Smith) and *U. pugilator* (Bosc). *Trans. Wis. Acad. Sci., Arts Lett.* **17,** 791–802.

Pearson, J. C. (1948). Fluctuations in the abundance of the blue crab in Chesapeake Bay. *Fish Wildl. Serv. (U.S.), Res. Rep.* No. 14, 1–26.

Pearson, W. H., and Olla, B. L. (1977). Chemoreception in the blue crab, *Callinectes sapidus. Biol. Bull. (Woods Hole, Mass.)* **153**, 346–354.

Petrocelli, S. R., Anderson, J. W., and Hanks, A. R. (1974). DDT and dieldrin residues in selected biota from San Antonio Bay, Texas, 1972. *Pestic. Monit. J.* **8**, 167–172.

Petrocelli, S. R., Anderson, J. W., and Hanks, A. R. (1975). Biomagnification of dieldrin residues by food-chain transfer from clams to blue crabs under controlled conditions. *Bull. Environ. Contam. Toxicol.* **13**, 108–116.

Pinschmidt, W. C. (1964). Distribution of crab larvae in relation to some environmental conditions in the Newport River estuary, North Carolina. *Diss. Abstr.* **24**, 4883.

Poole, R. L. (1966). A description of laboratory-reared zoeae of *Cancer magister* Dana, and megalopae taken under natural conditions (Decapoda, Brachyura). *Crustaceana (Leiden)* **11**, 83–97.

Poole, R. L. (1967). Preliminary results of the age and growth study of the market crab (*Cancer magister*) in California: The age and growth of *Cancer magister* in Bodega Bay. *In* ''Proceedings of the Symposium on Crustacea held at Ernakulam, Jan. 12–15, 1965,'' Part II, pp. 553–567. Mar. Biol. Assoc. India, Mar. Fish. P. O., Mandapam Camp.

Porter, H. J. (1960). Zoeal stages of the stone crab, *Menippe mercenaria* Say. *Chesapeake Sci.* **1**, 168–177.

Powell, E. H., Jr., and Gunter, G. (1968). Observations on the stone crab *Menippe mercenaria* Say, in the vicinity of Port Aransas, Texas. *Gulf Res. Rep.* **2**, 285–299.

Rasmussen, E. (1959). Behaviour of sacculinized shore crabs (*Carcinus maenas* Pennant). *Nature (London)* **183**, 479–480.

Rasmussen, E. (1973). Systematics and ecology of the Isefjord marine fauna (Denmark). *Ophelia* **11**, 1–495.

Rathbun, M. J. (1918). The grapsoid crabs of America. *U.S. Natl. Mus., Bull.* **97**, 1–461.

Rathbun, M. J. (1930). The cancroid crabs of America of the families Euryalidae, Portunidae, Atelecyclidae, Cancridae and Xanthidae. *U.S. Natl. Mus., Bull.* **152**, 1–609.

Reed, P. H. (1969). Culture methods and effects of temperature and salinity on survival and growth of Dungeness crab (*Cancer magister*) larvae in the laboratory. *J. Fish. Res. Board Can.* **26**, 389–397.

Rees, G. H. (1962). Effects of gamma radiation on two decapod crustaceans, *Palaemonetes pugio* and *Uca pugnax. Chesapeake Sci.* **3**, 29–34.

Rice, A. L., and Ingle, R. W. (1975). The larval development of *Carcinus maenas* (L.) and *C. mediterraneus* Czerniavsky (Crustacea, Brachyura, Portunidae) reared in the laboratory. *Bull. Br. Mus. (Nat. Hist.), Zool.* **28**, 103–119.

Robinson, L. A., ed. (1977). ''Fisheries Statistics of the United States, 1976,'' Curr. Fish. Statist. No. 7200. Natl. Oceanic Atmo. Adm., Natl. Mar. Fish. Serv., U.S. Dept. of Commerce, Washington, D.C. 96 pp.

Ropes, J. W. (1968). The feeding habits of the green crab, *Carcinus maenas* (L.). *Fish. Bull.* **67**, 183–203.

Ryan, E. P. (1956). Observations on the life histories and the distribution of the Xanthidae (mud crabs) of Chesapeake Bay. *Am. Midl. Nat.* **56**, 138–162.

Salmon, M., and Atsaides, S. P. (1968). Behavioral, morphological and ecological evidence for two new species of fiddler crabs (genus *Uca*) from the Gulf coast of the United States. *Proc. Biol. Soc. Wash.* **81**, 275–290.

Salmon, M., and Horch, K. W. (1972). Acoustic signalling and detection by semiterrestrial crabs of the family Ocypodidae. *In* ''Behavior of Marine Animals, Current Perspectives in Research'' (H. E. Winn and B. L. Olla, eds.), Vol. I, pp. 60–96. Plenum, New York.

Sandifer, P. A. (1973). Distribution and abundance of decapod crustacean larvae in the York River estuary and adjacent lower Chesapeake Bay, Virginia, 1968-1969. *Chesapeake Sci.* **14**, 235-257.

Sandifer, P. A. (1975). The role of pelagic larvae in recruitment to populations of adult decapod crustaceans in the York River estuary and adjacent lower Chesapeake Bay, Virginia. *Estuarine Coastal Mar. Sci.* **3**, 269-279.

Sastry, A. N. (1977). The larval development of the rock crab, *Cancer irroratus* Say, 1817, under laboratory conditions (Decapoda Brachyura). *Crustaceana (Leiden)* **32**, 155-168.

Savage, T. (1971). Mating of the stone crab, *Menippe mercenaria* (Say) (Decapoda, Brachyura). *Crustaceana (Leiden)* **20**, 315-316.

Scarratt, D. J., and Lowe, R. (1972). Biology of rock crab (*Cancer irroratus*) in Northumberland Strait. *J. Fish. Res. Board Can.* **29**, 161-166.

Schimmel, S. C., Patrick, J. M., Faas, L. F., Oglesby, J. L., and Wilson, A. J., Jr. (1979). Kepone®: Toxicity to and bioaccumulation by blue crabs. *Estuaries* **2**, 9-15.

Schoor, W. P. (1974). Accumulation of mirex-14C in the adult blue crab (*Callinectes sapidus*). *Bull. Environ. Contam. Toxicol.* **12**, 136-137.

Schwartz, B., and Safir, S. R. (1915). The natural history and behavior of the fiddler crab. *Cold Spring Harbor Monogr.* **8**, 1-24.

Seiple, W. (1979). Distribution, habitat preferences and breeding periods in the crustaceans *Sesarma cinereum* and *S. reticulatum* (Brachyura: Decapoda Grapsidae). *Mar. Biol.* **52**, 77-86.

Sheridan, P. F. (1975). Uptake, metabolism, and distribution of DDT in organs of the blue crab, *Callinectes sapidus. Chesapeake Sci.* **16**, 20-26.

Sikora, W. B., Heard, R. W., and Dahlberg, M. D. (1972). The occurrence and food habits of two species of hake, *Urophycis regius* and *U. floridanus* in Georgia estuaries. *Trans. Am. Fish. Soc.* **101**, 513-525.

Sloan, J. P., Thompson, J. A. J., and Larkin, P. A. (1974). Biological half-life of inorganic mercury in the Dungeness crab (*Cancer magister*). *J. Fish. Res. Board Can.* **31**, 1571-1576.

Smith, R. I. (1967). Osmotic regulation and adaptive reduction of water-permeability in brackish-water crab, *Rhithropanopeus harrisi* (Brachyura, Xanthidae). *Biol. Bull. (Woods Hole, Mass.)* **133**, 643-658.

Snow, C. D., and Nielsen, J. R. (1966). Premating and mating behavior of the Dungeness crab (*Cancer magister* Dana). *J. Fish. Res. Board Can.* **23**, 1319-1323.

Squires, H. J. (1965). Decapod crustaceans of Newfoundland, Labrador and the Canadian eastern Arctic. *Fish. Res. Board Can., Rep. Ser. (Biol.)* No. 810, pp. 1-212.

Subrahmanyam, C. B., Kruczynski, W. L., and Drake, S. H. (1976). Studies on the animal communities in two north Florida salt marshes. Part II. Macroinvertebrate communities. *Bull. Mar. Sci.* **26**, 172-195.

Sulkin, S. D., and Epifanio, C. E. (1975). Comparison of rotifers and other diets for rearing early larvae of the blue crab, *Callinectes sapidus* Rathbun. *Estuarine Coastal Mar. Sci.* **3**, 109-113.

Sulkin, S. D., and Norman, K. (1976). A comparison of two diets in the laboratory culture of the zoeal stages of the brachyuran crabs *Rhithropanopeus harrisii* and *Neopanope* sp. *Helgol. Wiss. Meeresunters.* **28**, 183-190.

Swartz, R. C. (1972). Postlarval growth and reproductive biology of the xanthid crab, *Neopanope texana sayi. Diss. Abstr.* **33**, 7150-7151B.

Tagatz, M. E. (1968). Biology of the blue crab, *Callinectes sapidus* Rathbun, in the St. Johns River, Florida. *Fish. Bull.* **67**, 17-33.

Tagatz, M. E. (1969). Some relations of temperature acclimation and salinity to thermal tolerance of the blue crab, *Callinectes sapidus. Trans. Am. Fish. Soc.* **98**, 713-716.

Tagatz, M. E., and Hall, A. B. (1971). Annotated bibliography on the fishing industry and biology of the blue crab, *Callinectes sapidus, NOAA Tech. Rep., NMFS SSRF* No. 640, 1-94.

Tagatz, M. E., Borthwick, P. W., Cook, G. H., and Coppage, D. L. (1974). Effects of ground applications of malathion on salt-marsh environments in northwestern Florida. *Mosq. News* **34**, 309–315.

Tagatz, M. E., Borthwick, P. W., and Forester, J. (1975). Seasonal effects of leached mirex on selected estuarine animals. *Arch. Environ. Contam. Toxicol.* **3**, 371–383.

Takahashi, F. T., and Kittredge, J. S. (1973). Sublethal effects of the water soluble component of oil: Chemical communication in the marine environment. *La. State Univ., Cent. Wetland Resour., Publ.* **LSU-SG-73-01**, 259–264.

Teal, J. M. (1958). Distribution of fiddler crabs in Georgia salt marshes. *Ecology* **39**, 185–193.

Teal, J. M. (1959). Respiration of crabs in Georgia salt marshes and its relation to their ecology. *Physiol. Zool.* **32**, 1–14.

Teal, J. M., and Carey, F. G. (1967). The metabolism of marsh crabs under conditions of reduced oxygen pressure. *Physiol. Zool.* **40**, 83–91.

Tennant, D. A., and Forster, W. O. (1969). Seasonal variation and distribution of ^{65}Zn, ^{54}Mn and ^{51}Cr in tissues of the crab *Cancer magister* Dana. *Health Phys.* **18**, 649–657.

Thurberg, F. P., Dawson, M. A., and Collier, R. S. (1973). Effects of copper and cadmium on osmoregulation and oxygen consumption in two species of estuarine crabs. *Mar. Biol.* **23**, 171–175.

Todd, M.-E., and Dehnel, P. A. (1960). Effect of temperature and salinity on heat tolerance in two grapsoid crabs, *Hemigrapsus nudus* and *Hemigrapsus oregonensis. Biol. Bull. (Woods Hole, Mass.)* **118**, 150–172.

Toombs, G. L. (1965). Radiological survey of the lower Columbia River in Oregon, August 1963– July 1964. *Radiol. Health Data* **6**, 563–568.

Turner, H. J., Jr. (1953). The edible crab fishery of Massachusetts. *In* "Sixth Report on Investigations of the Shellfisheries of Massachusetts," pp. 25–28. Dept. Nat. Resour., Div. Mar. Fish., Woods Hole Oceanogr. Inst., Woods Hole, Massachusetts.

Turner, H. J., Jr. (1954). The edible crab fishery of Boston Harbor. *In* "Seventh Report on Investigations of the Shellfisheries of Massachusetts," pp. 7–16. Dept. Nat. Resour., Div. Mar. Fish., Woods Hole Oceanogr. Inst., Woods Hole, Massachusetts.

Valiela, I., Babiec, D. F., Atherton, W., Seitzinger, S., and Krebs, C. (1974). Some consequences of sexual dimorphism: Feeding in male and female fiddler crabs, *Uca pugnax* (Smith). *Biol. Bull. (Woods Hole, Mass.)* **147**, 652–660.

Van Engel, W. A. (1958). The blue crab and its fishery in Chesapeake Bay. *Commer. Fish. Rev.* **20**, 6–17.

Vernberg, F. J. (1959). Studies on the physiological variation between tropical and temperate zone fiddler crabs of the genus *Uca*. II. Oxygen consumption of whole organisms. *Biol. Bull. (Woods Hole, Mass.)* **117**, 163–184.

Vernberg, F. J., and Costlow, J. D., Jr. (1966). Studies on the physiological variation between tropical and temperate-zone fiddler crabs of the genus *Uca*. IV. Oxygen consumption of larvae and young crabs reared in the laboratory. *Physiol. Zool.* **39**, 36–52.

Vernberg, F. J., and Tashian, R. E. (1959). Studies on the physiological variation between tropical and temperate zone fiddler crabs of the genus *Uca*. I. Thermal death limits. *Ecology* **40**, 589–593.

Vernberg, F. J., and Vernberg, W. B. (1970). Lethal limits and the zoogeography of the faunal assemblages of coastal Carolina waters. *Mar. Biol.* **6**, 26–32.

Vernberg, W. B., and DeCoursey, P. J. (1977). Effects of sublethal metal pollutants on the fiddler crab, *Uca pugilator. U.S. Environ. Prot. Agency, Off. Res. Dev. [Rep.] EPA* **EPA-600/3-77-024**, 215–220.

Vernberg, W. B., and O'Hara, J. (1972). Temperature–salinity stress and mercury uptake in the fiddler crab, *Uca pugilator. J. Fish. Res. Board Can.* **29**, 1491–1494.

Vernberg, W. B., and Vernberg, F. J. (1972). Synergistic effects of temperature, salinity, and mercury on survival and metabolism of the adult fiddler crab, *Uca pugilator*. *Fish. Bull.* **70,** 415–420.

von Hagen, H. O. (1976). Review. Jocelyn Crane, "Fiddler Crabs of the World. Ocypodidae: Genus *Uca,*" ISBN 08102-6. Princeton University Press, Princeton, New Jersey, 1975. *Crustaceana (Leiden)* **31,** 221–224.

Waldron, K. D. (1958). The fishery and biology of the Dungeness crab (*Cancer magister* Dana) in Oregon waters. *Contrib., Fish Comm. Ore.* No. 24, 1–43.

Walton, E., and Williams, A. B. (1971). Reef populations of mud crabs and snapping shrimp. *In* "Structure and Functioning of Estuarine Ecosystems Exposed to Treated Sewage Wastes" (E. J. Kuenzler and A. F. Chestnut, eds.), Annu. Rep. for 1970–1971, Sea Grant No. GH 103, Proj. UNC-10, pp. 205–238, University of North Carolina, Chapel Hill.

Ward, D. V., and Howes, B. L. (1974). The effects of Abate, an organophosphorous insecticide on marsh fiddler crab populations. *Bull. Environ. Contam. Toxicol.* **12,** 694–697.

Ward, D. V., Howes, B. L., and Ludwig, D. F. (1976). Interactive effects of predation pressure and insecticide (Temefos) toxicity on populations of the marsh fiddler crab *Uca pugnax*. *Mar. Biol.* **35,** 119–126.

Waterman, T. H., ed. (1960). "The Physiology of Crustacea," Vol. 1. Academic Press, New York.

Waterman, T. H., ed. (1961). "The Physiology of Crustacea," Vol. 2. Academic Press, New York.

Webb, D. A. (1940). Ionic regulation in *Carcinus maenas*. *Proc. R. Soc. London, Ser. B* **129,** 107–136.

Weis, J. S., and Mantel, L. H. (1976). DDT as an accelerator of limb regeneration and molting in fiddler crabs. *Estuarine Coastal Mar. Sci.* **4,** 461–466.

Welch, W. R. (1968). Changes in abundance of the green crab, *Cacinus maenas* (L.), in relation to recent temperature changes. *Fish. Bull.* **67,** 337–345.

Williams, A. B. (1965). Marine decapod crustaceans of the Carolinas. *Fish. Bull.* **65,** 1–298.

Williams, A. B. (1971). A ten-year study of meroplankton in North Carolina estuaries: Annual occurrence of some brachyuran developmental stages. *Chesapeake Sci.* **12,** 53–61.

Williams, A. B. (1974). The swimming crabs of the genus *Callinectes* (Decapoda: Portunidae). *Fish. Bull.* **72,** 685–798.

Williams, A. B., and Deubler, E. E., Jr. (1968). Studies on macroplanktonic crustaceans and ichthyoplankton of the Pamlico Sound complex. *N. C., Dep. Conserv. Dev., Div. Commer. Sports Fish., Spec. Sci. Rep.* No. 13, pp. 1–103.

Williams, B. G. (1968). Laboratory rearing of the larval stages of *Carcinus maenas* (L.) (Crustacea: Decapoda). *J. Nat. Hist.* **2,** 121–126.

Winget, R. R., Maurer, D., and Seymour, H. (1974). Occurrence, size composition and sex ratio of the rock crab, *Cancer irroratus* Say and the spider crab, *Libinia emarginata* Leach in Delaware Bay. *J. Nat. Hist.* **8,** 199–205.

Yang, W. T. (1971). Preliminary report on the culture of the stone crab. *In* "Proceedings of the Second Annual Workshop, World Mariculture Society," pp. 53–54. Div. Continuing Educ., Louisiana State Univ., Baton Rouge.

Young, J. E. (1974). Variations in the timing of spermatogenesis in *Uca pugnax* (Smith) and possible ' effectors (Decapoda, Brachyura, Ocypodidae). *Crustaceana (Leiden)* **27,** 68–72.

CHAPTER 7

Shrimps (Arthropoda: Crustacea: Penaeidae)

JOHN A. COUCH

I. Introduction

Penaeid shrimps play a significant role in the middle of the trophic web in the estuaries and inner oceanic littoral zones of most temperate, subtropical, and tropical regions of the world. These shrimps have relatively complex life cycles that bring metamorphic stages into contact with a wide range of environmental conditions. If we select as a representative example of penaeid life cycles a species of *Penaeus* in the northern Gulf of Mexico, then the varied environments required are well illustrated in Fig. 1. Gravid females spawn offshore, and early metamorphosis (nauplii, protozoea, and mysis stages) occurs in high-salinity coastal or offshore waters. Postlarvae enter estuaries, become juveniles, and then young adults prior to returning offshore upon sexual maturity (Williams, 1965). The juvenile and young shrimp are intimately dependent upon the quality of estuarine environments. The adults and larvae are dependent indirectly upon quality of the estuaries and directly upon the quality of coastal and near-shore oceanic waters.

Shrimps of the family Penaeidae form the basis for one of the most valuable fisheries in the world. The highly esteemed quality and edibility of these

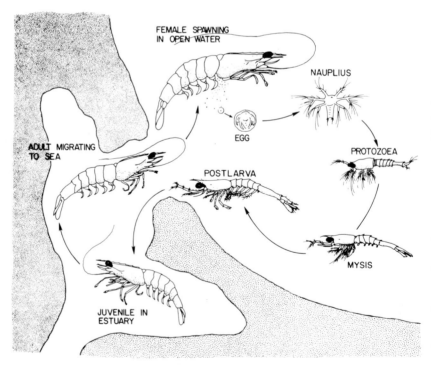

Fig. 1. Life cycle of the penaeid shrimp.

shrimps, their relatively large size, and their broad distribution worldwide have elicited a large investment of manpower and funds in the shrimp industry. In the western hemisphere, the penaeid shrimp fishery is an annual multimillion dollar enterprise; no accurate monetary value can be placed on its worth worldwide.

The purpose of this chapter is to review information available on the pollution ecology of representative species of penaeid shrimps. Data and information from both field and laboratory studies, as reported in the literature and from the author's research, will be used.

The majority of reports available about pollution and penaeid shrimps concerns studies involving the commercially valuable penaeid shrimps of the U.S. Atlantic states and Gulf Coast. Therefore, most of the information presented here will be related to the following three species: *Penaeus duorarum* (pink shrimp), *Penaeus aztecus* (brown shrimp), and *Penaeus setiferus* (white shrimp), all Atlantic and Gulf of Mexico species. Reference to other species of penaeid and some nonpenaeid crustacea will be made when specific studies contribute significantly to our understanding of pollution ecology of shrimps.

This chapter will cover the following pollutant categories and situations: organic chemicals other than petroleum, petroleum and related compounds, heavy metals, biological agents, and interactions of pollutants and other factors. Under each of these divisions toxicity and specific tissue, organismic, population, and ecological effects will be reviewed when known. Further, the uptake, transport, and fate of pollutants will be discussed as they may affect the ecology of penaeid shrimps.

II. Organic Chemicals Other Than Petroleum

A. INDUSTRIAL ORGANIC CHEMICALS

The last four decades have witnessed unprecedented release of synthetic organic chemicals into the natural environment. These so-called xenobiotics have numerous origins and uses; rarely are they innocuous upon entering natural waters. Because penaeid shrimps occupy estuaries during a significant portion of their life cycles, they are exposed to pollutants characteristic of industries associated with the estuaries. Relatively few industrial pollutants have been investigated specifically with regard to ecology of penaeid shrimps.

PCB's

Polychlorinated biphenyls (PCB's) are industrial pollutants to be considered with regard to their exemplary roles in penaeid shrimp ecology and aquatic ecology in general. Table I presents a summary of relationships of PCB's and penaeid shrimps.

TABLE I

POLYCHLORINATED BIPHENYLS AND PENAEID SHRIMPS

Compound	Toxicity	Physiological/histological effects	Ecological/environmental implications	References
Aroclor 1254	3 μg/liter 30-day exposure caused 50% mortality, pink shrimp	Lethargy; chromatophore changes; ultrastructural pathology in hepatopancreas cells.	Pink shrimp absorbed 1254 from sediments from polluted estuary; loss of burrowing activity–exposure to predators; shrimp could not avoid contaminated waters as some fish could. Long life of PCB's in sediments maintain pollution at chronic levels over several years.	Duke et al. (1970); Nimmo et al. (1971, 1975)
	0.9 μg/liter caused some mortality in 14 days, pink shrimp			Couch et al. (1974)
Aroclor 1016	10 μg/liter, 96-hour exposure caused 43% mortality	None reported	None reported	Hansen et al. (1974)
	0.9 μg/liter, 96-hour exposure caused 8% mortality, brown shrimp			

For many years these industrial chemicals have been at large in the aquatic environment as a result of waste effluents, disposal of dielectric fluids, and other industrial sources (Broadhurst, 1972). It is a well-established fact that certain fresh and marine bodies of water are contaminated with various compounds of PCB (Sodergren *et al.*, 1972; Nimmo *et al.*, 1971, 1975). In 1970, Duke *et al.* reported the PCB, Aroclor 1254*, in water, sediments, and tissue of animals (including penaeid shrimps) from Escambia Bay, near Pensacola, Florida.

Considerable research has been done on the effects of PCB's on estuarine species with emphasis on pink and brown shrimps. These two penaeids were killed in 2-week exposures to 0.9, 1.4, and 4.0 μg/liter Aroclor 1254 in flowing seawater (Nimmo *et al.*, U.S. Environmental Protection Agency, Gulf Breeze, Florida). The minimum concentration causing mortality was 0.9 μg/liter. Penaeid shrimps suffered greatest mortality when exposed during premolt (just before molting) and during molt. Most exposed shrimp became lethargic, stopped feeding, and did not dig into the substrate (digging is a normal activity for penaeids). Subtle to dramatic chromatophore changes in the cuticle of exposed shrimp were more frequent and obvious than in control shrimp.

On the light microscopical level, no lesions were consistently found that were indicative of PCB exposure in shrimp (Couch and Nimmo, 1974a). However, several interesting cytopathic changes were noted in exposed shrimp studied with electron microscopy.

Pink shrimp were exposed to 3 μg/liter Aroclor 1254 in flowing seawater for 30 to 52 days. During these exposures up to 50% of the animals died. Both living and dead shrimp were analyzed by gas chromatography and from 33 to 40 μg/gm Aroclor 1254 was found in their hepatopancreatic tissues. Aroclor uptake in hepatopancreas was linear with time (Couch and Nimmo, 1974a). Hepatopancreas was fixed and processed for electron microscopy. Hepatopancreatic absorptive cells from exposed shrimp revealed the following departures from those of controls: (1) an increase or proliferation in rough endoplasmic reticulum in 30–50% of cells; (2) production of membrane whorls with enclosed lipid droplets; and (3) nuclear degeneration characterized by the occurrence of vesicles in the nucleoplasm (20–50 nm and 100–700 nm in diameter).

The proliferation of smooth endoplasmic reticulum (ER) in hepatocytes of higher animals has been described as indicative of toxic responses to drugs or chemicals such as phenobarbitol, dilantin, dieldrin, and carbon tetrachloride. Proliferation of ER has been related to detoxification of poisons and may, in shrimp, represent an attempt, on the part of hepatopancreatic cells, to metabolize PCB absorbed from the lumen of hepatopancreatic ducts. If this is the case, cellular alterations at the ultrastructural level such as proliferation of ER may be

---

*Use of trade names does not imply endorsement of commercial products by the U.S. Environmental Protection Agency.

valuable as early indicators of sublethal effects of certain pollutants in penaeid shrimps.

Another PCB, Aroclor 1016, has been more recently introduced for limited use in the United States. This compound has been tested for toxicity in brown shrimp. Aroclor 1016 was found to have nearly the same toxicity for penaeid shrimp as Aroclor 1254: 0.9 μg/liter Aroclor 1016 in flowing seawater killed 8% of test shrimp in 96 hours; 10 μg/liter Aroclor 1016 killed 43% of test shrimp in 96 hours (Hansen *et al.*, 1974a).

It is apparent that PCB's as pollutants pose a threat to penaeid shrimps which show a high sensitivity to these compounds. In this regard, Nimmo *et al.* (1971) demonstrated that pink shrimp could absorb a PCB (Aroclor 1254) from sediments taken from a PCB-polluted estuary, Escambia Bay, Florida. Hansen *et al.* (1974b) found that some estuarine species could avoid waters contaminated with Aroclor 1254, but pink shrimp showed no avoidance reaction when given choices of clean or PCB-contaminated water. These and other data suggest that PCB's, as pollutants, could have influence on relative survival and abundance of penaeid shrimps in natural waters.

B. Pesticidal Chemicals

In the last 30 years, many kinds of chemical pesticides have been inadvertently or intentionally released into the environment. Aquatic life is exposed to these compounds because the aquatic portion of the biosphere often behaves as a "sink" or receptacle for these compounds because of runoff or fallout. Some of these pesticides, such as certain organochlorines or their metabolites, are refractory to breakdown, and thus tend to accumulate in various compartments of the aquatic environment. Shrimps have been found to accumulate certain pesticidal compounds in the laboratory, and feral shrimps have possessed detectable levels when taken directly from contaminated or apparently "clean" waters. The staff of the U. S. Environmental Protection Agency Laboratory, Gulf Breeze, Florida, has found, over several years of testing, that penaeid shrimps generally are far more sensitive to toxic and ecological effects of most pesticides than are fishes or mollusks (Couch, 1978). The effects of some of the well-known compounds will be reviewed here.

1. Organochlorines

The following organochlorine pesticides have been studied in regard to organismic and/or ecological effects on penaeid shrimps: chlordane, DDT, dieldrin, endrin, heptachlor, hexachlorobenzene, lindame, mirex, and toxaphene. Data on the effects are summarized in Table II for these pesticides.

White shrimp, which died as a result of DDT exposure, accumulated up to 40.4 μg/gm DDT and DDE in hepatopancreas after 18 days of exposure to 0.20

ORGANOCHLORINE PESTICIDES AND PENAEID SHRIMPS

Pesticide	Toxicity	Physiological/histological effects	Ecological implications	References
Chlordane	96-hour LC_{50}, 0.4 μg/liter, pink shrimp	None reported	Residues in estuaries from five southeastern states; effect in nature unknown	Parrish et al. (1976)
DDT	18 days LC_{100}, 0.1 μg/liter, white shrimp	Loss of cations (Na^+, K^+) in hepatopancreas; possible ATPase inhibitor	Death of exposed shrimp takes out possible step in trophic accumulation in nature	Butler et al. (1970), Nimmo et al. (1970)
Dieldrin	96-hour LC_{50}, 0.9 μg/liter, pink shrimp	None reported	Commonly found in estuaries; second most common pesticide in mollusks (1965–1972)	Parrish et al. (1973), Butler (1973)
Endrin	96-hour LC_{50}, 0.28 μg/liter, pink shrimp	None reported	None reported	Schimmel et al. (1974)
Heptachlor	96-hour LC_{50}, 0.11 μg/liter, pink shrimp	None reported	None reported	Schimmel et al. (1976)
Hexachlorobenzene	Lethargy in 25 μg/liter and 33% death at 96 hours, pink shrimp	Hepatopancreas grossly abnormally white	Accumulated by estuarine fishes. Exposed shrimp possibly unable to escape predation due to lethargy	Parrish et al. (1974)
Lindane	24-96 hour LC_{50}; range, 0.17–35 μg/liter; pink, brown, white shrimps	None reported	None reported	Chin and Allen (1957), Schimmel et al. (1977)
Mirex	7-day LC_{25} 1.0 μg/liter, pink shrimp	Delayed toxicity to pink shrimp 4 days after removal from mirex	Mirex applied as a particle bait in nature may render it unavailable to many organisms	Lowe et al. (1971), Tagatz et al. (1976), Markin et al. (1974)
Toxaphene	LC_{50} nauplius, 22 μg/liter; LC_{50} protozoea, 1.8 μg/liter; LC_{50} mysis, 1.4 μg/liter, pink shrimp	Differential toxic responses of different larval stages.	Temperature dissolved O_2 variations affected survival in exposed mysis	Courtenay and Roberts (1973)

μg/liter in flowing seawater (Nimmo *et al.*, 1970). Exposure to DDT concentrations greater than 0.10 μg/liter was lethal to pink shrimp in 28 days. A physiological effect of DDT exposure in pink and brown shrimp was loss of certain cations in the hepatopancreas (Nimmo and Blackman, 1972). Sodium and potassium concentrations in shrimp exposed to 0.05 μg/liter DDT for 20 days were lower than in those not exposed. Magnesium, however, was not significantly lowered. The significance of reduced cations in the hepatopancreas of shrimp for the pathophysiological behavior of shrimp is not known but a loss of ATPase activity in ion transport may be indicated. Blood protein levels also have been found to drop in shrimp exposed to DDT. There are no reports of histopathological changes in penaeids following exposure to DDT. In acute, high-concentration laboratory exposures, shrimp showed tremors, hyperkinetic behavior, and paralysis, classic signs of DDT poisoning in arthropods. After extended exposure to low concentrations of DDT, shrimp did not become paralyzed but sank into lethargy, refused food, and then died.

Pink shrimp were more sensitive to dieldrin than were grass shrimp (Palaemonidae) in test exposures. However, both species died when exposed to concentrations of dieldrin in the low parts per billion range.

Some juvenile pink and brown shrimp died after laboratory exposure to low concentrations of mirex (Table II). However, all survivors from this test died after 4 days in mirex-free seawater, demonstrating a delayed toxic effect of mirex (Lowe *et al.*, 1971). Mirex poisoning in shrimp produces loss of coordination, equilibrium, and finally signs of lethargy and paralysis.

I have examined both shrimp and blue crabs exposed to low concentrations of mirex for long periods (30 days or more) for histopathological effects. No pathologic effects at the tissue level were found in the animals that were examined. The organs studied were muscle, hepatopancreas, and gonads.

Markin *et al.* (1974) completed a detailed ecological monitoring study of the distribution of mirex in oysters, crabs, shrimps, and fishes from estuarine waters from the Gulf of Mexico to Delaware Bay. Sampling stations were in Mississippi Sound; Mobile Bay, Alabama; Tampa Bay, Florida; Jacksonville, Florida; Savannah, Georgia; Charleston, South Carolina; Morehead City, North Carolina; Chesapeake Bay; and Delaware Bay. Only shrimp from the Savannah, Georgia area had detectable concentrations of mirex (0.007 μg/gm in tissues). The Savannah, Georgia area has a long history of mirex usage. Mirex does not appear to be as widespread in estuarine regions as are PCB's and DDT (Markin *et al.*, 1974). Mirex applied usually as a particle bait poison would not be as directly available to many marine organisms as are broadcast liquids or powder formulations of other pesticides.

The finding of 0.007 μg/gm mirex in shrimp near Savannah probably reflects a considerable accumulation of mirex by shrimp over a long period of time from possibly extremely low concentrations in the environment. Lowe *et al.* (1971)

found (Table II) that 25% of test pink shrimp died in the laboratory after 7 days of exposure to only 1.0 μg/liter mirex. Therefore, the concentrations of mirex must be assumed to be considerably lower than 1.0 μg/liter in Georgia estuarine waters. If the aqueous concentrations were above 1.0 μg/liter few shrimp would probably survive in that area.

Toxaphene affects differentially the early metamorphic stages of pink shrimp (Table II) according to Courtenay and Roberts (1973). They found the mysis stage of pink shrimp most susceptible to toxaphene, particularly under various temperature and dissolved oxygen conditions. This might suggest that fluctuations in abundance in juvenile and young adult penaeid shrimps in toxaphene contaminated waters could result from impact of the pesticide on an early life-cycle stage during periods of changing temperature and/or during periods of dissolved oxygen fluctuations.

2. Organophosphates and Carbamates

Few organophosphate compounds have been tested on species of crustaceans. However, those tested have shown approximately 1000 times greater toxicity to shrimps than most other pesticides tested (Butler, 1966), and penaeid shrimps have shown greater sensitivity than fishes or mollusks to organophosphates.

Baytex (Bayer 29, 493) was toxic to penaeid shrimp (Butler and Springer, 1963) in the laboratory. Naled (1,2-dibromo-2,2-dichlroethyldimethyl phosphate) had little effect in field tests on shrimp. Fast dilution and instability without persistence of compounds may contribute to the lack of mortality of shrimps in field testing of this organophosphate. In the laboratory, Dibrom is lethal to postlarval brown shrimp at 2.0 μg/liter, and at 5.5 μg/liter (= LC_{50} for 48 hours) it is lethal to adult pink shrimp. Malathion at 14 μg/liter caused hyperactivity, paralysis, and death in penaeids. Parathion lethal concentration for 48 hours in pink shrimp was 0.2 μg/liter (D. Coppage, personal communication). No histopathogenesis has been reported for penaeids exposed to organophosphates.

Conte and Parker (1975) found malathion aerially applied to flooded marshes in Texas caused from 14 to 80% mortality in brown and white shrimps held in cages. They recommended that malathion not be applied to flooded marshes that maintained shrimp. Tagatz et al. (1974), however, found that caged pink shrimp were not killed in small areas receiving malathion via thermal fogging for mosquito control in salt marshes in northwest Florida.

Both organophosphates and carbamates are reported to be potent acetylcholinesterase (AChe) inhibitors in the vertebrates. Little evidence of early, presyndromic inhibition of AChe activity in ventral nerve cord of pink shrimp was found, but inhibition as high as 75% was found in moribund shrimp experimentally exposed to malathion (Coppage and Matthews, 1974).

Carbamate pesticides have not been tested extensively in regard to penaeid shrimps, but sevin may be lethal to other crustacea when applied to field sites in the marine environment (Haven *et al.*, 1966). J. Lowe (personal communication) has found carbaryl (sevin) to be quite toxic to penaeid shrimps in laboratory tests. Tagatz *et al.* (1979) report that brown shrimp have a 24- to 96-hour LC_{50} of 2.5 $\mu g/l$, the lowest concentration of sevin reported to kill any crustacean tested.

III. Petroleum

Charter *et al.* (1973) estimated that the annual spillage of oil and oil products into the ocean is approximately 3 million tons. Most of these spills occur in estuaries or near coastal regions of significant biological value. Of particular importance is the fact that many coastal areas impacted by oil spills or which are potential spill areas are also in penaeid shrimp-producing regions. Unfortunately, only a few detailed studies of ecological and physiological effects of oil on penaeid shrimps have been published.

Neff *et al.* (1976a) found that when penaeid shrimps were experimentally exposed to oil-contaminated seawater, they accumulated hydrocarbons in their tissues. Table III gives information on two forms of water-soluble petroleum products in relation to penaeid shrimps. Aromatic hydrocarbons were accumulated in greater amounts than were alkane hydrocarbons. Bioaccumulation of aromatics increased in proportion to the increase in their molecular weights up to, but not including, the heavier polycyclic aromatics (four to five rings). Brown shrimp (*Penaeus aztecus*) exposed to oil in seawater and then depurated in clean seawater released accumulated hydrocarbons more rapidly than clams (*Rangia cuneata*) and oysters (*Crassostrea virginica*). Shrimp can metabolize the hydrocarbons whereas the mollusks have limited, if any, capacity to do this. The higher molecular weight hydrocarbons are released from shrimp tissues more slowly than lower molecular weight hydrocarbons (Neff *et al.*, 1976a). Carcinogenic polycyclic hydrocarbons such as benzo[*a*]pyrene, therefore, would be retained longer, possibly to exert chronic effects or to be shunted into the human food chain.

Anderson *et al.* (1974a) tested water-soluble fractions and oil in water dispersions of four oils [(South Louisiana crude; Kuwait crude; refined No. 2 fuel; and Bunker C (residual)] against brown shrimp postlarvae for toxicity. The fractions of refined oils were generally more toxic to penaeid shrimps and to other aquatic species tested than were the fractions of crude oils. Crustacean species tested, including brown shrimp postlarvae, were more sensitive to oil fractions than fish species tested (*Cyprinodon, Menidia,* and *Fundulus*). The 24-hour median tolerance limit of juvenile brown shrimp exposed to components of No. 2 fuel oil (naphthalenes, methyl naphthalenes, and dimethyl naphthalenes) ranged from 0.77 to 2.51 ml/liter. The naphthalenes were the most toxic components of fuel oils (Anderson *et al.*, 1974b) to shrimp.

TABLE III

Petroleum and Penaeid Shrimps

Component	Toxicity	Physiological implications	Ecological/environmental implications	References
Water-soluble fraction of No. 2 fuel oil	2.9 mg/liter = 96-hour LC_{50}, brown shrimp juveniles; 1.0 mg/liter = 96-hour LC_{50}, white shrimp juveniles	Differential susceptibilities of postlarvae, early juveniles, and late juveniles of single species, i.e., brown shrimp. Postlarvae are more tolerant than juveniles	Different susceptibilities of different species in same genus, i.e., brown shrimp vs. white shrimp; might lead to changes in species composition in polluted regions	Neff *et al.* (1976)
Louisiana crude oil (water-soluble fraction)	19.8 mg/liter = 96-hour LC_{50}, brown shrimp postlarvae	May show chronic toxicity, but not as acutely toxic as refined product above	Brown shrimp postlarvae less sensitive than polychaetes and one fish (*Menidia* sp.); about equal in sensitivity with two fishes (*Fundulus* and *Cyprinodon*)	Neff *et al.* (1976)

Steed and Copeland (1967) and Neff *et al.* (1976b) observed considerable interspecies variation in oxygen consumption between different species of *Penaeus* and among different genera of marine animals when they were exposed to the same oil–seawater mixtures. Therefore, physiological responses to oil pollution may vary with test species used, and one cannot with certainty predict a priori which species will be more sensitive to oil (Neff *et al.*, 1976b).

Studies of detailed pathologic mechanisms of oil toxicity in shrimps have been reported only rarely. Yarbrough and Minchew (1975) reported nonspecific lesions in the cuticular chitin, the lining of the gastric mill, and the mouth region of penaeid shrimp exposed to sonified crude oil. They also found the proliferation of cells and necrosis in the basal portion of gill filaments of exposed shrimps.

The ecological effects of an oil spill depend to a large extent upon the environmental, meterological, and geographical variables surrounding the spill. Oil retained in enclosed coastal regions might be expected to affect juvenile shrimps more seriously than would spills occurring in open ocean areas. Oil spills in nature have not been studied specifically in relation to penaeid shrimp effects.

IV. Heavy Metals

A variety of heavy metals are found as both naturally and anthropogenically derived components of the estuarine environment. These metals may exist in several oxidation states with different reaction potentials depending on their specific chemistries (Wood, 1974). Certain heavy metals are pollutants generated by industry. Some may be acted upon by estuarine microbes to produce alkylmetallic compounds which can be accumulated by estuarine species and are potent toxicants.

The heavy metals that can be methylated, such as mercury, tin, palladium, platinum, gold, and thallium pose special threats as environmental pollutants and some should be monitored continuously (Wood, 1974). Other metals such as cadmium, lead, and zinc do not form stable alkyl-metals in aqueous solutions, but may have different modes of toxic action than do the alkyl-metals such as methylmercury, a neurotoxicant. Cadmium, which does not persist as an alkyl-metal in aquatic systems, but does as an ion, is a strong cytotoxicant to gill cells of crustacea (see below). Reports on effects of cadmium, copper, and mercury have been published concerning penaeid shrimps, and are reviewed below (see also Table IV).

A. CADMIUM

Unusually high levels of cadmium have been reported from certain estuaries in which penaeid shrimps commonly occur (i.e., Laguna Madre, Corpus Christi,

TABLE IV

HEAVY METALS AND PENAEID SHRIMPS

Metal	Biologically active state	Toxicity	Physiological/histological effects	Environmental implications	References
Cadmium	Ca^{2+}	718 μg/liter = 30-day LC_{50} pink shrimp	Destroys gill respiratory, osmo-regulatory tissues. Causes black gills	Salinity, respiratory stresses combined with cadmium toxicity causes mortality of shrimp	Couch (1977), Nimmo et al. (1977)
Copper	Cu^{2+}	0.05 mg/liter, lethal to larval stages of brown shrimp in seawater brine mixture	None reported	Copper tubing used in de-salination plants may contribute copper to estuaries	Mandelli (1971)
Mercury	Hg-CH_3, Hg^{2+}	17 μg/liter = 96-hour LC_{50} for postlarval white shrimp	Causes inhibition of brown shrimp's ability to regulate internal ion concentrations. Dysfunction in osmoregulation under salinity stress		Petrocelli et al. (1975), Green et al. (1976)

Texas). This metal is also a pollutant from several industrial effluents into aquatic systems.

Nimmo *et al.* (1977) observed that in a significant number of pink shrimp exposed to approximately 760 μg/liter cadmium (as $CdCl_2$) for 9 days or longer in flowing seawater, an unusual blackening of gills occurred. Control shrimp did not develop black gills. It was found that the LC_{50} of cadmium in 30 days was 718 μg/liter, and that during these tests many exposed shrimp developed the black gill syndrome prior to death.

I have completed light and electron microscopic studies of gill tissues from exposed blackened gills and control gills of surviving pink shrimp from the above tests (Couch, 1977). My findings indicate that the gross blackening of gills results from necrosis of subcuticular tissues (gill epithelial tissue). This necrosis stems from the death of cells in the distal gill filaments (smallest unit in gill of shrimp). Actual cell death occurs prior to gross blackening in tiny foci, followed by gradual involvement of the whole filament. Electron microscopy reveals polymorphic black deposits in the cytoplasm of moribund or necrotic gill cells (early around mitochondria, later throughout). A complete loss of structural and concomitant functional integrity of the gill soft tissue leads to organ necrosis. However, the cuticle and epicuticle remain intact at the ultrastructural level and hold the moribund or necrotic soft tissue within their boundaries. Apparent melanization of injured gill filaments may hasten the blackening syndrome. However, electron microscopy does not present evidence for the presence of melanosomes, melanocytes, or melanophores. An alternative possibility, that cell death and necrosis lead to the deposition of metal sulfides or other black deposits in necrotic tissues in the living animal, could account for the blackened gill syndrome. The interesting concept of specific cell and tissue death preceding organismic death is represented in the pink shrimp's response to cadmium exposure. Death of cells concerned with osmoregulation and respiration probably lead to dysfunction and eventual death of shrimp.

Bahner (1975) studied the uptake of cadmium in the tissue of pink shrimp. He found that between 1 and 10 μg/liter Cd in water elicited uptake by hepatopancreas, gills, and exoskeleton. Below concentrations of 1 μg/liter Cd, there was no accumulation of the metal in shrimp tissues.

B. COPPER

Mandelli (1971) found that a copper concentration of 0.05 mg/liter was lethal to nauplii, protozoea, and mysis of *Penaeus aztecus* and *P. duorarum* that were exposed in a seawater–brine mixture similar to mixtures derived from desalination plants. The larval stages were able to grow normally in seawater (35‰) containing 0.025 mg/liter copper.

C. MERCURY

Mercury as a metal has not been suspect in toxic effects on organisms; however, mercuric salts and methylated mercury, are extremely toxic with both short- and long-term chronic effects. Mercuric chloride is used in some histological fixative fluids because of its protein-precipitating effects in tissues. Enzymes of liver and other tissues may be bound and their activity may be altered by ionic mercury.

Petrocelli *et al.* (1975) studied the uptake and distribution of mercuric chloride in brown shrimp (*P. aztecus*), and also examined the effects of mercuric chloride exposure on ability of brown shrimp to adjust to salinity changes. They found that after 2 hours of exposure to 0.5 μg/liter mercuric chloride in seawater, accumulation of mercury in shrimp was 285 mg/gm with only 9% of the mercury in the meat (muscle) and 91% in the shell. This suggested a surface adsorptive process for mercury in brown shrimp exposed for brief periods. These authors also reported that shrimp obtained off Louisiana's Southwest Pass had natural levels of only 4.6 ng/gm mercury distributed as 64% in the muscle and 36% in the cuticle, suggesting that chronic exposure results in internal accumulation of mercury.

Green *et al.* (1976) found that 17 μg/liter ionic mercury is the LC_{50} (96 hours) for postlarval white shrimp (*P. setiferus*). Chronic exposure of postlarval white shrimp to 1.0 μg/liter for 60 days did not affect respiration, growth, and molting (Green *et al.*, 1976).

Brown shrimp are active regulators of blood chloride levels (ion regulators). Petrocelli *et al.* (1975) found that exposure of brown shrimp to mercury and to salinity changes resulted in interference with the shrimp's ability to regulate their internal ion levels to compensate for external salinity changes. Mercury could prove to be detrimental to penaeid shrimps if the form and amount were sufficient enough to prevent their adjustment to freshets or high saline conditions that result from rapid changes in estuaries or tidelands due to seasonal or meterological extremes.

To date, no reports have been published which suggest a specific mechanism for mercury toxicity in shrimps or other invertebrates. Investigations on the relative roles of different species of certain heavy metals need to be carried out to determine the behavior of specific metals in their different chemical states in estuaries.

V. Biological Agents

This section includes a discussion and review of biological agents which are being considered or developed for commercial, agricultural, and health purposes as biopesticides or biological control agents. Two major categories that may influence penaeid shrimps are biochemical agents (arthropod hormonal agents or their mimics) and infectious pathogenic agents for arthropods (viruses, bacteria, fungi, protozoa, helminths, and parasitoids).

To date, there is no strong evidence to assign any of the above roles as pollutant factors in penaeid shrimp ecology. There is concern, however, for the future use and safety of such agents in coastal regions because of their specific roles in arthropod ecology, physiology, and pathology (Couch, 1975). Biological control agents must be tested and evaluated regarding their safety for nontarget species such as valuable aquatic arthropoda (i.e., crustacea and some aquatic insecta). The effects of some insect growth regulators and insect pathogens, which are being used or developed as biocontrol agents, have been tested in nontarget crustacean species such as penaeid shrimps or related forms. The results of these studies are reviewed in the following sections.

A. Biochemical Agents

Naturally occurring insecticidal chemicals such as pyrethrum, nicotine, rotenone, hellebore, ryania, and sabadilla will not be discussed here because of lack of evidence that they are potential pollutants (Soloway, 1976). Insect growth regulators, however, have shown promise as synthetic insecticides and have the potential to be used on a wide scale thus resulting in possible environmental exposure of nontarget species.

The major example of an insect growth regulator that has been tested for effects on crustacea is methoprene (Altosid). This compound is the only registered (Environmental Protection Agency) commercially available insect growth regulator, and is listed as a control agent for floodwater mosquitoes. Persistence of methoprene in the aquatic environment is of short duration (4–7 days or less according to Siddall, 1976), and does not bioaccumulate. Methoprene is effective in many different insect species but has not shown significant toxicity to nontarget species (Miura and Takahashi, 1973). The LC_{50} of methoprene for white and pink shrimps (penaeids) was 100 ppm according to Siddall (1976) and Wright (1976). Freshwater crayfish also required relatively high concentrations ($LC_{50} = 100$ ppm) for toxic effects.

There are no reports available on sublethal physiological effects of an insect growth regulator in penaeid shrimps. Their mechanisms of effect in insects (interference with maturation and reproduction) should be examined as possible mechanisms in Crustacea because of relative close phylogenetic affinities of Crustacea and Insecta.

B. Infectious Pathological Agents

The major group of potential bioinsecticides studied in penaeid shrimps have been the entomopathogenic viruses known as nuclear polyhedrosis viruses

(*Baculovirus* group). These insect viruses are of particular interest because related viruses have recently been found in feral and cultured shrimps (Couch, 1974, 1975, 1978) and in feral crabs (Bazin *et al.*, 1974; Johnson, 1976) as natural pathogens.

Couch (1975), Lightner *et al.* (1973), Summers *et al.* (1975), and Heimpel and Buchanan (1967) have discussed the potential relationships between insect Baculoviruses (being developed as biological control agents) and nontarget species from shrimps to man. Summers *et al.* (1975) has pointed out the need to understand better the physical and biochemical nature of these viruses in different host systems. Couch (1975) suggested a need to evaluate regularly and experimentally new viral insecticides in nontarget crustacean species such as penaeid shrimps, with an emphasis on potentially susceptible larval stages of shrimps. Lightner *et al.* (1973) reported that tests of insect Baculoviruses in penaeid shrimps revealed no susceptibility of shrimps for the then-available insect viral insecticidal formulations. Postlarval, early, and late juvenile stages of brown and white shrimps were inoculated with free virus and were fed a diet containing virus. Over a 30-day test period, no mortality attributable to virus exposure occurred in test shrimps. These types of cross-infection experiments (i.e., insect virus in shrimp) should include careful ultrastructural, biochemical, and genetic monitoring to ensure that sublethal infections establishing the virus in a new host do not occur.

The modes of behavior in the estuary of viral, fungal, bacterial, and protozoan biological control agents developed as commercial products and for use in coastal agriculture are not known. The future need to understand the fates and effects of novel agricultural agents in estuaries is predictable, particularly regarding the development of integrated pest control methods which may utilize combinations of chemical and biological control agents. Mutagens in the form of chemicals such as fertilizer components, pesticides, or preservatives may interact with biological control agents in ways not presently anticipated, resulting in mutants with expansionist potential. Viruses, bacteria, and fungi should be studied in this regard.

VI. Interactions of Pollutants, Pollutant Mixes, and Other Factors in Penaeid Shrimp Ecology

Perhaps the most important area for future research in pollution ecology is the study of interactions of specific pollutants and pollutant complexes with physical, chemical, and biological factors in the environment. Estuaries are prime examples of multivariate and complex ecosystems where no single variable is altered without influencing some other component of the system. Penaeid shrimps are very dependent upon varying physical, chemical, and biological factors for their survival because of the wide range of factors that they must contend with in their life cycle, ranging from the open ocean to the tidal estuary.

Chief among the natural factors influencing penaeid shrimps are salinity, temperature, oxygen concentration (Venkataramiah *et al.*, 1977), bottom types or substrate, nutrition, and infectious diseases (Couch, 1978). A severe or sudden perturbation of any of these factors, combined with pollutant stress, may affect total stress or injury in penaeid shrimps. Some two-factor interactions will be reviewed.

A. Pollutant–Salinity Stress Interactions

Few reports have been published on the interaction of pollutant and salinity stress in penaeid shrimps. Nimmo and Bahner (1974) found that when brown shrimp were exposed to 3.0 μg/liter of the polychlorinated biphenyl Aroclor 1254 and to a gradual decrease in salinity over an 8-hour period, mortality of exposed shrimp was greater than in controls exposed to salinity change only or PCB only. The 3.0-μg/liter concentration of PCB was found previously to be sublethal in an 8-hour period. Therefore, the deciding factor that induced mortality was the combination of PCB exposure and salinity change. The authors also found that most major ions in the blood of dying shrimp became significantly less as the ambient salinity decreased. However, osmotic pressure overall showed no corresponding loss in the PCB–salinity stressed shrimp, thus complicating any interpretation of physiological mechanism for the lethal effect.

Jones (1975) reported that cadmium, zinc, and lead toxicity was enhanced in marine and estuarine isopods under stressful salinity changes in experimental exposures. Hemolymph osmotic concentrations were altered significantly in isopods exposed to cadmium, zinc, and mercury. Since it is known that cadmium destroys the soft gill tissues of penaeid shrimps (see Section IV and Couch, 1977), it is probable that the ion and water regulatory mechanism at the epithelial and cell membrane levels are structurally and functionally affected in crustacea exposed to certain heavy metals. Thus when salinity changes occur, the osmoregulatory mechanisms in metal-exposed crusacea fail to respond or are inhibited.

B. Pollutant–Temperature Stress Interactions

Penaeid shrimps have definite upper and lower temperature tolerance limits; they are shrimps of the temperate, subtropic, and tropic zones. Below water temperatures of about 5°–8°C penaeid shrimps become listless, and at lower temperatures they die. Above temperatures of 33°–35°C penaeids do not do well. Any pollutant that has a high oxygen demand would probably, at higher water temperatures in warm months, contribute to lower oxygen tension in water and thus to asphyxiation of penaeids that are sensitive to marginal or low ambient oxygen concentrations. No reports were found of studies explicitly concerned with the interactive effects of pollutants and varying temperature in penaeid shrimps.

The synergistic effects of temperature, salinity, and heavy metals in the fiddler crab *Uca pugilator* are discussed by Vernberg and Vernberg (1972). No three-factor interaction reports involving penaeid shrimps were found in the literature.

C. POLLUTANT–POLLUTANT INTERACTIONS

Most specific pollutants enter the estuary as components of complex mixtures of pollutant fallout, runoff, or effluents. In many cases estuarine species are exposed to several pollutants at once or to varying concentrations of two or more pollutants. Therefore, it is important to know what the combined and single effects of pollutants may be in estuaries. Bahner and Nimmo (1976) have examined the effects of combinations of toxicants in experimentally exposed pink shrimp (*Penaeus duorarum*). They studied the effects of the following toxicant combinations: cadmium–malathion, cadmium–methoxychlor, and cadmium–methoxychlor–Aroclor 1254. Essentially they found, with their particular mixtures, that the toxicities of the combinations reflected no dramatic interactions when compared to the toxicities of each single component of the mixtures. They stated that the toxicity of each component was independent and additive and no synergism was detected. More studies utilizing a broader range of defined pollutant mixes should be done to explore more fully the possibility of pollutant-pollutant interaction and possible synergism.

D. POLLUTANT–NATURAL DISEASE INTERACTION

Pollutant–disease interaction is one of the least studied phenomena in estuarine pollution ecology. There are numerous examples in man and other vertebrates of pollutant stress leading to increased susceptibility to natural pathogens such as viruses, bacteria, and noninfectious diseases such as neoplasia and respiratory dysfunction. There is growing evidence that some fish diseases are enhanced by certain pollutants (Couch *et al.*, 1974b).

Penaeid shrimps have a virus disease caused by a *Baculovirus* that is enhanced by a chemical pollutant, Aroclor 1254 (Couch, 1974; Couch and Courtney, 1977). This virus–shrimp system has been studied in order to test its suitability as a model indicator system for determining the degree of influence that sublethal concentrations of pollutant chemicals may have on a natural virus–host complex. The pink shrimp (*P. duorarum*) has proven to be a natural host for the shrimp-specific *Baculovirus* originally described as a result of findings made on experimentally stressed penaeid shrimps (Couch, 1974). A captive population of pink shrimp with relatively low initial virus prevalence stressed by 1–3 μg/liter Aroclor 1254 (PCB) for up to 30 days experienced a significantly more rapid spread and increased prevalence of the *Baculovirus* than did a control, nonstressed population with equal, initial viral prevalence (Couch and Courtney, 1977). The mortality in the Aroclor–*Baculovirus*-stressed population was greater than in the control population. At the end of the test, prevalence of virus infections in the Aroclor-stressed population was approximately 50% more than in the control population and intensity of individual virus infections was greater in the Aroclor-stressed shrimp.

The ecological implications for such findings are suggestive of phenomena in nature that to date have not been evaluated. If population immunity or resistance to natural pathogens were reduced in dense populations of Crustacea by sublethal exposures to specific pollutants, then natural disease, aided and abetted by human pollutant activity, could cause subtle but high mortality over longer periods of time and gradually contribute to an alteration of the ecological balance among certain susceptible species. Whether or not this happens in nature is unknown, but field studies should be conducted to compare disease prevalences in aquatic species in polluted versus nonpolluted regions.

VII. Discussion

Penaeid shrimps are valuable, desirable component species of estuarine ecosystems. The well-being of penaeids could reflect the general quality of their immediate estuarine or offshore environments. Penaeids, for example, are generally more sensitive to some chemical pesticides than are mollusks or most fishes. The fact that species of penaeids are found world-wide in the temperate, subtropic, and tropic regions indicates that they could be studied in these regions as indicators of estuarine health.

There are several characteristics of penaeids, however, that must be taken into consideration before they are used as choice indicator species in pollutant studies. These characteristics are (1) a limited north–south distribution beyond the temperate zones; (2) open ocean littoral and estuarine habitat requirements of life cycle could confuse identity of source of pollutant effects if population fluctuations were considered as criteria for effect; (3) culture of penaeids is difficult and, therefore, laboratory egg to egg or life cycle studies are difficult; and (4) penaeids are less sensitive to some specific pollutants than are other estuarine fishes and invertebrates.

The need to better understand the interaction of physical, chemical, and biological factors and pollutants in the ecology of penaeid shrimps and other estuarine species should be emphasized. Mortalities of penaeid shrimps and most other crustacean species are rarely detected in nature, except occasionally after-the-fact, as reflected in poorer commercial catches or in population changes. There is a need to develop monitoring and sentinel methods that will permit early detection of injury or damage to estuarine species or alteration of species composition. Without these early-warning methods little can be done to predict and then ameliorate or control the impact of estuarine pollution short of complete halting of pollutant discharge, which seems unlikely with present technology.

References

Anderson, J. W., Neff, J. M., Cox, B. A., Tatem, H. E., and Hightower, G. M. (1974a). Characteristics of dispersions and water-soluble extracts of crude and refined oils and their toxicity to estuarine crustaceans and fish. *Mar. Biol.* **27,** 75–88.

Anderson, J. W., Neff, J. M., Cox, B. A., Tatem, H. E., and Hightower, G. M. (1974b). The effects of oil on estuarine animals: Toxicity, uptake, depuration, and respiration. *In* "Pollution and Physiology of Marine Organisms" (F. J. Vernberg and W. B. Vernberg, eds.), pp. 285–310. Academic Press, New York.

Bahner, L. H. (1975). Mobilization of cadmium in the tissue of pink shrimp, *Penaeus duorarum*. *In* "Program of the First Workshop on the Pathology and Toxicology of Penaeid Shrimps," p. 8. U.S. EPA, Gulf Breeze, Florida.

Bahner, L. H., and Nimmo, D. R. (1976). Metals, pesticides, and PCB's: Toxicities to shrimp singly and in combination. *In* "Estuarine Processes" (M. Wiley, ed.), Vol. 1, pp. 523–532. Academic Press, New York.

Bazin, F., Monsarrat, P., Bonami, J. R., Croizier, G., Meynadier, G., Quiot, J. M., and Vago, C. (1974). Particules virales de type baculovirus observées chez le crabe *Carcinus maenas*. *Rev. Trav. Inst. Peches Marit.* **38,** 205–208.

Broadhurst, M. G. (1972). Use and replaceability of PCB's. *Environ. Health Perspect.* **2,** 81–102.

Butler, P. A. (1966). The problems of pesticides in estuaries. *In* "A Symposium on Estuarine Fisheries," Publ. No. 3 pp. 110–115. Am. Fish. Soc.

Butler, P. A. (1973). Organochlorine residues in estuarine molluscs, 1965–1972. National Pesticide Monitoring Program. *Pestic. Monit. J.* **6,** 238–362.

Butler, P. A., and Springer, P. F. (1963). Pesticides—a new factor in coastal environments. *Trans. North Am. Wildl. Nat. Resour. Conf.* **28,** 378–390.

Butler, P. A., Childress, R., and Wilson, A. J., Jr. (1970). The association of DDT residues with losses in marine productivity. *FAO Tech. Conf. Mar. Pollut. Effects Living Resour. Fishing (Rome)* pp. 1–13.

Charter, D. B., Sutherland, R. A., and Porricelli, J. D. (1973). Quantitative estimates of petroleum to the oceans. *In* "Workshop on Inputs, Fates and Effects of Petroleum in the Marine Environment," Vol. 1, pp. 7–30. Ocean Affairs Board, Natl. Acad. Sci.—Natl. Res. Counc., Washington, D.C.

Chin, E., and Allen, D. M. (1957). Toxicity of an insecticide to two species of shrimp *Penaeus aztecus* and *Penaeus setiferus*. *Tex. J. Sci.* **9,** 270.

Conte, F. S., and Parker, J. C. (1975). Effects of aerially-applied malathion on juvenile brown and white shrimp *Penaeus aztecus* and *P. setiferus*. *Trans. Am. Fish. Soc.* **104,** 793–799.

Coppage, D. L., and Matthews, E. (1974). Short term effects of organophosphate pesticides on cholinesterases of estuarine fishes and pink shrimp. *Bull. Environ. Contam. Toxicol.* **11,** 483–488.

Couch, J. A. (1974). An enzootic nuclear polyhedrosis virus of pink shrimp: Ultrastructure, prevalence, and enhancement. *J. Invertebr. Pathol.* **24,** 311–331.

Couch, J. A. (1975). Evaluation of the exposure of fish and wildlife to nuclear polyhedrosis and granulosis viruses (Discussant). *In* "Baculoviruses for Insect Pest Control: Safety Considerations" (M. D. Summers *et al.*, eds.), pp. 111-113. Am. Soc. Microbiol., Washington, D.C.

Couch, J. A. (1977). Ultrastructural study of lesions in gills of a marine shrimp exposed to cadmium. *J. Invertebr. Pathol.* **29**, 267-288.

Couch, J. A. (1978). Diseases, parasites, and toxic responses of commercial penaeid shrimps of the Gulf of Mexico and South Atlantic Coasts of North America. *Fish. Bull.* **76**, 1-44.

Couch, J. A., and Courtney, L. (1977). Interaction of chemical pollutants and a virus in a crustacean: A novel bioassay system. *Ann. N.Y. Acad. Sci.* **298**, 497-504.

Couch, J. A., and Nimmo, D. R. (1974a). Ultrastructural studies of shrimp exposed to the pollutant chemical polychlorinated biphenyl (Aroclor 1254). *Bull. Soc. Pharm. Ecol. Pathol.* **2**, 17-20.

Couch, J. A., and Nimmo, D. R. (1974b). Detection of interactions between natural pathogens and pollutants in aquatic animals. *Proc. Gulf Coast Reg. Symp. Dis. Aquatic Anim.* LSU Publ. LSU-SG-74-05, pp. 261-268.

Courtenay, W. R., and Roberts, M. H., Jr. (1973). Environmental effects on toxaphene toxicity to selected fishes and crustaceans. *U.S. Environ. Prot. Agency, Off. Res. Dev.* [Rep.] EPA **EPA-R3-73-035.**

Duke, T., Lowe, J. I., and Wilson, A. J., Jr. (1970). A polychlorinated biphenyl (Aroclor 1254) in the water, sediment, and biota of Escambia Bay, Florida. *Bull. Environ. Contam. Toxicol.* **5**, 171-180.

Green, F. A., Anderson, J. W., Petrocelli, S. R., Presley, B. J., and Sims, R. (1976). Effect of mercury on the survival, respiration, and growth of postlarval white shrimp, *Penaeus setiferus*. *Mar. Biol.* **37**, 75-81.

Hansen, D., Parrish, P. R., and Forester, J. (1974a). Aroclor 1016: Toxicity to and uptake by estuarine animals. *Environ. Res.* **7**, 363-373.

Hansen, D., Schimmel, S. C., and Matthews, E. (1974b). Avoidance of Aroclor 1254 by shrimp and fishes. *Bull. Environ. Contam. Toxicol.* **12**, 253-256.

Haven, D., Castagna, M., Chanley, P., Wass, M., and Whitcomb, J. (1966). Effects of the treatment of an oyster bed with polystream and Sevin. *Chesapeake Sci.* **7**, 179-188.

Heimpel, A. M., and Buchanan, L. K. (1967). Human feeding tests using a nuclear-polyhedrosis virus of *Heliothis zea*. *J. Invertebr. Pathol.* **9**, 55-57.

Johnson, P. T. (1976). A *baculovirus* from the blue crab, *Callinectes sapidus*. *Proc. Int. Colloq. Invertebr. Pathol., 1st, 1976*, p. 24.

Jones, M. B. (1975). Synergistic effects of salinity, temperature, and heavy metals on mortality and osmoregulation in marine and estuarine isopods (crustacea). *Mar. Biol.* **30**, 13-20.

Lightner, D. V., Proctor, R. R., Sparks, A. K., Adams, J. R., and Heimpel, A. M. (1973). Testing penaeid shrimp for susceptibility to an insect nuclear polyhedrosis virus. *Environ. Entomol.* **2**, 611-613.

Lowe, J. I., Parrish, P. R., Wilson, A. J., Jr., Wilson, P. D., and Duke, T. W. (1971). Effects of mirex on selected estuarine organisms. *Trans. North Am. Wildl. Nat. Resour. Conf.* **36**, 171-186.

Mandelli, E. F. (1971). A study of the effect of desalination plant effluents on marine benthic organisms. *U.S. Off. Saline Water, Res. Dev. Prog. Rep.* **803.**

Markin, G. P., Hawthorne, J. C., Collins, H. L., and Ford, J. H. (1974). Levels of mirex and some other organochlorine residues in seafood from Atlantic and Gulf coastal states. *Pestic. Monit. J.* **7**, 139-143.

Miura, T., and Takahashi, R. M. (1973). Insect developmental inhibitors. Effects on non-target aquatic organisms. *J. Econ. Entomol.* **66**, 917.

Neff, J. M., Cox, B. A., Dixit, D., and Anderson, J. W. (1976a). Accumulation and release of petroleum-derived aromatic hydrocarbons by four species of marine animals. *Mar. Biol.* **38**, 279-289.

Neff, J. M., Anderson, J. W., Cox, B. A., Laughlin, R. B., Jr., Rossi, S. S., and Tatem, H. E. (1976b). Effects of petroleum on survival, respiration, and growth of marine animals. *In* "Sources, Effects, and Sinks of Hydrocarbons in the Aquatic Environment," pp. 515–540. American University, Washington, D.C.

Nimmo, D. R., and Bahner, L. H. (1974). Some physiological consequences of polychlorinated biphenyl and salinity stress in penaeid shrimp. *In* "Pollution and Physiology of Marine Organisms" (F. J. Vernberg and W. B. Vernberg, eds.), p. 427. Academic Press, New York.

Nimmo, D. R., and Blackman, R. R. (1972). Effects of DDT on cations in the hepatopancreas of penaeid shrimp. *Trans. Am. Fish. Soc.* **101**, 547–549.

Nimmo, D. R., Wilson, A. J., Jr., and Blackman, R. R. (1970). Localization of DDT in body organs of pink and white shrimp. *Bull. Environ. Contam. Toxicol.* **5**, 333–341.

Nimmo, D. R., Wilson, P. D., Blackman, R. R., and Wilson, A. J., Jr. (1971). Polychlorinated biphenyls absorbed from sediments by fiddler crabs and pink shrimp. *Nature (London)* **231**, 50–52.

Nimmo, D. R., Hansen, D. J., Couch, J. A., Cooley, N. R., Parrish, P. R., and Lowe, J. T. (1975). Toxicity of Aroclor 1254 and its physiological activity in several estuarine organisms. *Arch. Environ. Contam. Toxicol.* **3**, 22–39.

Nimmo, D. R., Lightner, D. V., and Bahner, L. H. (1977). Effects of cadmium on the shrimps *Penaeus duorarum, Palaeomonetes pugio*, and *Paleomonetes vulgaris*. *In* "Physiological Responses of Marine Biota to Pollutants" (F. J. Vernberg *et al.,* eds.), p. 131. Academic Press, New York.

Parrish, P. R., Couch, J. A., Forester, J., Patrick, J. M., Jr., and Cook, G. H. (1973). Dieldrin: Effects on several estuarine organisms. *Proc. 27th Annu. Conf. Southeast. Assoc. Game Fish Comm.* pp. 427–434.

Parrish, P. R., Cook, G. H., and Patrick, J. M., Jr. (1974). Hexachlorobenzene: Effects on several estuarine animals. *Proc. 28th Annu. Conf. Southeast. Assoc. Game Fish Comm.* pp. 179–187.

Parrish, P. R., Schimmel, S. C., Hansen, D. J., Patrick, J. M., Jr., and Forester, J. (1976). Chlordane: Effects on several estuarine organisms. *J. Toxicol. Environ. Health* **1**, 485–494.

Petrocelli, S. R., Roseijada, G., Anderson, J. W., Presley, B. J., and Sims, R. (1975). Brown shrimp exposed to inorganic mercury in the field. *In* "Program of the First Workshop on the Pathology and Toxicology of Penaeid Shrimps," p. 1. U.S. EPA, Gulf Breeze, Florida.

Schimmel, S. C., Parrish, P. R., Hansen, D. J., Patrick, J. M., Jr., and Forester, J. (1974). Endrin: Effects on several estuarine organisms. *Proc. 28th Annu. Conf. Southeast. Assoc. Game Fish Comm.* pp. 187–194.

Schimmel, S. C., Patrick, J. M., Jr., and Forester, J. (1976). Heptachlor: Toxicity to and uptake by several estuarine organisms. *J. Toxicol. Environ. Health* **1**, 955–965.

Schimmel, S. C., Patrick, J. M., and Forrester, J. (1977). Toxicity and bioconcentration of BHC and Lindane in selected estuarine animals. *Arch. Environ. Contam. toxicol.* **6**, 355–363.

Siddall, J. B. (1976). Insect growth regulators and insect control: A critical appraisal. *Environ. Health Perspect.* **14**, 119–126.

Sodergren, A., Svensson, B. J., and Ulfstrand, S. (1972). DDT and PCB in south swedish streams. *Environ. Pollut.* **3**, 25–36.

Soloway, S. B. (1976). Naturally occurring insecticides. *Environ. Health Perspect.* **14**, 109–118.

Steed, D. L., and Copeland, B. J. (1967). Metabolic responses of some estuarine organisms to an industrial effluent. *Contrib. Mar. Sci.* **12**, 143.

Summers, M. D., Engler, R., Falcon, L. A., and Vail, P. (1975). Baculoviruses for Insect Pest Control: Safety Considerations." Am. Soc. Microbiol., Washington, D.C.

Tagatz, M. E., Borthwick, P. W., Cook, G. H., and Coppage, D. L. (1974). Effects of ground applications of malathion on salt-marsh environments in Northwestern Florida. *Mosq. News* **34**, 309–315.

Tagatz, M. E., Borthwick, P. W., Ivey, J. M., and Knight, J. (1976). Effects of leached mirex on experimental communities of estuarine animals. *Arch. Environ. Contam. Toxicol.* **4,** 435–442.

Tagatz, M. E., Ivey, J. M., and Lehman, H. K. (1979). Effects of sevin on development of experimental estuarine communities. *J. Toxicol. Environ. Health* **5,** 643–651.

Venkatarmiah, A., Lakshmi, G. J., Biesiot, P., Valleau, J. D., and Gunter, G. (1977). "Studies on the Time Course of Salinity and Temperature Adaptation in the Commercial Brown Shrimp *Penaeus aztecus* Ives. U. S. Army Corps of Engineers, Contract No. DACW 39- 73C-0115. U. S. Govt. Printing Office, Washington, D. C.

Vernberg, W. B., and Vernberg, J. (1972). The synergistic effects of temperature, salinity, and mercury on survival and metabolism of the adult fiddler crab, *Uca pugilator. Fish. Bull.* **70,** 415–420.

Williams, A. (1965). Marine decapod crustaceans of the Carolinas. *Fish. Bull.* **65,** 1–298.

Wood, J. M. (1974). Biological cycles for toxic elements in the environment. *Science* **183,** 1049–1052.

Wright, J. E. (1976). Environmental and toxicological aspects of insect growth regulators. *Environ. Health Perspect.* **14,** 127–132.

Yarbrough, J. D., and Minchew, D. (1975). Histological changes in the shrimp related to chronic exposure to crude oil. *In* "Program of the First Workshop on the Pathology and Toxicology of Penaeid Shrimps," p. 5. U.S. EPA, Gulf Breeze, Florida (abstr.).

CHAPTER 8

Larval Decapods (Arthropoda: Crustacea: Decapoda)

CHARLES E. EPIFANIO

I. Introduction

During the past two decades there have been many studies of effects of pollutants upon marine organisms. Results of these experiments have been important in furthering understanding of the consequences of environmental perturbation and have led to restriction in the use of many biologically active compounds, e.g., chlorinated hydrocarbons. More recently, however, scientists have become aware that conclusions concerning effects of pollutants on marine ecosystems have been based largely upon studies of juvenile and adult organisms. This has led in the last 5 years to an increasing number of studies of larval forms.

Work involving larval forms of decapod crustaceans has been especially fruitful because of highly developed techniques for culture and well-defined phases of development, i.e., discrete molt stages. The number of studies of effects of pollutants such as pesticides, petroleum oil, and heavy metals is great enough that some generalized conclusions can be drawn.

A. GENERAL BIOLOGY—HISTORICAL

There has been only one comprehensive review of the general biology of decapod larvae, and that was written over 25 years ago (Gurney, 1942). In that work Gurney brought together considerable information on the morphology, taxonomy, and systematics of these larval forms. He pointed out that a large amount of confusion surrounded very early studies. In fact, the most notable zoologists of the early nineteenth century steadfastly refused to conclude any connection between adult decapods and what are now known to be larval decapods. The terms ''zoea'' and ''megalopa,'' presently descriptors of larval stages of brachyuran crabs, were considered genera during that period (*Zoea* attributed to Bosc in 1802 and *Megalopa* to Leach in 1813). This of course led to some systematic difficulty and, according to Gurney, the animals were classified variously as ''Oecapodes douteux'' by Milne-Edwards, branchiopods by Bosc and Lamarck, and podophthalmata by Leach.

This confusion is somewhat surprising as, again according to Gurney, Cavolini published a figure in 1787 ''showing quite clearly the general form of the Brachyuran Zoea at the moment before hatching.'' Furthermore, Thompson (1828, from Gurney, 1942) concluded that all species of the genus *Zoea* were truly larval brachyurans and that metamorphosis is the rule among decapods. This was strongly contested by Westwood (1835, from Gurney, 1942), who relied on evidences of direct development in freshwater crayfish, and by Rathke (1838, from Gurney, 1942), who eventually acknowledged his error publically.

The whole area became one of lively controversy and the Societe Hollandaise des Sciences de Haarlem in 1840 offered a prize for research aimed at determining whether metamorphosis existed among crabs (Gurney, 1942). Finally Couch (1845, from Gurney, 1942) observed hatching in several species of brachyruans and macrurans, carefully set forth figures, and effectively settled the matter.

Systematic and morphological studies of decapod larvae continued through the nineteenth century with important milestones, including (1) the discovery by Muller (1892) that penaeid shrimp passed through naupliar and protozoeal stages as well as zoeal stages; (2) the description by Faxon (1879) of a progression of zoeal stages of *Palaemonetes vulgaris* (Say) 1818 (Holthius, 1949); and (3) the comparative morphological work of Claus (1876, from Gurney, 1942). Notable morphological studies in the present century include the works of Williamson (1910), Caroli (1918), Sollaud (1924), and Lebour (1925).

B. General Biology—Life Cycles

Four fundamental larval forms are evidenced among the decapods: the nauplius, the protozoea, the zoea, and the postlarva. There appears to have been an evolutionary trend toward compression of the life cycle and only the more primitive decapod families pass through all four phases outside the egg (Gurney, 1942).

Furthermore, morphological expression of the zoeal and postlarval stages is not uniform among the decapod groups and while the zoeae of the natantins and the brachyuran and anomuran reptantians are morphologically similar, zoeal expression as mysis or phyllosoma larvae among the macruran reptantians represents a radical morphological departure.

Postlarval stages are seen in all decapod groups with glaucothea and megalopa of the anomurans and brachyurans most distinctive and the postlarvae of the natantins and macrurans less morphologically distinct. Table I summarizes typical life cycles of the decapod taxa.

C. Culture Techniques

Although techniques for culturing larvae of lobsters of the genus *Homarus* were well developed by the beginning of the twentieth century (Barnes, 1911), the lack of suitable foods made culture of most other decapod larvae very difficult. Early work at Plymouth with natural plankton diets yielded chance successes but it was not until the discovery that commercially available nauplii of the branchiopod *Artemia salina* (Linne) were an excellent food that much progress was made in the area (Gurney, 1942).

Culture of larval decapods has been brought to a fine art in the laboratory of J. D. Costlow, Jr., and C. G. Bookhout (Costlow and Bookhout, 1959), and virtually all present techniques are based upon their work. Most commonly used methods involve holding gravid females in clean, filtered seawater under specified conditions of salinity and temperature until hatching occurs. [Alternatively, eggs can be removed from the female and cultured *in vitro* until hatching (Costlow and Bookhout, 1960).] Larvae are then placed in finger bowls (0.6 larvae per milliliter) and fed freshly hatched *Artemia salina* nauplii at (5 nauplii per milliliter) (Mootz and Epifanio, 1974). Water is changed daily, fresh *A. salina* added, and the number of living and newly molted larvae determined. Dead larvae are removed. This procedure, of course, allows tabulation of

TABLE I

Types of Postembryonic Development and Larvae in Decapods[a]

Group	Postembryonic development	Larvae
Suborder Natantia		
Family Penaeidae	Slightly metamorphic	Nauplius → protozoea → mysis → mastigopus (zoea) (postlarva)
Family Sergestidae	Metamorphic	Nauplius → elaphocaris → acanthosoma → mastigopus (protozoea) (zoea) (postlarva)
Section Caridea	Metamorphic	Protozoaea → zoea → parva (protozoea) (postlarva)
Section Stenopodidea	Metamorphic	Protozoea → zoea → postlarva
Suborder Reptantia		
Section Macrura		
Superfamily Scyllaridea	Metamorphic	Phyllosoma → puerulus, nisto, or pseudibaccus (zoea) (postlarva)
Superfamily Nephropsidea	Slightly metamorphic	Mysis → postlarva (zoea)
Section Anomura	Metamorphic	Zoea → glaucothöe in pagurids, grimothea (postlarva)
Section Brachyura	Metamorphic	Zoea → megalopa (postlarva)

[a] From Waterman and Chase (1960), reprinted by permission.

mortality and molt-frequency data. Upon molting to megalopa, larvae are separated to individual containers and fed freshly hatched *A. salina* nauplii until molting to first crab.

Under conditions when detailed observations of individual larvae are necessary, larvae are reared individually in plastic boxes divided into compartments. This allows collection of exuvia from newly molted larvae for morphological studies.

Recent innovations in culture technology include utilization of alternate foods for species whose early larvae are too small to ingest *Artemia salina* nauplii (Sulkin and Epifanio, 1975) and development of devices for automated culture of large numbers of larvae (Rice and Williamson, 1970).

II. Effects of Pollutants

A. PESTICIDES

1. General

The earliest insecticides were organic materials of natural origin. For example, there are records of use of a pyrethrum-like material in China during the Middle Ages and the compound is said to have been brought to Europe in the thirteenth century by Marco Polo. By the nineteenth century, organic compounds such as various highly refined oils and ground tobacco were commonly used as insecticides in Europe. Inorganic compounds such as lime washes, HCN, and various salts were also in common use. Synthetic organic pesticides began to be developed in the late nineteenth century. Dinitrophenols were used in Germany as early as 1892, and naphthalene, paradichlorobenzene, thiocyanate, and cyclohexylamines were developed in the early twentieth century. The use of pesticides up to that point in history, however, was of such small magnitude as to give no broad ecological effects (U.S. Department of Health, Education and Welfare, 1969).

Discovery of the insecticidal properties of DDT in Switzerland in 1939 marked the beginning of an era of extremely widespread use of pesticides. Competition in the area of production of effective pesticides has led to existence of an extremely large number of compounds. (There were over 900 active pesticidal compounds registered in 1969 and these were formulated into over 60,000 preparations.) The majority can be classified as organochlorides, organophosphates, or carbamates. Organochlorides are often toxic to a wide variety of insects, are relatively cheap, and are refractory to chemical and physical degradation in the environment. It is this last characteristic, combined with a high solubility in lipids, that has led to accumulation of organochlorides in food webs. Organophosphates and carbamates are susceptible to hydrolytic degradation and are reduced to less toxic compounds in relatively short periods of time in the environment. While it is certain that all of the above classes of pesticides are neurotoxins, the exact mode of toxicological action of organochlorides is not known (O'Brien, 1967). Organophosphates and carbamates are cholinesterase inhibitors.

Recently there has been interest in development of compounds that mimic the action of juvenile hormones of insects. The insecticidal action of such compounds stems from the fact that insects exposed to the material are unable to metamorphose. These insecticides have not experienced widespread use as yet.

2. Chlorinated Hydrocarbons

The first study of effects of a chlorinated hydrocarbon on crab larvae was that of Bookhout and Costlow (1970), who showed that larvae of four species of crabs displayed reduced survival and developmental anomalies when fed *Artemia salina* nauplii from Great Salt Lake. Larvae of the same species developed normally when fed *A. salina* from brine ponds near San Francisco Bay. Subsequent gas chromatographic analyses indicated that nauplii from Great Salt Lake contained whole-body residues of 7.05 ppm DDT, while those from California contained only 2.3 ppm. The authors concluded that reduced survival and developmental abnormalities among crab larvae fed Utah *A. salina* may have been due to the high content of DDT (Table II).

A subsequent series of experiments (Epifanio, 1971, 1972, 1973) investigated relative toxicities of a chlorinated hydrocarbon, dieldrin (Table II), dissolved in seawater and in the food of crab larvae. The experiments were unique in that they encompassed the entire period of larval development of two species of crabs, *Leptodius floridanus* (Gibbes)* (Rathbun, 1930) and *Panopeus herbstii* Milne-Edwards (Rathbun, 1930). One experiment (Epifanio, 1971) was designed to test the effect of dieldrin on individual zoeal stages of *L. floridanus*. One group was continuously exposed to dieldrin in seawater (1 ppb) from hatching to the postlarval stage (megalopa), while a second group was not exposed until they reached zoea stage II. A third group was not exposed until zoea stage III and fourth zoea stage IV. All groups were reared to the megalopal stage.

Survival was affected only in treatment groups that were first exposed to dieldrin in either zoea stages I or II. Dieldrin did not affect survival in groups not exposed to the pesticide until stages III or IV. It was concluded that there appear to be certain larvae that are genetically sensitive to dieldrin, and if these larvae are exposed to 1 ppb dieldrin during zoea stage I, they are unable to molt to zoea

---

Leptodius floridanus (Gibbes, 1850) is the type-species of a new genus, *Cataleptodius* Guinot, 1968 created for reception of five Atlantic–east Pacific species (see Guinot, 1968). *Cataleptodius* is used by Williams and Duke in Chapter 6 of this volume.

TABLE II

PESTICIDES USED IN STUDIES WITH DECAPOD LARVAE

Common name	Chemical name	Study
DDT	Dichlorodiphenyltrichloroethane	Bookhout and Costlow (1970)
Dieldrin	Hexachloroepoxyoctahydroendo, exo-dimethanonaphthalene	Epifanio (1971, 1972, 1973)
Mirex	Dodecachlorooctahydro-1,3,4-metheno-14-cyclobuta[cd]methalene	Bookhout *et al.* (1972), Bookhout and Costlow (1975)
Methoxychlor	1,1,1-Trichloro-2,2-bis-(p-methoxyphenyl)-ethane	Bookhout *et al.* (1976)
Sevin	1 Naphthyl-N-methylcarbamate	Buchanan *et al.* (1970)
Malathion	O,O-Dimethyl phosphorodithionate of diethyl mercapto succinate	Bookhout and Monroe (1977)
Fenitrothion	O,O-Dimethyl O-(4-nitro-m-tolyl) phosphorothioate	McLeese (1974)
Methoprene	Isopropyl 11-methoxy-3,7,11-trimethyl-dodeca-2,4-dienoate	Christiansen *et al.* (1977a)
Hydroprene	Ethyl 3,7,11-trimethyldodeca-2,4-dienoate	Christiansen *et al.* (1977b)

stage II, while those larvae that are not highly sensitive to dieldrin survive as well as control larvae. The toxicity of dieldrin to *Leptodius floridanus* larvae appears more dependent on stage of development at exposure than on the length of exposure (Table III).

Larvae of *Panopeus herbstii* were also most sensitive during zoea stages I and II, but survivorship was not affected until concentrations reached 5 ppb. Neither species was able to complete development at concentrations of 10 ppb or above.

Another aspect of the toxicity to dieldrin of *Leptodius floridanus* and *Panopeus herbstii* was the increase in the duration on intermolt periods of larvae exposed to the pesticide. This effect was seen even at concentrations that did not cause elevated mortality among the larvae, e.g., 0.5 ppb with *L. floridanus* and 1.0 ppb with *P. herbstii*. It is interesting that this same response had been observed in the earlier study of Buchanan *et al.* (1970) with sevin and *Cancer magister* Dana (Rathbun, 1930) larvae. Those authors suggested that the toxicity of sevin to *C. magister* was associated with the molting process. Since molting in crab larvae is known to be a neuroendocrine function (Costlow, 1963, 1966a,b) and since both dieldrin and sevin are neurotoxins, it might be expected that one aspect of the sublethal toxicity of these compounds would be interference with the molting process. Costlow *et al.* (1960, 1962), however, have shown that stress salinities delay molting in larvae of *Sesarma cinerum* (Bosc) (Rathbun, 1918), *Panopeus herbstii*, and *Rithropanopeus harrisii* (Gould) (Rathbun, 1930).

More recent studies have shown that many types of pesticides, various petroleum hydrocarbons, and some heavy metals also affect the molting process in

TABLE III

Summary of Effects of Three Chlorinated Hydrocarbons on Larvae of Five Species of Crabs[a]

Pesticide	Concentration (ppb)	Species	Sublethal	Acute	Sensitive stage	Delay of development
Dieldrin[b]	0.5	*Leptodius floridanus*		Insignificant mortality		X
Dieldrin	1.0	*Leptodius floridanus*	X		I	X
Dieldrin	5.0	*Leptodius floridanus*		X	I	—[f]
Dieldrin	1.0	*Panopeus herbstii*		Insignificant mortality		X
Dieldrin	5.0	*Panopeus herbstii*		X	I	—[f]
Mirex[c]	0.01	*Rhithropanopeus harrisii*	X		III, IV	X
Mirex	0.1	*Rhithropanopeus harrisii*	X		III, IV	X
Mirex	1.0	*Rhithropanopeus harrisii*		X	I, II	X
Mirex	10.0	*Rhithropanopeus harrisii*		X	I, II	X
Mirex	0.01	*Menippe mercenaria*	X		III, IV	—[g]
Mirex	0.1	*Menippe mercenaria*	X		III, IV	—[g]
Mirex	1.0	*Menippe mercenaria*		X	I, II	—[g]
Mirex	10.0	*Menippe mercenaria*		X	I, II	—[g]
Mirex[d]	0.01	*Callinectes sapidus*	X		III	—[g]

266

Compound		Species				—[g]
Mirex	0.1	*Callinectes sapidus*	X		III, VII, M	X
Mirex	1.0	*Callinectes sapidus*		X	II, III	X
Mirex	10.0	*Callinectes sapidus*		X	I, II	X
Methoxychlor[e]	1.0	*Rhithropanopeus harrisii*		Insignificant mortality		X
Methoxychlor	2.5	*Rhithropanopeus harrisii*	X		I	X
Methoxychlor	4.0	*Rhithropanopeus harrisii*	X		I, II, III	X
Methoxychlor	5.5	*Rhithropanopeus harrisii*	X		I, II	X
Methoxychlor	7.0	*Rhithropanopeus harrisii*		X	I, II	X
Methoxychlor	0.7	*Callinectes sapidus*	X		II, III	X
Methoxychlor	1.0	*Callinectes sapidus*	X		II, III	X
Methoxychlor	1.3	*Callinectes sapidus*		X	I, II	X
Methoxychlor	1.6	*Callinectes sapidus*		X	I, II	X
Methoxychlor	1.9	*Callinectes sapidus*		X	I, II	X

[a] Roman numerals refer to zoeal stages, and M refers to megalopa. The symbol X is used as a positive indicator, e.g., an X in the Acute column means that the toxicity was acute.

[b] Epifanio (1971).

[c] Bookhout *et al.* (1972).

[d] Bookhout and Costlow (1975).

[e] Bookhout *et al.* (1976).

[f] No survival to second larval stage.

[g] No effect on duration of intermolt periods, but increased occurrence of extra zoeal stage.

267

crab larvae (see below). Possibly this delay in molting is a generalized response of crab larvae to any environmental stress. In any event, the response allows a very sensitive bioassay for sublethal effects of a variety of pollutants.

In a later study (Epifanio, 1972), I compared toxicity of dieldrin-contaminated food with that already shown for dieldrin in seawater. Contaminated food was produced by hatching *Artemia salina* eggs in several concentrations of dieldrin in seawater, allowing the nauplii to remain in the contaminated suspension for 36 hours, and washing the nauplii in filtered seawater before use as food. There was no effect on *Leptodius floridanus* when fed nauplii containing concentrations of 0.15 or 2.9 ppm dieldrin, but the larvae exhibited an increase in intermolt period when fed nauplii containing concentrations of 5.49 ppm dieldrin. No *L. floridanus* larvae completed development when fed nauplii containing 33.0 ppm dieldrin.

Clearly, then, dieldrin dissolved in seawater is toxic at much lower concentrations (parts per billion) than when in the food (parts per million) of crab larvae. I hypothesized that this is related to the relative pumping and feeding rates of the larvae. The pumping rate (mass H_2O per day) of a crab larvae is undoubtedly much higher than the feeding rate (mass food per day) and so the absolute amount of dieldrin available to larvae from minute concentrations in seawater is greater than from higher concentrations in food. This hypothesis was corroborated by another study (Epifanio, 1973), where I showed that the rate of uptake of [^{14}C]dieldrin from seawater was much higher than uptake from food (Fig. 1).

Subsequent studies have emphasized the effects of pesticides dissolved in seawater. Bookhout *et al.* (1972) studied the sublethal toxicity of the chlorinated hydrocarbon, mirex (Table II), to larvae of *Rhithropanopeus harrisii* and *Menippe mercenaria* (Say) (Rathbun, 1930). The criterion for sublethal toxicity was a concentration in which there is a differential survival in relation to controls but where at least 10% of the larvae reach crab stage I; those concentrations yielding less than 10% survival to crab stage I were considered lethal (Epifanio, 1971). Bookhout *et al.* showed that concentrations of 0.01 and 0.1 ppb mirex were sublethal to both species while no *R. harrisii* larvae survived beyond megalopa in 10.0 ppb. No *M. mercenaria* larvae developed beyond the megalopa in 1.0 ppb or beyond the zoea stage III in 10.0 ppb mirex.

As in earlier studies with pesticides, mirex increased the zoeal intermolt periods as well as total time of development to crab stage in *Rhithropanopeus harrissi,* but surprisingly there were no differences in duration of developmental stages in relation to mirex with *Menippe mercenaria*. Instead, an increased number of extra zoeal stages were noted. *M. mercenaria* generally passes through five zoeal stages before molting to the megalopal stage. The occurrence of zoea stage VI in control conditions was rare (2.5%) but increased to 65% in 0.01 ppm mirex, 66% in 0.1 ppb, and 90% in 1.0 ppb. The majority of stage VI zoea died before reaching the megalopal stage.

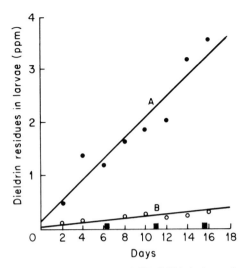

Fig. 1. *Leptodius floridanus.* Rates of uptake of [^{14}C]dieldrin by larvae from 0.5 ppb dieldrin in water (A), 0.213 ppm in food (B), from time of their hatching (Time 0). Regression lines were calculated by least-squares analysis. Reference marks on X axis indicate mean days of zoeal molts. (From Epifanio, 1973, reproduced by permission.)

Bookhout and Costlow (1975) continued studies of mirex by observing effects of 0.01, 0.1, 1.0, and 10.0 ppb in seawater on development of the blue crab *Callinectes sapidus* Rathbun (Rathbun, 1896). Results were similar to those with *Rhithropanopeus harrisii* and *Menippe mercenaria* in that sublethal effects were noted below 1.0 ppb. In all concentrations mortality was not evident for the first 5 days of development. Thereafter, mortality increased markedly in 1.0 and 10.0 ppb and rapidly approached 100%. Differential mortality with respect to controls was observed after 10 to 15 days in 0.01 and 0.1 ppb mirex. Mortality was greatest in zoeal stages III and VII and megalopa in 0.1 ppb, stages II and III in 1.0 ppb, and stages I and II in 10.0 ppb.

In contrast to results with *Rhithropanopeus harrisii,* sublethal concentrations of mirex did not affect duration of larval development of *Callinectes sapidus* at 0.01 and 0.1 ppb. Very few *C. sapidus* larvae completed development in 1.0 ppb and none in 10.0 ppb. Significant numbers of larvae passed through several zoeal stages before dying, however, and it was noted that mirex did increase duration of zoeal intermolt periods at these concentrations.

Bookhout and Costlow also measured the amount of mirex taken from the seawater and stored by *Callinectes sapidus* larvae (Table IV). It can be seen that concentrations of mirex residues increased with time of exposure and concentration of the pesticide in the water. Biological magnification, however, decreased with increasing concentration of mirex.

TABLE IV

Determination of Mirex in Crabs and Crab Larvae[a]

Sample identification by number	Sample identification treatment	Mirex detected (ppb)
1	First day larvae, seawater	<5
2	5-day larvae, acetone control	<5
3	15-day larvae, acetone control	<5
4	Megalopa, acetone control	<5
5	Crabs, 1st and 2nd, acetone control	<5
6	5-day larvae, 0.01 ppb mirex	11
7	15-day larvae, 0.01 ppb mirex	30
8	Megalopa, 0.01 ppb mirex	20
9	Crabs, 1st and 2nd, 0.01 ppb mirex	9
10	5-day larvae, .10 ppb mirex	33
11	15-day larvae, 0.10 ppb mirex	70
12	Megalopa, 0.10 ppb mirex	65
13	Crabs, 1st and 2nd, 0.10 ppb mirex	77
14	5-day larvae, 1.0 ppb mirex	301
15	8-day larvae, 1.0 ppb mirex	406
16	5-day larvae, 10.0 ppb mirex	1620
17	8-day larvae, 10.0 ppb mirex	1370

[a] From Bookhout and Costlow (1975), reprinted by permission.

Bookhout *et al.* (1976) conducted further studies of chlorinated hydrocarbons in an investigation of effects of methoxychlor (Table II) on development of *Rhithropanopeus harrisii* and *Callinectes sapidus*. *Rhithropanopeus harrisii* were reared in five concentrations ranging from 1.0 to 7.0 ppb. Survival to the first crab stage was 95.0% in 1.0 ppb, 79.5% in 2.5 ppb, 53.0% in 4.0 ppb, 19.0% in 5.5 ppb, and 0.5% in 7 ppb. Mortality was greatest in the first two zoeal stages at concentrations of 2.5 ppb and above. Increased duration of larval development was observed at all concentrations and the effect was greater at higher concentrations. *Rhithropanopeus harrisii* larvae accumulated considerable amounts of methoxychlor in their tissues over the period of larval development (Table V). Biological magnification was again highest at lowest concentrations of the pesticide in water but actual residues were highest in larvae reared at higher concentrations.

Callinectes sapidus were reared in methoxychlor concentrations ranging from 0.7 to 1.9 ppb. Survival to crab stage I was 53.5% in control conditions, 43.5% in 0.7 ppb, 26.5% in 1.0 ppb, 8.0% in 1.3 ppb, 2.0% in 1.6 ppb, and 0.5% in 1.9 ppb. Mortality was greatest in zoeal stages II and III at concentrations of 1.3

ppb and below. Mortality was highest in stages I and II above 1.3 ppb. As with *R. harrisii* there was a significant effect on duration of larval development at all concentrations. Also, uptake of methoxychlor was greater among *C. sapidus* larvae than *R. harrisii* larvae; residues in 10-day-old *C. sapidus* larvae in 1.0 ppb methoxychlor reached 2.87 ppm (Table VI).

Table III allows comparison of major effects of the three chlorinated hydrocarbons on five species of crab larvae. It can be seen that in almost every case sublethal toxicity was coupled with an increase in duration of larval development. This delay in molting was even noted among *Leptodius floridanus* and *Panopeus herbstii* larvae exposed to low dieldrin concentrations that caused no differential mortality relative to controls and has been observed among crab larvae under stress conditions of temperature and salinity in the absence of pesticides. It has been hypothesized that increased duration of intermolt period is associated with conditions of generalized stress and is not directly related to the action of the chlorinated hydrocarbon pesticide (Epifanio, 1971).

It is difficult, then, to explain the lack of effect of low concentrations of mirex (0.01 and 0.1 ppb) on duration of larval development of *Callinectes sapidus*. This is especially perplexing because these low concentrations do cause differen-

TABLE V

METHOXYCHLOR RESIDUES AND BIOLOGICAL MAGNIFICATION IN DEVELOPMENTAL STAGES OF *Rithropanopeus harrisii*[a]

Sample identification and treatment	Methoxychlor (ppm)	Biological magnification
5-day larvae, 1.0 ppb methoxychlor	<0.1	
10-day larvae, 1.0 ppb methoxychlor	0.40	400.0×
Megalopa, 1.0 ppb methoxychlor	<0.1	
1st crab, 1.0 ppb methoxychlor	<0.1	
5-day larvae, 2.5 ppb methoxychlor	0.56	224.0×
10-day larvae, 2.5 ppb methoxychlor	0.51	204.0×
Megalopa,. 2.5 ppb methoxychlor	1.12	448.0×
1st crab, 2.5 ppb methoxychlor	0.23	92.0×
5-day larvae, 4.0 ppb methoxychlor	Lost	
10-day larvae, 4.0 ppb methoxychlor	0.59	147.5×
Megalopa, 4.0 ppb methoxychlor	0.55	137.5×
1st crab, 4.0 ppb methoxychlor	0.76	190.0×
5-day larvae, 5.5 ppb methoxychlor	1.28	232.7×
1st crab, 5.5 ppb methoxychlor	0.55	100.0×

[a]From Bookhout *et al.* (1976), reprinted by permission.

TABLE VI

METHOXYCHLOR RESIDUES AND BIOLOGICAL MAGNIFICATION IN DEVELOPMENTAL
STAGES OF *Callinectes sapidus*

Sample identification and treatment	Methoxychlor (ppm)	Biological magnification
5-day larvae, 0.7 ppb methoxychlor	<0.1	
10-day larvae, 0.7 ppb methoxychlor	<0.1	
Megalopa, 0.7 ppb methoxychlor	0.51	728.6×
1st crab, 0.7 ppb methoxychlor	<0.1	
5-day larvae, 1.0 ppb methoxychlor	0.52	520.0×
10-day larvae, 1.0 ppb methoxychlor	2.87	2870.0×
1st crab, 1.0 ppb methoxychlor	0.15	150.0×
5-day larvae, 1.3 ppb methoxychlor	0.76	584.6×
13-day larvae, 1.3 ppb methoxychlor	2.62	2015.4×
1st crab, 1.3 ppb methoxychlor	0.34	261.5×
4-day larvae, 1.6 ppb methoxychlor	1.25	781.2×
1st crab, 1.6 ppb methoxychlor	2.68	1675.0×
5-day larvae, 1.9 ppb methoxychlor	0.81	426.3×

[a]From Bookhout *et al.* (1976), reprinted by permission.

tial mortality relative to controls and a higher concentration (1.0 ppb) causes both mortality and delay of molting.

Equally puzzling is the occurrence of an extra and apparently abnormal larval stage among *Menippe mercenaria* exposed to mirex. This is interesting since the duration of intermolt in the normal stages is not longer than among larvae in control conditions. To add further to the confusion, mirex increases the duration of larval development of *Rhithropanopeus harrisii* at all concentrations tested.

The effect of dieldrin on larval development of *Leptodius floridanus* and *Panopeus herbstii* was clearly most pronounced in the first zoeal stage. In experiments with mirex, mortality of first stage larvae only occurred at the higher concentrations. Mortality at 0.01 and 0.1 ppb was generally delayed until the third zoeal stage or later. In the case of methoxychlor, there was significant mortality during the first zoeal stage with *R. harrisii* at all concentrations where a sublethal effect was delayed until the second or third stage among *C. sapidus* exposed to low concentrations. At higher concentrations first stage *C. sapidus* were most severely affected.

Reasons for variation in the toxicity of the three pesticides to different larval stages of different species are not clear. It was hypothesized with dieldrin

(Epifanio, 1971), that larvae of a given species may have different genetically based resistance to the pesticide and that those least resistant die soon after exposure, i.e., before molting to the next larval stage. Those larvae genetically resistant to the compound continue through larval development with the only apparent effect manifested as the increased duration of each intermolt period. It was shown clearly that mortality was not directly related to the amount of dieldrin absorbed by the larvae. First stage larvae that died had relatively low residues of dieldrin in their tissues while those surviving continued to absorb dieldrin throughout development (Epifanio, 1973). Results from Bookhout and Costlow (1975) and Bookhout *et al.* (1976), however, suggest that mortality among *Callinectes sapidus* exposed to mirex and methoxychlor is related to the magnitude of pesticide residue in the larval tissue. Mortality at low concentrations of either pesticide does not occur until after the first 10–15 days of development when the larvae have accumulated considerable quantities of pesticide in their tissues. Mortality occurs during the first 10 days of development at higher concentrations, possibly because the larvae accumulate high body burdens of the pesticides from higher concentrations in seawater.

Mortality among *Rhithropanopeus harrisii* larvae exposed to methoxychlor is more similar to that seen in *Leptodius floridanus* and *Panopeus herbstii* exposed to dieldrin. Mortality occurred early in development and did not appear related to the magnitude of pesticide residue in the larval tissues.

Conversely, mortality among *Rhithropanopeus harrisii* and *Menippe mercenaria* low concentrations of mirex did not occur until the third and fourth larval stages (Bookhout *et al.*, 1972), presumably after the animals had accumulated considerable amounts of mirex in their tissues (uptake was not measured in this study).

3. Other Pesticides

There have been relatively few studies of pesticide effects other than chlorinated hydrocarbons on decapod larvae. Buchanan *et al.* (1970) investigated the toxicity of the carbamate, sevin (Table II), on development of *Cancer magister*. These workers exposed zoeae to nominal concentrations of sevin in seawater ranging from 0.1 to 10.0 ppb for a period of 25 days from hatching. They reported mortality of up to 40% at the higher concentrations and noted a delay in the onset of molting to second zoeal stage among experimental larvae. Furthermore, there was a direct relationship between the concentration of pesticide in the water and the degree of delay of molting (Table VII).

Bookhout and Monroe (1977) studied effects of the organophosphate, malathion (Table II), on development of *Rhithropanopeus harrisii* and *Callinectes sapidus*. The most apparent effect of malathion was its relatively low toxicity. It was not until concentrations of 0.011 and 0.014 ppm were reached that sublethal toxicity to *Rhithropanopeus harrisii* was noted. Concentrations of 0.017 and

TABLE VII

COMPARISON OF EFFECTS OF SEVIN AND MALATHION ON DEVELOPMENT OF CRAB LARVAE[a]

Pesticide	Species	Concentration (ppb)	Sublethal	Acute	Sensitive stage	Delay of development
Sevin[b]	Cancer magister	0.1	X		I	X
	Cancer magister	0.3	X		I	X
	Cancer magister	1.0	X		I	X
	Cancer magister	3.2	X		I	X
	Cancer magister	10.0		X	I	—[d]
Malathion[c]	Rhithropanopeus harrisii	11.0	X		II, M	X
	Rhithropanopeus harrisii	14.0	X		II, IV, M	X
	Rhithropanopeus harrisii	17.0		X	II, M	X
	Rhithropanopeus harrisii	20.0		X	II, III	X
	Rhithropanopeus harrisii	50.0		X	I	—[d]
	Callinectes sapidus	20.0	X		II, III	X
	Callinectes sapidus	50.0	X		II, III	X
	Callinectes sapidus	80.0		X	I, II, III	X
	Callinectes sapidus	110.0		X	I, II	X

[a]Roman numerals refer to zoeal stages; M refers to megalopa. The symbol X is used as a positive indicator, e. g., an X in the Acute column means that the toxicity was acute.

[b]Buchanan et al., 1970.

[c]Bookhout and Monroe, 1977.

[d]No survival beyond stage II.

0.02 ppm were acutely toxic and larvae reared in 0.05 ppm did not survive beyond the second zoeal stage. Duration of larval development was increased at both sublethal concentrations compared with the control situation. (Another sublethal effect was increased autotomy of legs of larvae in megalopal stage. The authors noted this in the text but did not quantify the effect.)

Mortality was generally delayed until the second zoeal stage and only larvae in 0.05 ppm experienced high mortality during the first stage. There was an additional increase in mortality rate at stage IV among larvae reared in 0.014 ppm. There was a similar increase in mortality among stage III larvae reared in 0.02 ppm and among megalopa at 0.011, 0.014, and 0.017 ppm.

Callinectes sapidus was considerably more resistant to malathion than *Rhithropanopeus harrisii*. Both 0.02 and 0.05 ppm were sublethal to *C. sapidus* but acutely toxic to *R. harrisii*. This is surprising because *C. sapidus* larvae are relatively difficult to culture and are considered very sensitive to general environmental stress, e.g., *C. sapidus* larvae were more sensitive to methoxychlor than *R. harrisii* (Bookhout *et al.*, 1976). Increased mortality among *C. sapidus* at 0.02 and 0.05 ppm occurred in stages II and III with no appreciable increases thereafter.

There is only one other report of effects of an organophosphate on development of decapod larvae. McLeese (1974) conducted studies of effects of fenitrothion (Table II) on lobster larvae *Homarus americanus* Milne-Edwards (Herrick, 1896). He noted no effects at concentrations below 0.1 ppb in seawater, but observed abnormal swimming behavior and expansion of chromatophores at 0.1 ppb and above. He calculated a 96-hour LC_{50} of approximately 1.0 ppb.

Two recent studies have investigated effects of juvenile hormone mimics, methoprene and hydroprene, on development of *Rhithropanopeus harrisii* (Christiansen *et al.*, 1977a,b). The first study, based on preliminary experiments

TABLE VIII

Rhithropanopeus harrisii; Concentrations in Parts per Million of the Juvenile Hormone Mimic Methoprene (ZR-515) Used to Test Larval Development in Five Salinities and Three Cycles of Temperature[a]

Temperature (°C)	Salinity (ppt)				
	5.0	12.5	20.0	27.5	35.0
20–25 25–30 30–35	Control, 0.01, 0.1, 1.0	Control 0.01, 0.1	Control, 0.01, 0.1, 1.0	Control 0.01, 0.1	Control 0.1, 1.0

[a]From Christiansen *et al.* (1977a), reprinted by permission.

of Costlow (1977), involved culture of larvae at combinations of three concentrations of methoprene, five salinities, and three cycles of temperature (Table VIII). The toxicity of methoprene was not significantly affected by salinity but did increase at higher concentrations. Highest mortality occurred among pesticide-exposed groups at 30°–35°C and was lowest at 20°–25°C. Acute toxicity was found in all salinity–temperature combinations at 1.0 ppm methoprene.

Stage I zoeae were most sensitive to methoprene but there was no evidence that any later stages were affected by the compound in concentrations below 1.0 ppm (Table IX). Duration of intermolt periods increased in groups exposed to methoprene and total time from hatching to megalopa increased with increasing concentrations. An additional effect of methoprene was an increase in the percentage of abnormal megalopa. At optimum salinity and temperature (20‰, 20°–25°C), the percentage of abnormal megalopa among control groups was less than 5%. Among some groups reared at stress salinities and temperatures in the presence of methoprene, 100% of the megalopa were abnormal. This abnormal-

TABLE IX

Summary of Effects of Methoprene on Development of *Rhithropanopeus harrisii* Larvae[a]

Concentration (ppm)	Salinity (ppt)	Temperature (°C)	Sublethal	Acute	Sensitive stage	Delay of development
0.01	5.0–35.0	20–25	X		I	X[b]
	5.0–35.0	25–30	X		I	X[b]
	5.0–35.0	30–35	X		I	X[b]
0.1	5.0	20–25		X	I	X[b]
	5.0	25–30		X	I	X[b]
	5.0	30–35		X	I	X[b]
	12.5–27.5	20–25	X		I	X[b]
	12.5–27.5	25–30	X		I	X[b]
	12.5–27.5	30–35		X	I	X[b]
	35.0	20–25	X		I	X[b]
	35.0	25–30		X	I	X[b]
	35.0	30–35		X	I	X[b]
1.0	5.0–35.0	20–25		X	I	—[c]
	5.0–35.0	25–30		X	I	—[c]
	5.0–35.0	30–35		X	I	—[c]

[a]Roman numerals refer to zoeal stages; M refers to megalopa. The symbol X is used as a positive indicator, e.g., an X in the Acute column means that the toxicity was acute. From Christiansen *et al.* (1977b), reprinted by permission.

[b]Increased number of morphologically abnormal megalopa.

[c]No survival beyond stage I.

TABLE X

Summary of Effects of Hydroprene on Development of *Rhithropanopeus harrisii*[a]

Concentration (ppm)	Salinity (ppt)	Temperature (°C)	Sublethal	Acute	Sensitive stage	Delay of development
0.01	20	20–25	X		I	—[b]
0.01	20	25–30	X		I	
0.01	20	30–35	X		I	
0.1	20	20–25	X		I, II	
0.1	20	25–30	X		I, II	—[b]
0.1	20	30–35	X		I, II	
0.5	20	20–25		X	I, II, M	X[b]
0.5	20	25–30		X	I, II, M	X
0.5	20	30–35		X	I, II, M	

[a]Roman numerals refer to zoeal stages; M refers to megalopa. The symbol X is used as a positive indicator, e.g., an X in the Acute column means that the toxicity was acute. From Christiansen *et al.* (1977b), reprinted by permission.

[b]Occurrence of extra zoeal stage.

ity consisted of deformations of the legs, but the majority of these abnormal individuals successfully completed molting to first crab stage.

Effects of a second juvenile hormone mimic, hydroprene, were determined in a similar experiment (Table X). In this study, based on earlier work by Bookhout and Costlow (1974), *Rhithropanopeus harrisii* larvae were reared at three cyclical temperatures and three concentrations of pesticide, but salinity was a constant 20‰. Sublethal effects were noted at all temperatures in concentrations of 0.01 and 0.1 ppm and an acute toxicity was observed in 0.5 ppm. It was further noted that larvae from some female crabs were more susceptible to hydroprene than larvae from other females, and large variations in survival between some replicates resulted.

Stage I larvae were most sensitive to hydroprene but there was also a significant increase in mortality rate among stage II larvae in 0.1 and 0.5 ppm and among megalopa in 0.5 ppm. Duration of zoeal development was significantly increased only among larvae reared at the two lower temperature cycles in 0.5 ppm. An extra zoeal stage was observed a few times in the hydroprene concentrations and these larvae died without molting to megalopa.

4. Summary

While only a few of the immense number of pesticides in use have been tested with larvae of a small number of species of decapods, some generalizations can be made. It appears that there is no one most sensitive stage. Different concentrations of a given pesticide may first yield increased mortality at various stages in the development of a species. For example, low concentrations of mirex did not cause increased mortality among *Callinectes sapidus* zoeae until stage III while higher concentrations caused increased mortality as early as stage II (Bookhout and Costlow, 1975). This would imply some relationship between mortality and the amount of mirex taken up and stored in body tissue. If one considers the work with dieldrin, however, it can be seen that toxicity is always manifested in stage I regardless of concentration and regardless of the amount taken up and stored in body tissue (Epifanio, 1971, 1972, 1973).

In the majority of studies of toxicity of pesticides to decapod larvae, zoeae are exposed to a given concentration from hatching to metamorphosis and increased mortality in each stage is statistically verified. The problem here is that there is no way to account for the possibility of selection for more resistant larvae as development progress. For example, if increased mortality is first observed in stage II and again in megalopa, there is no way to infer how those larvae that did not survive stage II might have responded in stages III or IV. So while this type of experiment allows strong inference concerning stages where increased mortality *did* occur, inferences concerning stages where increased mortality *did not* occur are less certain.

An alternate technique (Epifanio, 1971) involves mass culture of a large number of crab larvae in the absence of a pesticide with initial exposure as the larvae reach a given developmental stage (Fig. 2). With this technique it is possible to assess mortality in any stage without the complication of exposure to the pesticide during a previous stage. The obvious drawback is the larger number of larvae that have to be cultured for each experiment.

A relatively consistent effect of exposure to pesticides, even among resistant larvae, however, is an increase in the duration of larval development. This effect was seen with every pesticide tested and with all species except *Menippe mercenaria*. In that case, *M. mercenaria* exposed to various concentrations of mirex showed an increase in the appearance of an additional zoeal stage. Occurrence of an extra zoeal stage is normally rare and, while increased frequency was noted with other species and other pesticides, it was always associated with a general increase in duration of normal stages.

Among the pesticides tested, the chlorinated hydrocarbons and the one carbamate, sevin, appear most toxic. Sublethal effects were seen with mirex at concentrations as low as 0.01 ppb in seawater and acute toxicities with all organochlorides were noted at concentrations between 1.0 and 10.0 ppb. Sevin showed a similar toxicity. Generalizing from studies of dieldrin-contaminated food, it appears that the effective toxicity of waterborne chlorinated hydrocarbons is greater than that of the compounds found in the food of crab larvae.

The toxicity of malathion is somewhat less than that of the chlorinated hyd-

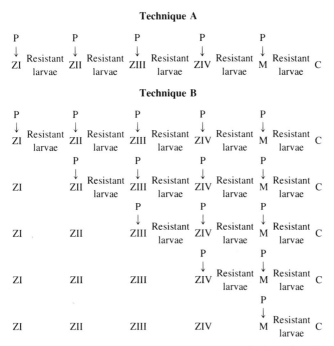

Fig. 2. Comparison of two bioassay techniques for assessment of effects of a pollutant on crab larvae. With technique A, larvae are continuously exposed to the pollutant (P) from hatching to metamorphosis. With technique B, each stage is exposed independently. ZI–ZV is zoea stage I–IV, M is megalopa, and C is crab stage 1.

rocarbons or carbamate. Sublethal effects were not apparent until larvae were exposed to concentrations of at least 10.0 ppb in seawater and acute effects did not occur with one species until exposure to 80.0 ppb. The other organophosphate, fenitrothion, appeared more toxic but the limited nature of the experiments in this case make comparison difficult.

Juvenile hormone mimics appear least toxic of all. Even under extreme conditions of salinity and temperature, sublethal effects were not observed below 10 ppb in seawater and under more favorable environmental conditions, sublethal effects were not noted until exposure to 100 ppb. Since juvenile hormone mimics are designed to interfere with metamorphosis in insec/s, it is interesting that no effects of this nature were observed among crab larvae exposed to the compounds. There was an increase in the occurrence of an extra zoeal stage with hydroprene, however, and increased number of morphologically abnormal megalopa with methoprene.

B. Petroleum Hydrocarbons

1. General

Ehrhardt and Blumer (1972) estimated that 5 to 10 million metric tons of oil are introduced annually into the aquatic environment through oil tanker spills, ballasting and bilge cleaning, transferring of oil, and several other types of mechanical and industrial failures. Various types of refined and unrefined oils have rather different compositions and the extent of biological damage from oil spillage is related to the type of oil involved. Specifically, aromatic hydrocarbons are the most toxic fraction. Those aromatics with low boiling points are most toxic of all but because of their low boiling points they volatilize rapidly in natural situations. The intermediate and high boiling aromatics are somewhat less toxic but are more persistent following a spill and hence have greater environmental effects.

Sublethal effects of these compounds have been observed with adult marine invertebrates at concentrations ranging from 10 to 100 ppb in seawater while acute effects are generally restricted to the range of 1.0 to 100 ppm (Moore and Dwyer, 1974). Some controversy exists, however, concerning the value of testing effects of individual components of petroleum in the absence of others and there has been an equally large number of studies of effects of various crude and refined oils on marine organisms (Wilson and Hunt, 1975).

2. Effects on Decapod Larvae

Three have been rather few studies of effects of petroleum oils upon larval decapods. The first reported laboratory study was that of Mironov (1969), who found acute toxicity of various oil fractions to larvae of several species of crabs and shrimp ranged from 0.001 to 0.1 ml/liter. These values are somewhat difficult to compare to later results, which are generally reported as micrograms per liter or milligrams per liter (parts per billion or parts per million).

Wells (1972) reported effects of Venezuelan crude oil dispersed in seawater on larval development of the lobster *Homarus americanus*. He noted acute and sublethal toxicity in terms of milliliters of an oil–water emulsion per milliliter of clean seawater. Wells and Sprague (1976) extended this work with a study of the effects of Venezuelan crude oil on mortality, rate of development, and feeding rate. Oil was dispersed in seawater by stirring and ultrasonic vibration, thereby, according to the authors, simulating dispersal at sea. Test media were then prepared by drawing the oil–water emulsion from beneath the residual oil and diluting the emulsion with clean seawater. Actual amounts of oil in test media were assayed by uv absorption (Zitko and Carson, 1970). Acute toxicities were measured as 4-day LC_{50} and 30-day LC_{50}. First stage larvae were most sensitive (4-day LC_{50} = 0.86 ppm) with stages III and IV more resistant (4-day LC_{50}, 4.9 ppm). The 30-day LC_{50} was 0.14 ppm and duration of development was increased in larvae that survived to the postlarval stage. An increased frequency of an additional larval stage was noted at all concentrations of oil. A final effect of oil was a lowered rate of food consumption. In three trials involving 30 larvae, control larvae ate 1.3, 1.7, and 1.9 times as many brine shrimp as did larvae in test media (0.19 and 0.74 ppm).

Katz (1973), also working with Venezuelan crude oil, found stage I *Neopanope texana sayi* (Smith) (Rathbun, 1930) more sensitive to 4 ppm in seawater than later stages and noted an increase in duration of development when compared to controls. Similar results were reported by Caldwell *et al.* (1977), who found increased duration of larval development in *Cancer magister* reared in the water-soluble fraction of Alaskan crude oil (0.22 ppm as dissolved aromatics) and napthalene (0.13 ppm).

Since different crude oils may have rather different compositions, it is difficult to compare results of studies using both different oils and different test species. More recent experiments have been based on use of four standard oils (South Louisiana crude, Kuwait crude, No. 2 fuel oil, and Venezuelan Bunker C) provided by the American Petroleum Institute (API) (Table XI).

Another problem is comparing results of different studies is the various procedures used in preparing oil–water solutions or emulsions. Wells and Sprague (1976) found that a combination of stirring and ultrasonic vibration yielded higher concentrations of oil in test media than simple stirring. Furthermore, they found that ultrafiltration changed the toxicity of a given test medium as the filtered medium was more toxic per unit hydrocarbon than the unfiltered medium.

Anderson *et al.* (1975), however, suggested a stirring technique that has been used in the majority of studies thereafter. Here, the test medium is prepared by placing 1 part oil over 9 parts seawater in a glass bottle and slowly stirring for 20

TABLE XI

AROMATIC AND *n*-PARAFFIN CONTENT OF THE FOUR TEST OILS, THEIR WATER-SOLUBLE FRACTIONS (WSF), AND CALCULATED AROMATIC ENRICHMENT FACTORS (AEF)[a]

Oil	Aromatics		*n*-Paraffins		Aromatics: *n*-Paraffins		AEF:WSF
	Oil (%)	WSF (ppm)	Oil (%)	WSF (ppm)	Oil	WSF	Oil
South Louisiana	0.94	0.305	3.98	0.089	0.24	3.43	14.29
Kuwait	0.60	0.075	4.00	0.004	0.15	18.75	125.00
No. 2 fuel oil	9.18	2.002	7.38	0.047	1.24	42.60	34.35
Bunker C	4.31	0.935	1.26	0.012	3.42	77.92	22.78

[a]From Anderson *et al.* (1975), reprinted by permission.

hours at 20°C. The bottle is capped to minimize evaporation; after mixing, the oil and water phases are allowed to separate for 1-6 hours before the water phase is siphoned from beneath the remaining oil. Oil in the water phase is termed the water-soluble fraction (WSF).

By using this technique, Anderson *et al.* (1975) determined TL_m (LC_{50}) values for the postlarval stage of the shrimp *Penaeus aztecus* Ives (Burkenroad, 1939) exposed to three of the API standard oils. The 4-day TL_m to South Louisiana Crude was 19.8 ppm while refined oils were more toxic: TL_m to No. 2 fuel oil was 9.4 ppm and that for Bunker C was 1.9 ppm.

Laughlin *et al.* (1978) used Anderson's technique for preparation of test media and studied long-term effects of WSF of No. 2 fuel oil on larval and juvenile development of *Rhithropanopeus harrisii*. The aqueous phase siphoned from beneath the oil was termed 100% WSF and test media consisted of various dilutions of this with clean seawater. The actual concentrations of No. 2 fuel oil in test media were measured as napthalene (Neff and Anderson, 1975), where 2.5% WSF was 0.054 ppm, 5.0% WSF was 0.099 ppm, 10% WSF was 0.070 ppm, 15% WSF was 0.277 ppm, and 20% WSF was 0.357 ppm.

Acute toxicity was observed only at 20% WSF but sublethal effects were seen at all lower concentrations. Highest mortality occurred during stage I at 10, 15, and 20% WSF and mean duration to megalopa increased with concentrations. Upon reaching megalopa, however, effects of oil were less pronounced. There was no effect of increasing oil concentration upon mortality of megalopa and while duration of megalopa stage was greater at 2.5, 5.0, and 10.0% WSF than under control conditions, it decreased at 15 and 20% WSF.

Those crabs that successfully metamorphosed from megalopa to crab stage 1 were cultured in their respective test media for an additional 6 months and mortality and rate of molting and growth were monitored. There was no effect on WSF on molt frequency but there was a striking effect on rate of growth. For a given molt stage, size (carapace width) of controls was always greater than that of crabs in 2.5, 5.0, and 10.0% WSF. The mean size of crabs in 15 and 20% WSF, however, was greater than controls at each molt stage. The reasons for this paradoxical result are not clear.

Cucci (1978) conducted a similar study of long-term effects of WSF of Kuwait crude and No. 2 fuel oil on development of *Eurypanopeus depressus* (Smith) (Rathbun, 1930). Test media were again prepared according to Anderson *et al.* (1975) and WSF was reported as a percentage of the initial aqueous phase. Actual amounts of dissolved hydrocarbons were not measured but were calculated from Anderson *et al.* (Table XII).

Both short- and long-term toxicities were tested. In the former, each larval stage was exposed separately to various concentrations of both oils and 48-hour TL_m was calculated. Number 2 fuel oil was considerably more toxic than Kuwait

TABLE XII

Total Dissolved Hydrocarbons and Aromatics at Acute and Sublethal Levels[a]

Oil	Concentrations	Total dissolved hydrocarbons[b] (ppm)	Total dissolved aromatics (ppm)
No. 2 fuel oil	100% WSF	6.28	5.74
	Acute level (2.4%)	0.15	0.14
	Sublethal levels (1.0%)	0.06	0.06
	(2.0%)	0.13	0.11
	(3.0%)	0.19	0.16
Kuwait crude oil	100% WSF	21.65	10.03
	Acute level (47.3%)	10.24	4.74
	Sublethal levels (20.0%)	4.33	2.01
	(40.0%)	8.66	4.02

[a] Calculated from results of Anderson *et al.* (1975). Analyses were performed utilizing gas chromatographic methods. All concentrations are in parts per million. (From Cucci, 1978).

[b] Alkyl benzenes (boiling points between 176° and 218°C), polycyclic aromatics greater than C_{16} and alkanes (C_8–C_{11} and higher than C_{24}) were not analyzed.

crude and stage I larvae were most sensitive to both oils. The 48-hour TL_m of stage I larvae to Kuwait crude was 47.3% WSF while the 48-hour TL_m to No. 2 fuel oil was 2.4% WSF.

Investigation of long-term effects involved exposure of one group from hatching through larval development, a second group from stage II, another from stage III, and a final group from stage IV (Fig. 2). Those exposed to 20 and 40% WSF Kuwait crude suffered substantial mortality during stage I and significantly fewer survived to megalopa than in a control group. Those first exposed to Kuwait crude at stage II showed increased mortality only at 40% WSF.

Surprisingly, mortality was not greater among groups exposed to 1, 2, or 3% WSF No. 2 fuel oil than in the control group. Cucci attributed this to the fact that larvae used in the short-term experiments defining 48-hour TL_m to No. 2 fuel oil were hatched at the very end of the spawning season while those used in the long-term experiments were from midspawning season during the following year. He hypothesized a generally increased ability to cope with environmental stress among larvae hatched in midseason.

Increased duration of larval development was observed at both 20 and 40% WSF Kuwait crude and the effect was greatest among larvae first exposed during stage I. A similar effect was seen in larvae first exposed to all three concentrations of No. 2 fuel in stage I, but not in groups first exposed at stage II or later.

Megalopa successfully molting to crab stage 1 were cultured in their respective test media through four intermolt periods. As opposed to the results of Laughlin *et al.* (1978), those crabs first exposed to oil in zoea stage I continued to exhibit slower rates of molting during juvenile development. However, juvenile crabs first exposed during zoea stage II or later molted at rates similar to crabs under control conditions.

A comparison of carapace width and dry weight of crabs successfully molting to crab stage 5 indicated generally smaller size among those exposed to the various oil concentrations. Crabs first exposed to 2 and 3% WSF No. 2 fuel during zoea stages III and IV were considerably larger than controls at crab stage 5, lending credence to similar results of Laughlin *et al.*

3. Summary

It is clear, then, that petroleum hydrocarbons are generally less toxic to decapod larvae than pesticides. Among petroleum oils, refined products with their increased percentage of medium-boiling-point aromatics are considerably more toxic than crude oils. Early stage larvae are more sensitive than later stage zoeae or megalopa and sublethal toxicity is almost always associated with an increase in duration of larval development. The increased growth rate of juvenile crabs exposed to low levels of petroleum hydrocarbons is difficult to explain and it is not known whether this might have occurred with pesticides had juvenile crabs been exposed.

C. MISCELLANEOUS POLLUTANTS

In addition to pesticides and petroleum hydrocarbons, there have been studies of effects of various other pollutants (heavy metals, detergents, polychlorinated biphenyls, polychlorinated naphthalenes, phthalate esters, and dredge spoils) on development of decapod larvae.

1. Heavy Metals

Studies of heavy metals began with the work of DeCoursey and Vernberg (1972), who investigated effects of mercury on survival, behavior, and metabolism of *Uca pugilator* (Bosc) (Rathbun, 1918) larvae. They determined that a concentration of 118 ppb mercury (as $HgCl_2$) in seawater was acutely toxic to all larval stages tested and while concentrations of 18 and 1.8 ppb were sublethal, they did affect metabolic and swimming rates. Generally, advanced stages were more sensitive than early stages.

In a following study, Vernberg *et al.* (1973) reared *Uca pugilator* larvae in 1.8 ppb mercury at four combinations of salinity and temperature. Mercury increased mortality at stress conditions of salinity and temperature, with a particularly pronounced effect at low temperature, low salinity (20°C–20‰). The effect of mercury on metabolism varied; at high temperature (25 and 30°C), rates were depressed while at 20°C rates were elevated. The effect of mercury on swimming rate was also temperature dependent with reduced rates at combinations of mercury and low temperature.

Rosenberg and Costlow (1976) carried out a related study involving effects of cadmium at various salinity and temperature combinations. In this investigation, the megalopa and three juvenile crab stages of *Callinectes sapidus* were exposed to 12 combinations of cadmium in seawater (0, 50, and 150 ppb) and salinity (10, 20, 30, and 40‰) at 25°C. In another part of the study, larvae of *Rhithropanopeus harrisii* were exposed to from hatching to crab stage 1 to 63 combinations of cadmium (0, 50, and 150 ppb), salinity (10, 20, and 30‰), constant temperature (20, 25, 30, and 35°C), and cycling temperature (20°–25°C, 25°C–30°C, and 30°–35°C).

Cadmium was most toxic to *Callinectes sapidus* megalopa at low salinity, where no individuals survived to crab stage 1 at 150 ppb cadmium. At higher salinities survival was always lowest in 150 ppb and highest in controls, and there was a significant increase in duration of development time from megalopa stage to crab stage 3 at all concentrations of cadmium.

Survival of zoeal stages of *Rhithropanopeus harrisii* zoeae was reduced by cadmium at all combinations of salinity and temperature but reported results do not indicate which stages are most sensitive. A combination of 150 ppb cadmium at 10‰ was lethal regardless of temperature. Mortality among those larvae surviving to megalopa was not increased by cadmium, but this may have been attributable to selection for less sensitive individuals during zoeal development. Effects of cadmium were generally less at cycling temperatures than at either of the comparable constant temperatures.

When larvae survived in 150 ppb, duration of development was significantly increased compared to 50 ppb or controls at all salinities and temperatures. This was similar to results of Benijts and Benijts (1975), who exposed *Rhithropanopeus harrisii* larvae to 50 ppb lead and 50 ppb zinc. Effects of 50 ppb cadmium were dependent on temperature and salinity with increases in duration of development at all temperatures in 10‰ but at only 25° and 30°C in 20 and 30‰.

2. Detergents

There has been very little study of effects of detergents on decapod larvae. Czyzewska (1976) exposed larvae of *Rhithropanopeus harrisii* (Gould) to a detergent consisting of 19.40% ethoxylate alkylophenol, 4.85% ethoxylate deacetylate alcohol, 7.25% triethanolamine salt of alkyl benzene sulfonic acid, 67.80% water, and 0.70% inert ingredients. Concentrations of 10 ppm and above were acutely toxic to the larvae and none survived to megalopa. Sublethal effects were noted at 1.0 and 0.1 ppm where duration of zoeal development was increased. In all cases, stage I zoeae were most sensitive to the detergent.

3. *Polychlorinated Biphenyls, Polychlorinated Naphthalenes, and Phthalate Esters*

Polychlorinated biphenyls are a group of chlorinated polycyclic compounds of industrial importance. It is well known that they are widely disseminated in marine and estuarine environments and are toxic to a variety of organisms (Peakall, 1975). Polychlorinated naphthalenes are related industrial compounds that, along with phthalate plasticizers, have only recently been found in marine environments. Relatively little is known of the toxicity of these compounds.

Initial work with decapod larvae and polychlorinated biphenyls was that of Roesijadi *et al.* (1976), who investigated effects of Aroclor on larvae of the caridean shrimp *Palaemonetes pugio* Holthius (Holthius, 1949). (Various polychlorinated biphenyls are produced by Monsanto Company* under the trade name, Aroclor.) Sublethal effects were observed at concentrations less than 100 ppb.

Neff *et al.* (1977) compared effects of Aroclor 1016 with those of two polychlorinated naphthalenes, Halowax 1000 and Halowax 1099, and two phthalate ester plasticizers, dimethylphthalate and dibutylphthalate. (Several polychlorinated naphthalenes are produced by Koppers Company† under the trade name, Halowax. The plasticizers were produced by Monsanto Company.)

A concentration of 80 ppb Aroclor 1016 in seawater was acutely toxic to *Rhithropanopeus harrisii* larvae as only 10% of the larvae survived to megalopa. Initial mortality was associated with the molt from zoea stage I to stage II, but the highest rate of mortality occurred during stage III.

Significant increases in mortality were not observed until concentrations of Halowax 1099 and 1000 reached 100 and 300 ppb, respectively. The two plasticizers were even less toxic as no increased mortality was noted in either case at 1000 ppb.

Duration of larval development was not significantly different from controls at 20 and 40 ppb Aroclor 1016, 50 and 150 ppb Halowax 1000, or 20 and 50 ppb Halowax 1099. However, duration of development was increased at higher concentrations of each of these compounds. Rates of development of larvae exposed to the plasticizers were not significantly different from controls.

One novel aspect of the Neff *et al.* study was the correlation between increasing concentrations of polychlorinated biphenyls and polychlorinated napthalenes and a decrease in the size of megalopa exposed to the compounds. They concluded

---

*Monsanto Company, 800 North Lindbergh Boulevard, St. Louis, Missouri 61366.

†Koppers Company, Inc., 1420 Koppers Building, Pittsburgh, Pennsylvania 15219.

that this was the most sensitive index of sublethal toxicity as it was observed at concentrations which did not affect duration of larval development.

4. Dredge Spoils

DeCoursey and Vernberg (1975) studied the effects of water from dredge spoil dikes receiving spoil from the Cooper River, a tributary of Charleston Harbor, South Carolina. Although they conducted no analyses of pollutants in the water, their review of the literature indicated that it should contain elevated levels of pesticides, polychlorinated biphenyls, and heavy metals.

Palaemonetes pugio larvae were exposed to water from the dredging site, 200 m downstream of the dredging site, and the outflow of the spoil dike. Generally, water from the outflow was most toxic and cumulative mortality of larvae reared in this water was 43% greater than controls after 10 days. Furthermore, swimming speed of larvae reared in water from the outflow was only 56% of controls.

III. Conclusions

A. SYNOPSIS

Two major criticisms apply to virtually all the studies reviewed here. The first is that most experiments were conducted in static culture where a toxin is introduced into the water at periodic intervals, e.g., with petroleum hydrocarbons or polychlorinated biphenyls, this results in volatile loss of considerable portions of the toxin each day (Neff *et al.*, 1977). Furthermore, the test organisms themselves may accumulate toxic compounds in their tissues, and in a static culture where toxins are replenished periodically, this can result in additional loss of compounds from the test medium. The implicit assumption in work with larvae is that the volume of test medium is so great relative to the larvae's ability to remove a toxin from water, that the change in concentration over a 24-hour period is negligible. There are little data to support this assumption, however.

A flow-through test system where small amounts of a toxin are added continuously is preferable but very difficult to use with larval forms. Only Czyzewska (1976) has reported using such a system with decapod larvae and he did not report results of analyses of amounts of toxin in the test medium over time.

A second criticism is not restricted to studies of decapod larvae but applies to any laboratory-based study. It concerns the limited ecological inferences that can be drawn from laboratory toxicology. In many cases, the concentration of toxin shown to cause even sublethal effects is considerably higher than that in all but the most polluted of natural environments. One could then conclude that few of the toxins tested affect decapod larvae in natural systems. Alternatively, one could suppose synergistic effects of the many pollutants in natural systems or maintain that sublethal effects more subtle than those observed in the laboratory occur at lower, more environmentally realistic levels of a toxin. This is not a satisfying argument, but more extensive testing of environmentally realistic concentrations (alone and synergistically) would be very expensive. This is the very same dilemma experienced by toxicologists attempting to ascertain the safety of human food and drugs.

Nevertheless, laboratory toxicology studies are an important tool in environmental management. Without the results of such studies, there would be no basis whatsoever for limiting amounts of various compounds in the environment. Recent studies with larval forms have been particularly important in this respect since the majority of marine invertebrates undergo larval development and since it is generally held (though not always correctly) that larval forms are more sensitive to environmental perturbation than adults. In many ways decapod larvae are ideal test organisms. The fact that they share a phylum with the insects would make them prime candidates as sensitive, nontarget organisms of the immense variety and quantity of pesticides released into the environment. The fact that they undergo a discrete number of stages via a set number of ecdyses and that these stages are morphologically distinct makes quantification of effects of toxins relatively easy. The developmental pattern typically includes planktonic larval stages and semibenthic postlarval forms which adds to the value of decapod larvae as test organisms.

Decapod larvae do appear most sensitive to insecticides with sublethal effects for several of the compounds tested observed at concentrations below 1.0 ppb. Heavy metals (cadmium, lead, mercury, and zinc) are somewhat less toxic with sublethal effects between 10 and 100 ppb (between 1.0 and 10 ppb for mercury). Polychlorinated biphenyls are chemically similar to organochloride insecticides but are considerably less toxic with sublethal toxicities between 10 and 100 ppb. Polychlorinated naphthalenes appear less toxic (sublethal effects between 100 and 500 ppb) and phthalate plasticizers are virtually nontoxic. Sublethal effects of water-soluble fractions of various crude oils are first observed as concentrations approach 1.0 ppm while similar effects of refined oils occur at half that concentration.

The sensitivity of a given larval stage appears to depend upon species and the particular toxin tested. The toxicity of some pesticides, for example, is always greatest in early stage larvae and is apparently independent of accumulation of the compound by the larvae. On the other hand, several cases were cited where mortality occurred during later stages and could be correlated with accumulation of the toxin in larval tissues. Determination of sensitivity of later larval stages was complicated, however, by selection for more resistant individuals during larval development in presence of a toxin. In only a few cases was toxicity determined independently for each stage (Fig. 2).

An increase in duration of larval development is a sensitive and almost universal response by decapod larvae to sublethal concentrations of a toxin. The response is apparently a reaction to generalized stress since it is independent of the nature of the toxin and also occurs under other conditions of stress unrelated to toxins. The physiological basis of the effect is not known, but it is almost certainly an indirect effect of a toxin on the molting system, i.e., a compound need not be a neuroendocrine toxin to elicit the response. In some instances an increased duration of larval development is associated with an abnormal occurrence of additional larval stages or with morphological abnormalities in the larvae, but the relationship of the effects is not clear.

Recent studies (Cucci, 1978; Laughlin *et al.*, 1978) are interesting in that they followed growth and development of individuals exposed to petroleum as larvae through metamorphosis and juvenile development. Their results differed in that Laughlin *et al.* (1978) did not find continued effects at low concentrations while Cucci observed effects on rates of molting and growth through crab stage 5. A surprising increase in growth rate of juveniles was observed in both studies at relatively high concentrations, however. Reasons for this are not at all clear.

B. PROGNOSIS

1. Bioassays

Future studies might concentrate on one species which is readily available, easily cultured, and would allow easy comparison of data concerning various toxins. *Rhithropanopeus harrisii* fits the above criteria and has, in fact, already been used in more toxicology studies than any other decapod. Larvae of this species are relatively large, making metabolic and behavioral studies easier, exceptionally easy to culture, available all along the Atlantic Coast of the United States and in Europe, and considerable information concerning their basic physiology exists.

2. Long-Term Studies

The success of Laughlin *et al.* and Cucci with studies through metamorphosis and into the juvenile stages opens interesting possibilities. Experiments involving several generations reared in presence of low levels of toxins might yield exceptionally valuable information on previously unknown sublethal effects on reproduction and development. It would also be interesting to investigate the apparent stimulatory effect of toxins on the growth of juveniles. The major problem in such studies, of course, is the amount of time and effort involved in long-term culture of the organisms. Such studies are not out of the question, though, when it is considered that it is possible to rear species such as *Rhithropanopeus harrisii* from egg to mature adult in 6 months.

References

Anderson, J. W., Neff, J. M., Cox, B. A., Tatem, H. E., and Hightown, G. M. (1975). Characteristics of dispersions and water-soluble extracts of crude and refined oils and their toxicity to estuarine crustaceans and fish. *Mar. Biol.* **27**, 75–88.

Barnes, E. W. (1911). Revised edition of the methods of protecting and propagating the lobster, with a brief outline of its natural history. *Rep. R. I. Comm. Inland Fish.* **41**, 83–127.

Benijts, C. C., and Benijts, F. (1975). The effect of low lead and zinc concentrations on the larval development of the mud crab *Rhithropanopeus harrisii* Gould. *In* "Sublethal Effects of Toxic Chemicals on Aquatic Chemicals" (J. H. Koeman and J. J. T. W. A. Stric, eds.), pp. 43–52. Elsevier, Amsterdam.

Bookhout, C. G., and Costlow, J. D., Jr. (1970). Nutritional effects of *Artemia* from different locations on larval development of crabs. *Helgol. Wiss. Meeresunters.* **20**, 435–442.

Bookhout, C. G., and Costlow, J. D., Jr. (1974). Crab development and effects of pollutants. *Thalassia Jugosl.* **10**, 77–87.

Bookhout, C. G., and Costlow, J. D., Jr. (1975). Effects of mirex on the larval development of blue crab. *Water, Air, Soil Pollut.* **4**, 113–126.

Bookhout, C. G., and Monroe, R. (1977). Effects of Malathion on the development of crabs. *In* "Physiological Responses of Marine Biota to Pollutants" (F. J. Vernberg *et al.*, eds.), pp. 3–19. Academic Press, New York.

Bookhout, C. G., Wilson, A. J., Jr., Duke, T. W., and Lowe, J. I. (1972). Effects of mirex on the larval development of two crabs. *Water, Air, Soil Pollut.* **1**, 165–180.

Bookhout, C. G., Costlow, J. D., Jr., and Monroe, R. (1976). Effects of methoxychlor on larval development of mud-crab and blue crab. *Water, Air, Soil Pollut.* **5**, 349–365.

Buchanan, D. V., Millman, R. E., and Stewart, N. E. (1970). Effects of the insecticide sevin on various stages of the Dungeness crab, *Cancer magister*. *J. Fish. Res. Board Can.* **27**, 93–104.

Burkenroad, M. D. (1939). Further observations on penaeidae of the northern Gulf of Mexico. *Bull. Bingham Oceanogr. Collec.* **6**, 1–62.

Caldwell, R. S., Calderane, E. M., and Mallon, M. H. (1977). Effects of a seawater-soluble fraction of Cook Inlet crude oil and its major aromatic components on larval stages of the Dungeness crab, *Cancer magister* Dona. *In* "Fates and Effects of Petroleum Hydrocarbons in Marine Organisms and Ecosystems" (D. A. Wolfe, ed.), pp. 210–220. Pergamon, Oxford.

Caroli, E. (1918). *Miersia clavigera*, Chum, Stadio misidiforme di *Lysmata seticaudata*. *Boll. Soc. Nat. Napoli* **38**, 161–166.

Christiansen, M. E., Costlow, J. D., Jr., and Monroe, R. (1977a). Effects of the juvenile hormone mimic ZR-515 (Altosid) on larval development of the mud crab *Rhithropanopeus harrisii* in various salinities and cyclic temperatures. *Mar. Biol.* **39**, 269–279.

Christiansen, M. E., Costlow, J. D., Jr., and Monroe, R. (1977b). Effects of the juvenile hormone mimic ZR 512 (Altozar) on larval development of the mud crab *Rhithropanopeus harrisii* at various cyclic temperatures. *Mar. Biol.* **39**, 281–288.

Costlow, J. D., Jr. (1963). Effect of eyestalk extirpation on metamorphosis of megalops of the blue crab *Callinectes sapidus* Rathbun. *Gen. Comp. Endocrinol.* **3**, 120–130.

Costlow, J. D., Jr. (1966a). The effect of eyestalk extirpation on larval development of the crab, *Sesarma reticulatum* Say. *In* "Some Contemporary Studies in Marine Science" (H. Barnes, ed.), pp. 209–224. Allen & Unwin, London.

Costlow, J. D., Jr. (1966b). The effect of eyestalk extirpation on larval development of the mud crab *Rhithropanopeus harrisii* (Gould). *Gen. Comp. Endocrinol.* **7**, 255–274.

Costlow, J. D., Jr. (1977). The effect of juvenile hormone mimics on the development of the mud crab *Rhithropanopeus harrisii* (Gould). *In* "Physiological Responses of Marine Biota to Pollutants" (F. J. Vernberg *et al.*, eds.), pp. 211–220. Academic Press, New York.

Costlow, J. D., Jr., and Bookhout, C. G. (1959). The larval development of *Callinectes sapidus* Rathbun reared in the laboratory. *Biol. Bull. (Woods Hole, Mass.)* **116**, 373–396.

Costlow, J. D., Jr., and Bookhout, C. G. (1960). A method for developing brachyuran eggs *in vitro*. *Limnol. Oceanogr.* **5**, 212–215.

Costlow, J. D., Jr., Bookhout, C. G., and Monroe, R. (1960). The effect of salinity and temperature on larval development of *Sesarma cinereum* (Bosc) reared in the laboratory. *Biol. Bull. (Woods Hole, Mass.)* **118**, 183–202.

Costlow, J. D., Jr., Bookhout, C. G., and Monroe, R. (1962). Salinity-temperature effects on larval development of the crab *Panopeus herbstii* Milne-Edwards, reared in the laboratory. *Physiol. Zool.* **35**, 79–93.

Cucci, T. L. (1978). Effects of the water-soluble fractions of Kuwait crude and No. 2 fuel oil on the larval and juvenile development of *Eurypanopeus depressus* (Brachyura: Xanthidae). M.S. Thesis, University of Delaware, Newark.

Czyzewska, K. (1976). The effects of detergents on larval development of crab. *Mar. Pollut. Bull.* **7**, 108–112.

DeCoursey, P. J., and Vernberg, W. B. (1972). Effect of mercury on survival, metabolism, and behavior of larval *Uca pugilator*. *Oikos* **23**, 241–247.

DeCoursey, P. J., and Vernberg, W. B. (1975). The effect of dredging in a polluted estuary on the physiology of larval zooplankton. *Water Res.* **9**, 149–154.

Ehrhardt, M., and Blumer, M. (1972). The source identification of marine hydrocarbons by gas chromatography. *Environ. Pollut.* **3**, 179–194.

Epifanio, C. E. (1971). Effects of dieldrin in seawater on the development of two species of crab larvae, *Leptodius floridanus* and *Panopeus herbstii*. *Mar. Biol.* **11**, 356–362.

Epifanio, C. E. (1972). Effects of dieldrin-contaminated food on the development of *Leptodius floridanus* larvae. *Mar. Biol.* **13**, 292–297.

Epifanio, C. E. (1973). Dieldrin uptake by larvae of the crab *Leptodius floridanus*. *Mar. Biol.* **19**, 320–322.

Faxon, W. (1879). On the development of *Palaemonetes vulgaris*. *Bull. Mus. Comp. Zool.* **5**, 303–330.

Guinot, D. (1968). Recherches préliminaires sur les groupements naturels chez les Crustacés Décapodes Brachyoures. IV. Observations sur quelques genres de Xanthidae. *Bull. Mus. Hist. Nat., Paris* [2] **39**, 704–708.

Gurney, R. (1942). "Larvae of the Decapod Crustacea." Ray Society, London.

Herrick, F. H. (1896). The american lobster. *Bull. U.S. Fish. Comm.* **15**, 1–252.

Holthius, L. B. (1949). Notes on the species (Crustacea: decapoda) found in the United States of America. *Proc. K. Ned. Akad. Wet.* **52**, 87–95.

Katz, L. M. (1973). The effects of water soluble fraction of crude oil on larvae of the decapod crustacean *Neopanope texona sayi*. *Environ. Pollut.* **5**, 199–204.

Laughlin, R. B., Jr., Young, L. G. L., and Neff, J. M. (1978). A long-term study of the effects of water-soluble fractions of No. 2 fuel oil on the survival, development rate, and growth of the mud crab *Rhithropanopeus harrisii*. *Mar. Biol.* **47**, 87–95.

Lebour, M. V. (1925). Young anglers in captivity and some of their enemies. A study in a plunger jar. *J. Mar. Biol. Assoc. U. K.* **13**, 721–743.

McLeese, D. W. (1974). Olfactory response and fenitrothion toxicity in american lobsters (*Homarus americanus*). *J. Fish. Res. Board Can.* **31**, 1127–1131.

Mironov, O. G. (1969). On the influence of oil pollution on some representatives of Black Sea zooplankton. *Zool. Zh.* **48**, 1734–1736.

Moore, S. F., and Dwyer, R. L. (1974). Effects of oil on marine organisms: A critical assessment of published data. *Water Res.* **8**, 819–827.

Mootz, C. A., and Epifanio, C. E. (1974). An energy budget for *Menippe mercenaria* larvae fed *Artemia nauplii*. *Biol. Bull. (Woods Hole, Mass.)* **146**, 44–55.

Muller, F. (1892). O camarao preto, *Palaemon potiuna. Arg. Mus. Nac., Rio de Janeiro* **8**, 179–206.

Neff, J. M., and Anderson, J. W. (1975). Effects of polychlorinated biphenyls, polychlorinated naphthelenes and phthalate esters on larval development of the mud crab *Rhithropanopeus harrisii. Bull. Environ. Contam. Toxicol.* **14**, 122–128.

Neff, J. M., Laughlin, R. B., Jr., and Giam, C. S. (1977). Effects of polychlorinated biphenyls, polychlorinated naphthalenes and phthalate esters on larval development of the mud crab *Rhithropanopeus harrisii. In* "Pollutant Effects on Marine Organisms" (C. S. Giam, ed.), pp. 95–110. Lexington Books, Lexington, Massachusetts.

O'Brien, R. D. (1967). "Insecticides: Action and Metabolism." Academic Press, New York.

Peakall, D. B. (1975). PCB's and their environmental effects. *Crit. Rev. Environ. Control* **5**, 469–508.

Rathbun, M. J. (1896). The genus *Callinectes. Proc. U.S. Nat. Mus.* **18**, 349–375.

Rathbun, M. J. (1918). The grapsoid crabs of *America. U.S., Natl. Mus., Bull.* **97**, 1–461.

Rathbun, M. J. (1930). The Concroid Crabs of America of the Families Euryalidae, Portunidae, Atelecyclidae, Cancridae, and Xanthidae." U.S. Natl. Mus., Washington, D.C.

Rice, A. L., and Williamson, D. J. (1970). Methods for rearing larval decapod crustacea. *Helgol. Wiss. Meeresunters.* **20**, 417–434.

Roesijadi, G., Petrocelli, S. R., Anderson, J. W., Giam, C. S., and Neff, G. E. (1976). Toxicity of polychlorinated biphenyls (Arochlor 1254) to adult, juvenile and larval stages of the shrimp *Palaemonetes pugio. Bull. Environ. Contam. Toxicol.* **15**, 297–304.

Rosenberg, R., and Costlow, J. D., Jr. (1976). Synergistic effects of cadmium and salinity combined with constant and cycling temperatures on the larval development of two estuarine crab species. *Mar. Biol.* **38**, 291–303.

Sollaud, E. (1924). Comparison of rotifers and orther diets for rearing larvae of the blue crab, *Callinectes sapidus* Rathbun. *C. R. Hebd. Seances Acad. Sci.* **178**, 125–128.

Sulkin, S. D., and Epifanio, C. E. (1975). Comparison of rotifers and other diets for early larvae of the blue crab *Callinectes sapidus* Rathbun. *Estuarine Coastal Mar. Sci.* **3**, 109–113.

U.S. Department of Health, Education, and Welfare. (1969). "Report of the Secretary's Commission to Environmental Health." US Govt. Printing Office, Washington, D.C.

Vernberg, W. B., DeCoursey, P. J., and Padgett, W. J. (1973). Synergistic effects of environmental variables on larvae of *Uca pugilator. Mar. Biol.* **22**, 307–312.

Waterman, T. H., and Chase, F. A., Jr. (1960). General crustacean biology. *In* "The Physiology of Crustacea" (T. H. Waterman, ed.), Vol. 1, pp. 1–30. Academic Press, New York.

Wells, P. G. (1972). Influence of Venezulan crude oil on lobster larvae. *Mar. Pollut. Bull.* **3**, 105–106.

Wells, P. G., and Sprague, J. B. (1976). Effects of crude oil on american lobster (*Homarus americanus*) larvae in the laboratory. *J. Fish. Res. Board Can.* **33**, 1604–1614.

Williamson, H. C. (1910). On the larval and later stages of *Portunus holsatus, Portunus puber, Portunus depurator, Hyas araneus, Eupagurus berhardus, Galathea dispersa, Crangon trispinosus, Cancer pagurus. Fish. Scotland Sci. Invest., 1909* Vol. 1, pp. 1–20.

Wilson, E. B., and Hunt, J. M. (1975). "Petroleum in the Marine Environment." Natl. Acad. Sci. Washington, D.C.

Zitko, V., and Carson, W. V. (1970). PCB and other industrial halogenated hydrocarbons in the environment. *Fish. Res. Board Can., Tech. Rep.* **217**, 1–29.

CHAPTER 9

Cyathura (Arthropoda: Crustacea: Isopoda: Anthuridae)

W. D. BURBANCK and MADELINE P. BURBANCK

I. Introduction

It is possible that after studying species of a single genus of estuarine isopod, *Cyathura,* for 30 years and after collecting cyathurans in at least 100 localities in 17 Atlantic and Gulf of Mexico states of the United States and six foreign countries, that we shall tell you more about these crustaceans than you really want to know. Nevertheless, our prolonged interest reflects the continuing challenges these animals have made and our unflagging enthusiasm for them. Despite our unabashed pleasure in investigating each range extension and applauding each discovery, we trust that we have been able to maintain our scientific objectivity. We wish to share the information and insights we have accumulated over the years. Species of the genus *Cyathura* occur on both sides of the Atlantic Ocean, and not only is *C. carinata* a "genuine brackish-water organism" of the

293

Baltic Sea (Segerstråle, 1957), but on the western side of the Atlantic Ocean, another species of the genus, *C. polita,* merits the same designation as a member of the estuarine infauna. Carriker (1967) has called such organisms which are limited to estuaries "true estuarine" species.

Since the scene that is most familiar to us today and one that has immediate and critical estuarine problems is in the United States, we shall deal primarily with *Cyathura polita* (Stimpson) (Miller and Burbanck, 1961) and to a lesser extent with *C. burbancki* Frankenberg (Frankenberg, 1965). From time to time we shall compare them with *C. carinata* (Kröyer), the European estuarine cyathuran, which was confused with the North American *C. polita* and *C. burbancki* in early descriptions and distribution reports, and still today appears erroneously in species lists of animals present in marshes and estuaries of the Atlantic coast of the United States.

The role of *Cyathura polita* in estuarine environments was first noted in 1948 during a study of the bottom fauna of Rand's Harbor on Cape Cod, Massachusetts (W. D. Burbanck *et al.,* 1956) when the conclusion was reached that this small estuary was a transition or tension zone between freshwater and marine communities and was, therefore, an ecotone with an assemblage of animals, *C. polita* included, able to live in such an environment. Later, after an analysis of many *C. polita* habitats (Burbanck, 1962a), we concluded that "—*C. polita* lives in an ecotone, and because of its fidelity for this environment, it may be considered to be an index species of an ecotonal area of the estuaries studied."

In 1974, Harman listed the following characteristics for organisms that might be useful as indicator species:

> (1) they should be easily recognized by researchers that are not specialists; (2) they should be abundant in their preferred habitat, throughout a large geographic region; (3) they should exhibit approximately the same degree of tolerance to a particular phenomenon, or be indicative of the same conditions, throughout their range; (4) they should possess a relatively long life span; (5) they should be comparatively sessile, or at least not easily able to avoid temporarily stressed environments by rapid migrations.

Applying Harman's rigorous criteria to *Cyathura polita,* our conviction is that when present as a reproducing population, *C. polita* is an indicator of a healthy estuarine ecotone. *C. burbancki* has been more recently identified and is less well known, but it may prove useful as an indicator species of more saline coastal habitats.

Specific application of Harman's criteria to the American cyathurans of the Atlantic and Gulf coasts reveals that their characteristics agree remarkably well. Considering these criteria in the same order as above, we find that: (1) by using taxonomic keys by Smith (1964), Menzies and Frankenberg (1966), or Schultz (1969), cyathurans and other members of the order Isopoda, family Anthuridae can be recognized easily by the nonspecialist (Fig. 1); (2) breeding populations of

Fig. 1. Living, 20-mm *Cyathura polita* (stroboscopic photograph by Charles Ray, Jr.).

Cyathura polita may reach densities as high as 4000/m² (Burbanck, 1962a) and 13,458/m² (Cammen, 1976b), but typically, population densities are 100–250/m² throughout the long, Atlantic-Gulf coast range of the species; populations of *C. burbancki* have lower numbers and only an Atlantic distribution; (3) with the exception of variations in osmoregulation, which will be discussed later, *C. polita* seems to have virtually the same tolerance to widely fluctuating conditions throughout its long range and can be characterized as a hardy, euryokous animal, while again, *C. burbancki* is less well known but may be even more consistent than *C. polita*; (4) generally, individuals of *C. polita* live at least 2 years, and there is evidence (W. D. Burbanck and Burbanck, 1975) that some may live 3 years; the life-span of *C. burbancki* is not known; and (5) both *C. polita* and *C. burbancki* live interstitially as juveniles and are tubicolous as adults, and therefore, though not actually attached or sessile, their mobility is limited (W. D. Burbanck *et al.*, 1964); breeding populations of *C. polita* are known to have maintained the same locations and densities over a period of 20 years (M. P. Burbanck and Burbanck, 1975).

In addition to fulfilling Harman's criteria for an indicator species, *Cyathura polita* possesses two other valuable characteristics. It has been observed by us and by E. L. Bousfield (personal communication) that through much of its range there is a recognizable "*C. polita* association" consisting of certain species of gammarids, polychaetes, and bivalve mollusks living in the estuarine ecotone with *C. polita* (Table I). Second, cyathurans are large enough to be collected and handled easily, but are not commercially valuable and hence not subject to harvesting by commercial fishermen or to laws regulating their collection.

From our field and laboratory observations, we believe that the American Atlantic coast and Gulf of Mexico cyathurans are hardy representatives of a truly brackish water infauna; that the presence of breeding populations of *Cyathura* is indicative of a biotope with consistent and recognizable ecological characteristics; and that their absence from apparently suitable environments is indicative of stressed conditions deleterious to their existence. When cyathurans are used as monitors of environmental change, however, an evaluation of habitat conditions should include consideration of current perturbations, those of the immediate past, and even long-range geological changes. Moreover, cyathuran populations are dynamic, not static, biological entities, and, therefore, capable of undergoing adaptations in response to long-range geological and climatic changes.

II. Distribution and Habitat

Populations of *Cyathura polita* are found along the Atlantic coast from Saint John River, New Brunswick, Canada (M. P. Burbanck *et al.*, 1979) to Flamingo at the southern tip of Florida (Tabb and Manning, 1961), and from there along the west coast of Florida and the southern shores of Alabama and Mississippi to Pointe au Chien in Lake Pontchartrain, Louisiana (Miller and Burbanck, 1961; Burbanck, 1962b; W. D. Burbanck and G. P. Burbanck, 1967). Repeated attempts to find cyathurans north of their recently located Canadian habitat and west of Lake Pontchartrain in western Louisiana and Texas have failed. *Cyathura burbancki* populations are known from off the coasts of northern Florida (D. Frankenberg, private communication) to Long Island Sound (T. E. Bowman, private communication). It is also possible that some of the cyathurans described by Verrill and Smith (1874) as *Anthura brunnea* from Vineyard Sound and those discovered on buoys in Vineyard Sound off Cape Cod, Massachusetts, by George Gray (M. B. Gray, personal communication, 1958) were actually *C. burbancki*.

Typical of isopods, cyathurans have no larvae and both adults and juveniles are feeble swimmers. Also, since *C. polita* is an obligate estuarine infaunal species, its distribution is disjunct although its range extends over thousands of miles from north temperate to subtropical zones. Populations of *C. burbancki* lie seaward of those of *C. polita,* often beyond the mouths of rivers (Frankenberg, 1965), and, therefore, may have a more continuous distribution throughout their range than does *C. polita.*

The overt conditions for the existence of *Cyathura polita* are now quite well known (Burbanck, 1959, 1962b; Miller and Burbanck, 1961), but data on those for *C. burbancki* are still incomplete. We believe that the following summary of habitat descriptions, with variations and/or exceptions noted, will serve for the total range of *C. polita* from New Brunswick, Canada to Louisiana.

It should be noted that the conditions described are for breeding populations that live well upstream and not for isolated individuals or scatterings of small nonbreeding individuals that are swept downstream after torrential rains or flooding storm tides. Even though breeding populations can withstand sudden and drastic fluctuations in environmental parameters, no populations occur along intermittent water courses, and individual cyathurans are intolerant to drying.

Specifically, the living conditions for *C. polita* are as follows:

1. Generally, they live only where fresh and salt water mix. Typically, the mixture is freshwater drainage from land mixed with seawater, but in the southeastern United States, e.g., St. John's River, Florida, fresh water is mixed with the runoff from salt springs resulting in salinities of usually less than 1‰. *C. polita* and a few other estuarine and marine animals are able to live upstream in this unique river beyond the limits of penetration of tidal salt water in waters made slightly salty by dissolved NaCl derived from underground salt deposits (Kelley and Burbanck, 1972).

2. The water covering the substratum where they live is never quiet for any appreciable length of time, but is kept in motion relative to the amount of runoff from the land, tides, or wave action. In such cyathuran habitats as Seine and Swan Ponds on Cape Cod, Massachusetts, the western shore of Chesapeake Bay, Maryland, and Lake Pontchartrain, Louisiana, wave action caused by prevailing winds produce pronounced ripple marks in the sand which give evidence to the constancy and direction of the winds.

3. They live in simple unlined burrows of their own construction, although it is possible that they may also use annelid worm tubes. Where the substratum is firm or compacted and the overlying water relatively calm, the openings of cyathuran burrows are visible (Burbanck and Payne, 1970), but in substrata beneath running water, the burrows are not open at the surface (W. D. Burbanck and G. P. Burbanck, 1967). Small juvenile cyathurans are interstitial rather than tubicolous.

4. The substratum in which they live must be stable and contains much or little sand with an admixture of plant debris, shell fragments, and clay, and particularly in New Hampshire and Maine, blue clay. In the northern part of their range, the substrata are stable because of a mixture of different sized sediment particles and organic compounds (M. P. Burbanck *et al.,* 1979) such that the upper 7.0 cm in which cyathurans live are neither too compacted for the isopods to penetrate nor so loosely aggregated that they cannot maintain a tubicolous existence. In the unglaciated, southeastern part of the United States, relatively wide and slow moving rivers, often heavily laden with silt, traverse extensive coastal marshes composed of a soft, peaty, or sandy soil with the result that southeastern estuaries are characterized by shifting, unstable bottoms. Cyathurans are unable to become established in such substrata but do occur in South Carolina and Georgia in sandy clay along tidal creeks where the banks have been stabilized by ballast at railroad and highway bridges, among compacted shells used to form ramps for launching boats, and at river mouths stabilized by rip-rap

jetties. In much of Florida and in the Gulf states, the fine sand bottoms of streams and estuaries usually form a satisfactory substratum for cyathuran populations only where the runners of the tape grass, *Vallisneria americana*, or dense growths of the "grass phase" of *Sagittaria* sp. maintain stable conditions. At Flamingo, Florida, a mixture of marl and sand is stable enough for *C. polita* populations (Tabb and Manning, 1961; W. D. Burbanck and G. P. Burbanck, 1967). In northern Florida within St. Marks National Wildlife Refuge, *C. polita* is more abundant among the matted roots and rhizomes of *Juncus roemerianus* in the lower and upper marsh zones than in the tidal streams which drain the area (Subrahmanyam *et al.*, 1976; Kruczynski and Subrahmanyam, 1978).

Although *Cyathura polita* does not usually occur in association with rooted plants in the northern part of its range, marsh grasses and reeds do stabilize the banks of the tidal creeks in the bottom of which the cyathurans live. After digging and sieving many tons of tidal creek bottoms searching for *C. polita*, it was noted that from South Carolina northward, the presence of *C. polita* was correlated with the distribution of two plant genera, *Typha* and *Spartina*. Breeding cyathuran populations were consistently found in tidal streams where the downstream limit of the less euryhaline cattail, *Typha latifolia*, coincided with the upstream penetration of the salt marsh grass, *Spartina alterniflora* (Burbanck, 1962a).

The depth of water covering substrata in which *C. polita* lives varies throughout the range of the species. Ordinarily, at low tide the water is only a few centimeters to knee deep. This depth may increase by 3–4 cm or by 1–4.5 m at high tide depending upon the local amplitude of the tidal change. Maximum and minimum water depths at cyathuran habitats are also influenced by prevailing winds, seasonal tides, and storms with winds of gale or hurricane force, particularly if they coincide with the time of high tides. Populations of *C. polita* also occur in substrata under water that is 6–12 m deep in the Saint John River estuary in New Brunswick, Canada (M. P. Burbanck *et al.*, 1979) and in the Patuxent River and Chesapeake Bay, Maryland at depths up to 10.5 m (J. A. Mihursky, personal communication; Burbanck, 1963; Hamilton and LaPlante, 1972). At all sites, both those with shallow and deep water, the previously described prerequisites of a mixture of fresh and salt water in motion and a suitable, stable substratum exist.

The habitat requirements of *C. burbancki* are less well known than are those of *C. polita*. Boesch (1973) has reported *C. burbancki* from Hampton Roads, Virginia, and both Drs. Marvin Wass in Virginia and Richard Heard in Georgia (private communications) have observed that *C. burbancki* lives in the substratum at the mouths of tidal rivers at salinities greater than 20‰. The original collections of *C. burbancki* were from depths of 15.5–17 m at a distance of 8–14 miles off the coast of Sapelo Island, Georgia, where salinity ranges from 34.6–

36.0‰ (Frankenberg, 1965). Scattered field observations of *C. burbancki* indicate that the juveniles live interstitially and adults are tubicolous in a bottom which appears to be a stable mixture of coarse and medium sand or a mixture of sand, shell, and clay similar to the typical substratum of *C. polita*.

III. Ecological Niche and Associated Organisms

According to a relatively simple but useful definition by Charles Elton (1927), an ecological niche of an animal is its place in the foodchain relative to its food and predators. *Cyathura polita* feeds on detritus and diatoms and also appears to be a facultative predator and scavenger. Setae of oligochaete worms were found in the fecal pellets of cyathurans living in almost fresh Lake Okeechobee, Florida (W. D. Burbanck, unpublished), while in the more saline habitat at Fort Meyers, polychaete setae were found in the alimentary tract of a cyathuran (Payne, 1969). While being maintained under laboratory conditions, *C. polita* has been observed to attack and partially eat gammarids and to feed on tissue of dead elvers, *Crangon septemspinosus,* small fish, and dead or injured cyathurans (Burbanck, 1959). Similar feeding habits have been observed for *C. carinata* (Jazdzewski, 1969).

As to predators, feeding records and stomach analyses indicate that a number of animals use cyathurans as food. Of the vertebrates listed as occurring in association with *C. polita* (Table I), all but three, *Anguilla, Gobionellus,* and *Symphura,* are known to feed upon *Cyathura,* sometimes consuming large numbers of the iospod. It has been observed (Kendall, 1974) that the blue crab (*Callinectes sapidus*) will eat *C. polita.*

On the basis of our own work and that of many others, we have listed in Table I the animals found by us or reported by others as living in association with *C. polita.* Only one reference is given for each species listed although more than one investigator may have reported the association. Our own data include only those animals living with reproducing populations of *C. polita,* and in using the literature, based on our experience, the same criterion was used. Additional species of animals reported as being with *C. polita* may merely be living with scattered cyathurans that are "transients" in suboptimal or uncharacteristic situations because of intraspecific pressure within a breeding population of *C. polita* or because of displacement by catastrophic disturbance of the substratum such as flooding by storm tides and heavy rains.

The letters N, C, S, and G have been used in Table I to designate the North, Central, South Atlantic, and Gulf of Mexico locations of the species associated with *C. polita.* The species marked with an asterisk are those which form a predictable association of animals usually found with *C. polita,* but none of these

TABLE I

Animals Living in Association with *Cyathura polita*

Classification	Organism	Distribution[a]	Reference
		Nemertina	
	Micrura leidyi	C	W. D. Burbanck et al. (1956)
		Annelida	
Oligochaeta			
	Branchiura sowerbyi	S	W. D. Burbanck and G. P. Burbanck (1967)
	Limnodrilus sp.	C	Hogan et al. (1973)
	Oligochaeta spp.	N, C, S, G	Sanders et al. (1965)
	Pristina osborni	S	W. D. Burbanck and G. P. Burbanck (1967)
Hirudinea			
	Glossiphonia complanata	N, C	Frankenberg and Burbanck (1963)
Polychaeta			
Nephtyidae			
	Ciglophamus sp.	N	M. P. Burbanck et al. (1979)
Nereidae			
	Laeonereis culveri	S, G	Frankenberg and Burbanck (1963)
	Nereis diversicolor	N	Burbanck (1959)
	N. succinea[b]	N, C, S, G	Burbanck (1961)
	N. virens	N, C	Burbanck (1959)
Capitellidae			
	Heteromastus filiformis	C, S	Burbanck (1962a)
Spionidae			
	Scolecolepides viridis[b]	N, C, S, G	Jones and Burbanck (1959)
	Streblospio benedictis	N, C, S	Sanders et al. (1965)
Orbinidae			
	Scolopolos fragilis[b]	N, C, S, G	Frankenberg and Burbanck (1963)
Pectinariidae			
	Pectinaria gouldii	C, S	Burbanck (1962a)
Ampharetidae			

300

Mollusca

Gastropoda

Species		Reference
Amnicola limnosa	N	M. P. Burbanck *et al.* (1979)
Goniobasis floridanus	S	W. D. Burbanck and G. P. Burbanck (1967)
Ilyanassa obsoleta	C, S	Burbanck (1961)
Littorina irrorata	S, G	Burbanck (1962b)
Nassarius trivittata	C	Burbanck (unpublished)
Neritina sp.	S, G	Burbanck (1962b)
Pomacea paludosa	S	Burbanck (unpublished)
Valvata tricarinata	N	M. P. Burbanck *et al.* (1979)
Viviparus georgianus	S	W. D. Burbanck and G. P. Burbanck (1967)

Pelecypoda

Species		Reference
Amygdalum papyrium	G	W. D. Burbanck and G. P. Burbanck (1967)
Anodonta imbecilis	S	Burbanck (unpublished)
A. implicata	N	Frankenberg and Burbanck (1963)
Brachidontes recurvus	S	Kendall (1974)
Carunculina minor	S	W. D. Burbanck and G. P. Burbanck (1967)
Crassostrea virginica	C, S	Burbanck (1961)
Elliptio buckleyi	S	W. D. Burbanck and G. P. Burbanck (1967)
Geukensia demissa	C	Burbanck (1962a)
Macoma balthica[b]	N, C, S	Burbanck (1959)
M. mitchelli	S, G	W. D. Burbanck and G. P. Burbanck (1967)
Mercenaria mercenaria	C, S	Burbanck (1961)
Modiolus modiolus	S	Burbanck (unpublished)
Tagelus plebius[b]	C, S, G	Payne (1969)
Mulinia lateralis[b]	C, S, G	W. D. Burbanck and G. P. Burbanck (1967)
Musculium transversum	S	W. D. Burbanck and G. P. Burbanck (1967)
Mya arenaria	N, C, S	Burbanck (1959)
Mytilopsis leucophoetus	S	Burbanck (unpublished)
Polymesoda caroliniana	S, G	Payne (1969)
Rangia cuneata	S, G	Burbanck (1962b)

(Continued)

TABLE I (*Continued*)

Classification	Organism	Distribution[a]	Reference
		Arthropoda	
Crustacea			
Ostracoda			
	Chlamodotheca unispinosa	S	W. D. Burbanck (unpublished)
Copepoda			
	Cithadius cyathurae[b]	N,[c] C, S, G[c]	Bowman (1972)
Cirripedia			
	Balanus eburneus	C, S	Burbanck (1961)
	B. improvisus	S	W. D. Burbanck (unpublished)
Cumacea			
	Almyracuma proximoculi	C	Jones and Burbanck (1959)
Tanaidacea			
	Hargeria rapax	C, S	Jones and Burbanck (1959)
	H. savignyi	C	Jones and Burbanck (1959)
	Tanais carolinii	S	W. D. Burbanck (unpublished)
Isopoda			
	Cassidinidea lunifrons	S	Kendall (1974)
	Chiridotea almyra[b]	C, S	Bowman (1955)
	C. nigrescens	S	Kendall (1974)
	Edotea montosa	S	Kendall (1974)
	E. triloba	C	Sanders *et al.* (1965)
	Exosphaeroma papillae	C	W. D. Burbanck (unpublished)
	Idothea balthica	C	W. D. Burbanck (unpublished)
	Sphaeroma destructor	S	W. D. Burbanck (unpublished)
	S. quadridentatum	C	W. D. Burbanck (unpublished)
Amphipoda			
	Ampelisca abdita[d]	G	W. D. Burbanck (unpublished)
	Ampithoe longimana[d]	G	W. D. Burbanck (unpublished)
	Carinogammarus mucronatus	C	W. D. Burbanck (unpublished)

	Species		Reference
	C. toalsianum	G	Subrahmanyam *et al.* (1976)
	Cymadusa compta	C	W. D. Burbanck (unpublished)
	Gammarus fasciatus	C	Crumb (1977)
	G. lawrencianus	N	M. P. Burbanck *et al.* (1979)
	G. tigrinus[b]	N, C, S, G[a]	Burbanck (1961)
	Grandidierella nottoni	S, G	Burbanck (1962b)
	Haustorius sp.	S, G	Frankenberg and Burbanck (1963)
	Hyalella azteca	N, S	M. P. Burbanck *et al.* (1979)
	Leptocheirus plumosus	C	Jones and Burbanck (1959)
	Monoculodes edwardsi[d]	S	W. D. Burbanck (unpublished)
	Parahaustorius longimanus	S	Kendall (1974)
	Synchelidium americanum	S, G	W. D. Burbanck (unpublished)
Decopoda			
Caridea	*Alpheus heterochaelis*	S	Kendall (1974)
	Cambarus fallax	S	W. D. Burbanck and G. P. Burbanck (1967)
	Crangon septemspinosa	N	Frankenberg and Burbanck (1963)
	Palaemonetes pugio	C, S	Kendall (1974)
Brachyura			
Portunidae	*Callinectes sapidus*[b]	C, S, G	Burbanck (1961)
Xanthidae	*Rhithropanopeus harrisii*	S, G	Burbanck (1962b)
Ocypodidae	*Uca pugilator*	S	Frankenberg and Burbanck (1963)
	U. pugnax	C, S	W. D. Burbanck *et al.* (1956)
Insecta	*Chironomid larvae*	N, C, S, G	Haefner *et al.* (1969)
	Dolichopodid larvae	S	Kendall (1974)
	Gomphoides williamsoni	S	W. D. Burbanck (unpublished)
	Hexagenia munda	S	W. D. Burbanck (unpublished)
	Orieanthus purpureus	S	W. D. Burbanck (unpublished)

(Continued)

303

TABLE I (*Continued*)

Classification	Organism	Distribution[a]	Reference
Arachnoidea			
Xiphosura			
	Limulus polyphemus (larvae)	C, S	Burbanck (1962a)
Vertebrata	Chordata		
Pisces			
	Acipenser brevirostra (Shortnose sturgeon)	N	Dadswell (1976)
	Anguilla rostrata (elvers) (American eel)	N, C, S, G	Burbanck (1962a)
	Bairdiella chrysura (Silver perch)	S	Kendall (1974)
	Dasyatis sabina (Stingaree)	S	Kendall (1974)
	Fundulus diaphanus (Banded killifish)	C	Hogan *et al.* (1973)
	Fundulus heteroclitus (Munnichog)	C	Hogan *et al.* (1973)
	Gobionellus oceanus (Goby)	S	Kendall (1974)
	Ictalurus catus (White catfish)	S	Kendall (1974)
	Lepomis sp. (Bluegill)	S	McLane (1947)
	Micropogon undulans (Common croaker)	S	Kendall (1974)
	Micropterus salmoides (Large-mouthed bass)	S	McLane (1947)
	Morone americana (White perch)	C	Hogan *et al.* (1973)
	Morone saxatilis (Striped bass)	N, S	Burbanck (1967)
	Pogonias cromia (Black drum)	S	Kendall (1974)
	Pseudopleuronectes americanus (Winter flounder)	N, C	Burbanck (1967)
	Salvelinus fontinalis (Brook trout)	N, C	Burbanck (1967)
	Symphura plagius (Blackcheeked tonguefish)	S	Kendall (1974)
	Trinectes maculatus (Hog choker)	S	Kendall (1974)
Aves			
	Anas rubripes (Black duck)	N, C, S, G	Burbanck (1967)

[a] N = North Atlantic (Saint John River, New Brunswick, Canada to Cape Cod, Massachusetts); C = Central Atlantic (Cape Cod, Massachusetts to Cape Hatteras, North Carolina); S = South Atlantic (Cape Hatteras, North Carolina to Flamingo, Florida); G = Gulf of Mexico (Flamingo, Florida, to Lake Pontchartrain, Louisiana).
[b] Most commonly found with *C. polita.*
[c] Occurrence probable.

are always present, suggesting that they may be somewhat less euryokous than *C. polita*. Although the cyathuran habitat is consistently an estuarine ecotone, there are some environmental parameters which vary along a latitudinal gradient and may account for differences in associated species. Warmer temperatures in the south than in the north is an obvious difference. In general, the estuarine ecotonal areas occupied by *C. polita* in the northern part of its range are narrower in extent and subject to a greater range of salinity and a higher maximum salinity than are the habitats of *C. polita* in the southern Atlantic and Gulf regions. An exception to this generalization is the population of *C. polita* at Flamingo at the southern tip of Florida where salinities are high and according to Tabb and Manning (1961) may reach 70‰ during the summer.

One of the 12 species most commonly found with *C. polita* has a very close association with the isopod. In 1972, Bowman described a harpacticod copepod, *Cithadius cyathurae,* from the upper reaches of Chesapeake Bay living as an ectosymbiont on *Cyathura polita*. He stated that "It seems unlikely that *Cithadius* is limited to Chesapeake Bay, and its distribution will probably be found to approximate that of *Cyathura polita.*" We have identified this copepod on specimens of *C. polita* from Massachusetts and Georgia, and therefore, we believe that its distribution may indeed coincide with that of its cyathuran host.

Too little is known about the assemblage of species living with *Cyathura burbancki* to include a discussion of them comparable to that given for *C. polita*. From what we know at present, *C. burbancki* associates with marine species that live on or in a stable, sandy bottom in coastal waters.

IV. Life Cycle and Latitudinal Variation

Knowledge of the life cycle of an animal is necessary both for the identification of the species and for the evaluation of its place within an ecosystem. Reproduction and growth of cyathurans follow a typical isopod pattern in general, but there are certain distinctive peculiarities, and some variations among populations appear to be related to the latitude of the habitat. The following discussion pertains to *Cyathura polita* unless reference is made specifically to other species of *Cyathura*.

As is characteristic of isopods, embryonic cyathurans are carried in a ventrally located marsupium made up of overlapping oostegites on the third, fourth, and fifth thoracic segments and extending posteriorly over the sixth segment. One cyathuran peculiarity is that copulation has never been observed nor have seminal receptacles been identified in the females. The possibility of parthenogenesis has been suggested (Burbanck, 1967), but Strömberg (1972) has stated that the presence of a vitelline membrane surrounding the eggs in the marsupium at their earliest stage of development suggests an ordinary fertilization. It is assumed that if fertilization does occur, it takes place within the body of the female because embryonic development has already begun in all eggs removed from marsupia for examination, and Strömberg (1972) found no evidence of a micropyle to allow sperm to penetrate the tough chorion membrane that surrounds each egg during the early stages in the marsupium.

Development of the cyathuran embryos is similar to that described for other isopods. In his description of *Limnoria lignorum,* Strömberg (1967) includes outline drawings that show the developmental stages of early cleavage, curved embryo within the two egg membranes, the elongated and straightened embryo enclosed by only the embryonic membrane, and finally the embryo almost fully formed, free of any enclosing membrane, and with only a small yolk mass. These stages are delineated more fully by Kelley and Burbanck (1976) in an outline of the developmental stages of *Cyathura polita.* Developing embryos remain in the marsupium for a period of 21–27 days (Kelley, 1969). Embryonic studies are facilitated by the fact that all embryos are in approximately the same stage of differentiation within a marsupium, and single embryos can be removed for examination without disturbing the remaining ones. When shed from the marsupium, individuals have already undergone one real ecdysis, have little or no yolk in the liver lobes, and are fully mobile and ready to begin an interstitial existence in the same substratum in which the tubicolous adults live.

Both the length of a young animal and the number of embryos within a marsupium vary among the widely distributed populations of *C. polita.* We have found that on Cape Cod, Massachusetts, embryos attain a length of 2.0 mm and that the greater the length of the female the greater the number of embryos in the marsupium. Females with marsupia range from 14 to 25 mm in length and a 19-mm female was observed to be carrying 100 embryos (unpublished). Kelley (1969) has also reported the number of embryos as being relative to the size of the female in studies of both Florida and Massachusetts cyathurans and reported 27 embryos in the brood sac of a 14-mm female from Green Cove Springs, Florida. Kruczynski and Subrahmanyam (1978), working in a marsh at St. Marks on the Gulf of Mexico side of Florida, found no obvious correlation between size of mother and number of young. A characteristic of the St. Marks population, however, is the small size of the cyathurans. Fully developed intramarsupial embryos are 1.0 mm long and gravid females range from 8.3 to 13.0 mm in length, a maximum which is less than the minimum given above for Cape Cod females. The mean number of embryos in the marsupia of St. Marks cyathurans was 14 with a maximum of 32.

Specimens of the European species, *Cyathura carinata,* are similar in size to those in the St. Marks population of *C. polita.* In a Polish population of *C. carinata* (Jazdzewski, 1969), the maximum number of embryos was 45 in a 9.5-mm female, and the longer the female, the greater the number of embryos at a comparable stage of development. The more advanced the stage of development, however, the smaller the number of embryos of *C. carinata*.

Based on monthly collections made for a period of 17 months at St. Marks, Florida, Kruczynski and Subrahmanyam (1978) outline a 2-year life cycle for *C. polita*. Young are produced from May to October with most of the young being shed in early summer. During the following winter, the young cyathurans become sexually mature when 7–10 mm in length, young are produced in the spring and early summer, and most of the individuals of this generation are dead by the end of their second summer having attained a size of 11–13 mm. A few 12–15 mm cyathurans may persist during a second winter. Growth occurs at a rate of about 1 mm increase in length per molt with a total of at least eight molts.

Other populations of *C. polita* in the southern states which have been sampled at random rather than on a regular monthly basis, appear to have cyathurans with life cycles similar to that of the St. Marks population, but the time of production of young may vary. Ovigerous females have been observed at the following times elsewhere in Florida: in January at Silver Glen Springs, in early March at Flamingo, Lake Poinsett, and Lake Okeechobee, in April at Sanford (W. D. Burbanck, unpublished). As stated by Strömberg (1972) regarding the population at Flamingo, breeding seems to occur simultaneously in a whole population. Data from other southern sites, however, indicate that breeding may occur more than once a year, and regardless of the latitudinal location, there may be a few individuals that mature and reproduce before or after the time of maximum shedding of young. Kelley (1967) found production of young cyathurans in May and again in October in a population at the runoff of Green Cove Springs into the St. John's River, Florida, and large specimens were collected in summer, autumn, and winter rather than being more frequent at a single season. Later (1969), he reported that water temperature might be an important factor in the time of reproduction since gravid females were found in February in the runoff at Green Cove Springs where the water was 19–22°C, but none of the cyathurans collected at the same time in the nearby river (water temperature 12°C) were gravid.

Data are not as complete on life cycles of northern cyathurans as on the St. Marks population, but as already indicated, young are larger when shed and attain a greater maximum size on Cape Cod, Massachusetts than at St. Marks, Florida. Available data (Burbanck, 1962a, 1967; W. D. Burbanck and Burbanck, 1975; W. D. Burbanck, unpublished) indicate that Cape Cod cyathurans are shed from the marsupia chiefly in late June and early July (with a few stragglers in August and early September), are 9–11 mm in length by early fall, produce young the following year, and although some may die during the second summer of their existence, others probably are at least 20 mm in length by the fall of their second year, survive a second winter, and are present as very large individuals

(25–27 mm) in the spring of the third calendar year of their life span. Hogan *et al.* (1973) have reported that reproduction times of *C. polita* in the lower Hudson River are comparable to those reported for Massachusetts, and other New England populations are assumed to have a similar life cycle with young being produced only during the summer months rather than throughout the year as in the southern states. There is some indication that cold weather may so shorten the summer season in Maine that occasionally young are produced only very late in the summer (Burbanck, 1967), thus perhaps jeopardizing their chance of survival during the winter.

Although cyathurans living in the northern part of their range are generally larger than those in the south, stressed or unusual environmental conditions may either intensify this regional difference or appear to negate the generalization. The cyathurans at St. Marks National Wildlife Refuge are so small and live in such a peculiar substratum that they may be a distinct ecotype of *C. polita* (Kruczynski and Subrahmanyam, 1978). Two examples of northern populations with small individuals and unusual physical environments are the terminal population in a deep-water estuary in Canada (largest animal a 15-mm male) (M. P. Burbanck *et al.*, 1979) and the population on the rocky shore of the Hudson River at Dennings Point, New York, where the cyathurans were about half the size of those on Cape Cod (Burbanck, 1962b). Conversely, the environmental conditions in the St. John's River system in Florida would appear to be particularly favorable since the cyathurans there approach the size of those in northern populations, commonly being 16–23 mm in length (Kelley, 1967; W. D. Burbanck, unpublished). Maximum sizes at other southern sites not part of the St. John's River system are 20 mm at Lakeport, Florida (W. D. Burbanck, unpublished), 20 mm at Fort Myers, Florida, and 21 mm at Brunswick, Georgia (Payne, 1969). The latitudinal difference of smaller cyathurans in the south than in the north was also reported by Frankenberg and Burbanck (1963).

Another latitudinal difference between cyathuran populations is a consistent variation in the dorsal chromatophoral pattern, specifically the central part of the pattern on the first thoracic segment (W. D. Burbanck and Burbanck, 1961). The difference is not absolute, but in a sample of 20 or more animals, northern and southern specimens of *C. polita* can be distinguished, with specimens from New Jersey being the most northern of the southern type. Variations of pattern in geographically separated populations of the European *Cyathura carinata* have also been observed (W. D. Burbanck and Burbanck, 1964).

The life cycles of cyathuran populations need further study because certain peculiarities of sexual differentiation were not understood when some studies were made on *Cyathura polita,* and there are still unsolved problems concerning the sexuality of cyathurans. As reported by Miller and Burbanck (1961) in the description of the species, *Cyathura polita* males are characterized by a brushlike flagellum on antenna 1 and an appendix masculina on the endopod of the second pleopod. Sperm duct openings occur on the ventral surface of the last thoracic segment (M. P. Burbanck and Burbanck, 1974). The only distinguishing characteristic of a female, aside from the absence of male characteristics, is the presence of a ventral marsupium. Oviduct openings appear on the fifth thoracic segment at the same time that the marsupium is formed but are concealed by the overlapping oostegites. A working method for sexing members of *C. polita* populations has been to determine the least size for a male in the geographic region being studied and to assume that all smaller individuals are juveniles, and all those which are larger than juveniles, but not males, are females. In any given population, the smallest gravid female observed has always been larger than the smallest male leading to the assumption that males mature first. Some reported minimum sizes for males include 12.0 mm for Cape Cod (W. D. Burbanck and Burbanck, 1975), 11.0 mm for Marshfield, Massachusetts and 9.0 mm for Sapelo Island, Georgia (Frankenberg, 1962), 7.3 mm for northwest Florida (Kruczynski and Subrahmanyam, 1978), and 13.0 for Green Cove Springs, Florida (Kelley, 1967).

A puzzling phenomenon to all who have studied populations of *C. polita* has been seasonal variability of sex ratios. In Cape Cod populations, no male cyathurans are found during late summer, but by winter and spring the sex ratio is approximately 1:1 among adult specimens (W. D. Burbanck, unpublished). Dead or dying males have been seen in Massachusetts in June simultaneously with the occurrence of gravid females (Burbanck, 1967) which could account for the lack of males in late summer, but fall collections include males too large to have been produced earlier that same year. In studying the population of *C. polita* at Green Cove Springs, Florida, Kelley (1967) found female to male ratio of 2.4:1 in January and February. If only large animals (19–23 mm) were considered, however, there was a 1:1 ratio in summer and autumn, and in winter all large animals were males. We have observed that on Cape Cod, Massachusetts, during winter and spring, the largest cyathurans were usually males.

The clue to the answers to some of the questions concerning sex ratios and sex determination in populations of *C. polita* is the phenomenon of protogynic hermaphroditism as described by Legrand and Juchault (1961, 1963), Juchault (1966), and Jazdzewski (1969) for the European estuarine species, *Cyathura carinata*. They determined that after functioning as a female and producing young, such cyathurans during subsequent molts could develop into males with typical tufted antennae and appendix masculina on the second pleopods. Only a small number of specimens have actually been observed to undergo this change because it is very difficult to maintain cyathurans under laboratory conditions for extended periods of time. The data on *C. carinata* from populations in France and from the Baltic coast of Poland, however, indicate that the transformation of functional females into males is a regular phenomenon in this European species and explains the observed fluctuations of sex ratios and antennal polymorphism reported by Cléret (1959, 1960) for *C. carinata*.

Observations of *C. polita* made since the publication of the paper by Jazdzewski (1969) and a reexamination of data, both published and unpublished, in the light of reports on reproduction and sexual differentiation in *C. carinata*, suggest that sex reversal from female to male is a widespread phenomenon in populations of *C. polita*. For example, when the concept of protogynic hermaphroditism is applied to data on life cycles of *C. polita* in Massachusetts, possible answers are supplied to unexplained phenomena, particularly the seasonal variation in sex ratios. If after producing young during the summer, some of the females became males during the fall and winter, functioned as males in the spring, and then died, this would explain the absence of males in late summer, since the young produced that year would not yet have become sexually mature. Size classes of males and examination of specimens collected in late summer in Massachusetts (W. D. Burbanck, unpublished) and in Canada (M. P. Burbanck *et al.*, 1979) suggest that some juvenile cyathurans become males without ever having been functional females, and that this sexual differentiation begins at the end of the first summer. During the winter months, then, males present would represent two generations, those produced that year and those produced a year earlier and which had undergone sex reversal from female to male.

As to the ultimate sex and life span of 3 to 4-month-old cyathurans which show no male characters during the winter months, further investigations need to be made. Based on what is known of other American populations and of the European *C. carinata*, it seems possible that on Cape Cod there are at least two options for these "females": (1) some will develop marsupia and produce young the following spring or summer having reached 15–16 mm in size; some of these can undoubtedly then transform to males, but some may die soon after shedding young, or it is even possible that some may continue to grow and molt as "females" without marsupia and produce a second brood the following spring since 25-mm females with marsupia have been collected occasionally in Cape Cod populations; and (2) sexual maturation may be delayed in some young cyathurans and not occur until the second fall and winter of their existence when they may acquire male characteristics or may function as females for the first time during the second spring of their existence. It is not uncommon to find gravid females 20–22 mm in length and it seems unlikely that such a size would be attained in 1 year.

As the complexities of the reproductive cycle of *C. polita* have multiplied, it has become evident that the categories of juvenile, male and female are not entirely adequate. Kruczynski and Subrahmanyam (1978) categorize only those animals possessing oostegites as females. They write of juveniles and adults, of

males and females, but use the term "non-reproductive" for all animals, reg
less of size, which do not possess external sexual characteristics. It is not po
ble, then, to compare their winter figure of an 100% male sex ratio with our
Cape Cod ratio or with Kelley's 2.4:1 ratio for Green Cove Springs, both
which include non-males of adult size as females. In describing the populatio
C. polita at Saint John River estuary in Canada (M. P. Burbanck *et al.*, 1979)
have used the terms "subadult" and "immature male" in addition to th
previously employed. The term juvenile is still used for all individuals sma
than the smallest sexually differentiated cyathuran in the population, with
realization that such an arbitrary figure is approximate rather than absolute. 1
term subadult is used for sexually undifferentiated individuals which are lar
than juveniles. Immature male is used to designate an individual which 1
partially developed appendix masculina but no sperm duct openings 1
brushlike flagella on the first antennae; mature males possess all three exten
male characteristics. Females may be subdivided into two categories, those w
young in their marsupia and those with empty marsupia. In the future, invo
tigators of the reproductive cycle and populations dynamics of *Cyathura* w
need to choose appropriate categories and carefully define the size classes a
explain on what basis sex determinations were made.

Little is known about the reproductive cycle of *C. burbancki* other than th
ovigerous females were present among the samples collected at the type locali
off the coast of Georgia in March and April. There are striking parallels betwee
the life cycles of *C. polita* and the European *C. carinata* including a longer li
span and larger size of individuals in the northern populations in Polar
(Jazdzewski, 1969) than in the more southern population at Arcachon, Franc
(Amanieu, 1969), where reproduction occurs earlier in the season (April) than
Puck Bay, Poland. Cléret (1962) also noted that reproduction occurs at least
weeks earlier at Arcachon than at Luc-sur-Mer on the English Channel. It seem
reasonable to assume that these three species have similar reproductive cycles bi
that the life span varies with latitudinal location of populations.

There is still much to be learned about the reproduction of cyathurans. Legran
and Juchault (1961) described the role of the androgenous gland in sex determi
nation but no similar work has been done on the American species. We hav
studied spermatogenesis and mitotic divisions in early embryos and determine
that for *C. polita* on Cape Cod the chromosome count is $n=22$ and $2n=44$ (W
D. Burbanck and M. P. Burbanck, 1967; M. P. Burbanck and Burbanck, 1970)
Shaw (1972) found that cyathurans could be sexed microscopically on the basi
of internal differentiation of gonads before there was any external sexual dif
ferentiation, and that there was a positive correlation between size and sexua
maturity. More long-term observations of laboratory cultures of maturing cyath
urans are needed to clarify the relationships of increasing age and size to sexua

differentiation. Until cyathurans are observed *in copula,* however, or the mystery of if or when and how fertilization occurs is solved, the story of cyathuran reproduction is incomplete.

V. Physiology

In the Introduction, we stated that *Cyathura polita* is euryokous, i.e., it tolerates a wide range of physical–chemical conditions as well as being able to live in association with many different species. In light of the extensive geographical distribution of the species, it is of interest to know if *C. polita* is equally euryokous throughout its range. The data on the ability of *C. polita* to osmoregulate are relatively complete, but toleration tests of other environmental parameters such as pH, temperature, salinity, and dissolved oxygen are meager and the field data are scattered and regional. With regard to *C. burbancki,* several keen observers (D. Frankenberg, M. B. Gray, R. Heard, and M. Wass, personal communications) independently have made careful observations regarding the location of this species relative to environmental conditions, particularly salinity. They agree that breeding populations of *C. burbancki* live in stable substrata at salinities greater than 20‰ and as high as 34.6–36.0‰ (Frankenberg, 1965). All investigators have observed that breeding populations of *C. polita* live shoreward and usually upstream of populations of *C. burbancki* at low tide salinities of less than 20‰. At present, we have neither personally nor in the literature found any exceptions to this generalization.

The results of some physiological tests that have been performed to test the euryokous nature of *C. polita* show that in general the species is able to tolerate extremes of pH, temperature, and salinity greater than it normally encounters under natural conditions. Tests run on Cape Cod cyathurans for 12 hours (tidal exchange time) gave the following results on limits of toleration: pH, 11.0 and 3.4; temperature, 45° and -8°C; salinity, 140 and 0‰ (Burbanck, 1961). Actually, cyathurans lived for over 2 weeks in distilled water. At low tide, the salinity of the estuary from which Cape Cod experimental animals were collected was 0.5–1.0‰. When using *C. polita* from Lake Okeechobee and the upper Caloosahatchee River, Florida (not part of the St. John's River system) where there was no measurable salinity, $LD_{50_{12}}$ tests showed salinity tolerance up to 85‰ (W. D. Burbanck, unpublished). Osmoregulation studies have shown that cyathurans living in certain parts of the St. John's River system at 1.22‰ can regulate to somewhat higher experimental salinities but break down at salinities lower than those found in their natural habitats (Segal and Burbanck, 1963). The respiration of *C. polita* is aerobic, resembles that of other isopods, and does not differ significantly between Georgia and Massachusetts populations tested, and oxygen consumption was remarkably consistent at salinities of 0, 10, 20, 30, and 40‰ (Frankenberg and Burbanck, 1963).

A series of studies on the osmoregulation of *C. polita* living at various latitudes and throughout its life cycle have been most illuminating. Both Frankenberg and Burbanck (1963) and Segal and Burbanck (1963) found that cyathurans from Silver Glen Springs, Florida; Sapelo Island, Georgia; Pocasset River (Cape Cod), Massachusetts; and Marshfield (north of Cape Cod), Massachusetts, are isosmatic at salinities between 28 and 42‰ and exhibit hyperosmotic regulation between 1.22 and 28‰. Below 1.22‰, there are differences in osmoregulatory ability among the populations. In Massachusetts, both the Marshfield and the Pocasset River populations of *C. polita* experience daily tidal changes in salinity from about 0.5 to 29.0‰, and at Sapelo Island, *C. polita* lives where at some times in the year tidal exchange barely alters the salinity (28.4‰), but spring rains can freshen the water to 21.2‰. Flood waters from the nearby mouth of the Altamaha River may sometimes cover this exposed Sapelo Island, mainland side, beach habitat of *C. polita* with essentially fresh water. Cyathurans from these three populations, Marshfield, Pocasset, and Sapelo, were able to osmoregulate in experimental salinities of less than 1.0‰. Cyathurans from Silver Glen Springs (St. John's River system), however, which live naturally at a constant salinity of about 1.22‰, were not able to osmoregulate, and one-third of the experimental animals died when subjected to 1.0‰ and to distilled water for 48 hours.

Furthermore, temperature can affect osmoregulation. The above experiments with cyathurans from Pocasset River and Silver Glen Springs were performed at 22°C (Segal and Burbanck, 1963). Over an experimental range of temperatures from 5°–32°C, the Pocasset animals continued to be able to osmoregulate when subjected to various salinities. At 32°C, however, the Silver Glen Springs animals had a breakdown of their osmoregulatory mechanisms at both high and low experimental salinities. Again, it was the population from a habitat where the parameter being tested had a naturally large amplitude (temperature range of 2–23°C at Pocasset) which was more euryokous than the Silver Glen Springs populations where the annual temperature range of the water is 20–23°C.

The ability of *C. polita* to tolerate higher salinities than those usually present in their estuarine habitats raises the question as to why they are not present in saline coastal regions or in the seaward ends of estuaries where salinities are as high as those tolerated by *C. polita* in the laboratory. One explanation might be that only adult cyathurans can tolerate high salinities. Adult and juvenile *C. polita* from Green Cove Springs, Florida (between the mouth of the St. John's River and Lake George) were tested for their ability to tolerate different salinities (Kelley and Burbanck, 1972). Within the range of 0–32‰, juveniles and adults are not different in their ability to osmoregulate, but at 40‰, adults regulate significantly better than juveniles, although both age groups are well within their range of tolerance. To test further the effects of stress, both adults and juveniles were exposed to 44,000 rads of gamma irradiation from a ^{137}Cs source. The radiation affected the ability of both age groups to osmoregulate; irradiated juveniles and adults at 40‰ had less ability to osmoregulate than nonirradiated cyathurans of comparable age, and of those irradiated, the adults osmoregulated significantly better than did the juveniles.

Perhaps one of the best correlations between physiological capacity and the distribution of *C. polita* is the study of the osmoregulatory mechanisms of developing embryos in relation to the location of breeding populations (Kelley and Burbanck, 1976). We have pointed out earlier that breeding populations of *C. polita* usually live in substrata covered by water at or less than 20‰, or sometimes somewhat higher. Kelley and Burbanck (1976) showed that in water of 0-20‰, embryonic membranes function normally in their osmotic capacities but do not function normally at 30 and 40‰. At these higher salinities, embryos die. How the prospective cyathuran mothers ''know'' where to remain or to go to water of proper salinity so their susceptible embryos held in marsupia of overlapping oostegites may develop normally is one of those mind boggling questions of nature.

Since we have regularly found cyathurans living in stable substrata covered by water moved by tides, wind, or gravity and usually well oxygenated, it was something of a surprise to find the dissolved O_2 concentration of the interstitial water of their substrata to be as low as 0.3-1.7 ml/liter in the southeastern states (W. D. Burbanck and G. P. Burbanck, 1967). A possible explanation is that, being tubicolous animals, cyathurans keep water in their burrows in motion by movement of their pleopods (further evidenced by colonies of sessile vorticellid ciliates living as ectocommensals on their pleopods) and pump in oxygenated water from the overlying water column. We also noted that cyathurans crowded in collection jars died rapidly, and in finger bowls with high bacterial growth they leave their substratum and try to climb out of the bowls. It was, therefore, of interest to find (W. D. Burbanck, unpublished) that Lakeport, Florida, cyathurans (18-20 mm) could survive a lowering of DO from 5.6 ml/liter to 0.13 ml/liter during an experimental interval of 14 hours. Apparently, adult *C. polita* can survive low DO's if not of too long duration.

Cyathurans are obligate aquatic animals. They survive for only a few minutes when air dried. In covered plastic containers with substratum and water from their habitats, they live for weeks or months. Under such conditions, some may produce thin-walled ''cocoons'' of cemented (?) sand grains in which they remain quiescent but alive (M. P. Burbanck, unpublished). Whether this is a reaction to generally suboptimal conditions or to a lack of water movement and/or to evaporation, we do not know.

Analyses of relative amounts of amino acids in mollusk shells have shown differences between families and between specialized and more primitive forms (Hare and Abelson, 1964), and Degens *et al.* (1967) have analyzed data derived from amino acid determinations of a number of shell-forming organisms to discover possible phylogenetic relationships and the effect of differing habitat temperatures on the amino acid composition of certain species. To study cyathuran relationships, the technique of ion exchange chromotography was used to compare species and populations of *Cyathura* (W. D. Burbanck *et al.,* 1969). In addition to specimens of *C. burbancki, C. carinata,* and four geographically distinct populations of *C. polita* which are Anthuridea (Isopoda), specimens of *Lirceus fontinalis* (Asellota), a freshwater isopod collected near Atlanta, Georgia, were included in the analysis. Percentages of 16 α-amino acids, ammonia, and nitrogen were determined and analyzed statistically. All of the anthurids were quite different from the single asellot species. The two Florida populations of *C. polita* on either side of Lake Okeechobee (Fort Myers and St. Lucie) were almost identical but were slightly different from the northern Pocasset, Massachusetts population. The results of the interspecific comparisons of *Cyathura* were somewhat surprising. The European estuarine species, *C. carinata,* differed significantly from *C. burbancki,* the American saltwater species, only in percentages of alamine and ammonia, while populations of *C. polita* differed significantly from *C. burbancki* in 29 combinations representing 10 amino acids, ammonia, and nitrogen. *C. polita* and *C. carinata* differed to about the same extent as the four populations of *C. polita* did among themselves. These data are suggestive but not conclusive, particularly since geographically separated populations of *C. polita* were analyzed but similar samples were not available for *C. burbancki* and *C. carinata* which are both wide-ranging species.

VI. Population Dynamics

Having established that *Cyathura polita* is a euryokous, benthic animal with a life cycle and physiological tolerations that enable it to live in the ecotonal part of an estuary, typically subjected to tide-related salinity changes, and with a geographical distribution that subjects representatives of the species to variations in temperature, tidal amplitude, type of substratum, and associated species of plants and animals, the final question to consider is its ability to maintain a viable population when subjected to external stresses, both extraordinary natural disturbances and man-made perturbations.

During the survey study of populations of *C. polita* on Cape Cod, Massachusetts (Burbanck, 1962a), it was noted that over a period of 1 to 5 years, the density and location of 40 populations remained relatively stable. Only one population suffered irreversible damage and disappeared because of filling, dredging, and building of boat docking facilities in Falmouth Inner Harbor. Nine populations showed decreases in numbers which were apparently related to natural causes and human disturbances. A reexamination of the 39 sites in 1972 (M. P. Burbanck and Burbanck, 1975) revealed that *C. polita* is a resilient species since 38 of the 39 populations were still in existence after 12 years of increasing development of Cape Cod as a resort area and as a home for Boston commuters. The 39th population of *C. polita* subsequently became reestablished (M. P. Burbanck, unpublished).

Consideration of the kinds of environmental changes to which Cape Cod populations of *C. polita* were subjected and survived helps to explain the occurrence of this cyathuran in so many estuarine habitats along the Atlantic coast. For example, in three Cape Cod tidal creeks where populations were well upstream in 1960, natural processes of succession such as siltation of the creek bottom and downstream invasion of the banks by cattails (*Typha latifolia*) with subsequent narrowing of the channel and less penetration of salt water changed the habitats so completely that cyathurans were no longer there in 1972 but did occur downstream where the habitat was more typically estuarine. Similarly, there were man-made disturbances that destroyed cyathurans at five upstream sites, but populations downstream of the 1960 sites were present in 1972. It is not known whether or not a "downstream" population existed at each of these sites at the time of the original study, so the disappearance at one place and appearance at another may have been either a downstream migration of a population or the destruction of the upstream portion of an extensive population. It is known that cyathurans can become established both upstream and downstream of a reproducing population (Burbanck, 1962a), and that in the longer tidal creeks of Cape Cod such as Pocasset River, cyathurans may occur for a considerable linear distance in the upper reaches of the creeks (Sanders *et al.*, 1965; W. D. Burbanck, unpublished).

At one readily observed location on Cape Cod, we are relatively certain that there was no population present downstream of a cyathuran habitat when it was altered and partially destroyed by human activities. When first examined, this cyathuran population was living in a tidal creek a short distance below the dam of a small, freshwater pond. Between 1955 and 1960, the creek bed was filled in and the stream diverted and straightened to join its original course at a point downstream of the cyathuran habitat. In 1960, there were no cyathurans either in the new channel or where it joined the original course of the stream, but a few cyathurans were collected in the swampy remains of the original channel where seepage and surface runoff into the area created a sluggish flow of water. Subsequently, two large motels were built on the filled area, and the swampy, original channel became a low, wet area with stagnant rather than flowing water. No cyathurans were found in this original area in 1972, but a population was present where the new, altered stream course joined the old, natural one. The conclusion is that as conditions worsened, the cyathuran population gradually moved to a more suitable habitat, in this case a distance of about 100–150 m downstream (M. P. Burbanck, unpublished).

Introduction of domestic and industrial wastes into rivers and estuaries may so alter the environment that benthic populations of animals are almost obliterated. When sampled in 1957 (Dean and Haskin, 1964), the 20-km Raritan River estuary between New Brunswick and Raritan Bay, New Jersey, contained no benthic macrofauna in its upper reaches. Following pollution abatement early in 1958, the estuary became repopulated by both freshwater and marine species. In 1957, *Cyathura polita* was present only at a river mouth site (13/m²), but in 1960 was collected from five additional upstream sites with a density of 140/m² at a distance of 19 km from the mouth of the Raritan River. The supposition is that as conditions improved, *Cyathura polita* became reestablished in its typical ecotonal estuarine habitat.

If a population is completely destroyed, however, natural recolonization of a suitable site can occur only if there is a reproducing population nearby. Small populations are particularly vulnerable to destruction. The population of *C. polita* which disappeared between 1960 and 1972 (M. P. Burbanck and Burbanck, 1975) had been a small population in the short runoff of a spring on the shore of Oyster Pond near Chatham, Massachusetts (Burbanck, 1962a). In the late 1960's, dredged material was spread over this portion of Oyster Pond, and when the site was visited in 1972, natural vegetation had become reestablished, but at low tide, there was no clearly defined runoff channel typical of cyathuran habitats, and no cyathurans could be found in the exposed mudflat. In 1975, there was an admixture of sand with the mud, some upwelling of spring water discernable, and specimens of *C. polita* were collected from an area of about 30 m² (W. D. Burbanck, unpublished). It is assumed that recolonization was effected by dispersal of young from other spring runoffs which had not been smothered by dredged material.

Another example of colonization of dredge spoil by cyathurans is given by Cammen (1976a,b) in his study of the colonization by macroinvertebrates of bare and planted spoil material in salt marshes in North Carolina. At the Snow's Cut study area in the Cape Fear estuary, colonization took place under conditions of substratum and salinity suitable for *C. polita* and where there was a population with a very high density of cyathurans in a nearby, natural marsh. At another Cammen study site where dredge spoil had been deposited on the landward side of a coastal barrier island, *Cyathura polita* did not colonize either the planted or the bare areas. This is not surprising since the summer salinity was near oceanic levels and there were no cyathurans in the adjacent natural *Spartina* marsh.

The ability of *Cyathura* to withstand stressed conditions can be studied by making experimental transplantations of cyathurans in addition to observing them under natural conditions. If there are no cyathurans in an apparently suitable habitat, specimens can be transplanted to the site to test its suitability. In 1957, 55 C. *polita* were moved from Pocasset River of the Buzzards Bay drainage of Cape Cod to a similar substratum and upstream location in Town Creek, Sandwich, Massachusetts, on the north side of Cape Cod, in which there were no cyathurans. Three years later, there was a reproducing population of C. *polita* occupying an area that extended 15 m upstream and 15 m downstream of the transplantation site (Burbanck, 1962a). The population has continued to increase in size and now occupies all suitable habitats in Town Creek (now called Mill Creek) (W. D. Burbanck and M. P. Burbanck, 1975; W. D. Burbanck, unpublished). The fauna of Mill Creek had apparently been destroyed by oil and sulfur spills at the northern end of Cape Cod canal during World War II (Burbanck, 1962a), and although the stream recovered biotically in other ways, no cyathurans occurred there naturally in 1957 because there were no reproducing populations near enough to colonize the stream.

Within tidal creeks, some dissemination undoubtedly takes place by tiny juvenile cyathurans being carried passively by tides and colonizing the depositing sides of meanders. There are some reports of cyathurans being able to swim short distances (Omer-Cooper, 1916; Jazdzewski, 1969). Since they are without a swimming larval stage, however, cyathurans have a limited ability for dispersal, and another type of transplantation has demonstrated that C. *polita* is a relatively sessile animal. To determine intramarsh movements, 300 cyathurans were removed from Pocasset River, Massachusetts, labeled with $Zinc_{65}$, and returned to the creek bed. Recovery of labeled cyathurans 12 months later indicated a maximum movement of about 1.1 m (W. D. Burbanck *et al.*, 1964).

Transplantations may also be made to areas where there are no cyathurans and where unfavorable conditions inimical to their survival are suspected. The stressed conditions in the well-known Eel Pond at Woods Hole, Massachusetts, where C. *polita* does not occur naturally, have been revealed by periodically introducing Cape Cod cyathurans at the mouth of a small stream where it enters the pond. Particularly in the summer, Eel Pond receives organic wastes, trash, and oil from the moored boats, boat traffic, and from the homes and laboratory buildings that surround this tidal pond. Over a period of 10 years, we have recovered a few cyathurans from near the site of introduction, indicating that conditions in the pond are suboptimal but not impossible for C. *polita*.

An attempt to transplant southern C. *polita* from Florida to a northern site on Cape Cod (Burbanck, 1962a) failed for several reasons. It was difficult to locate a Cape Cod site that was not already occupied by C. *polita,* the transplant site selected had very little freshwater flow, the bottom was peaty instead of sand and clay, and the marsh supported a large population of *Uca pugnax* which are known to eat cyathurans under laboratory conditions. A few Florida cyathurans survived in this suboptimal habitat but only for several weeks, but this cannot be considered a real test of the ability of *C. polita* to adapt to a latitudinally different habitat because conditions were so unfavorable.

In another attempt to test the ability of cyathurans from Florida to survive in a more northern habitat (W. D. Burbanck, unpublished), *C. polita* from Sanford, Florida were transplanted to the campus of Emory University in the suburbs of Atlanta, Georgia. The Sanford population of *C. polita* is located in fresh water with a high pH at the runoff of Lake Jessup into the St. John's River upstream of Lake Monroe. The site selected at the Emory University Field Station, approximately 500 km from the Atlantic Ocean, was in the runoff of a small spring. Substratum from Sanford, bags of oyster shell from Palatka, Florida, and *Vallisneria* plants were added to the Emory site. Under such conditions, in two independent experiments, cyathurans lived for 6 months and each time died during uncharacteristically cold winters without ever experiencing a mild Georgia winter. Temperatures dropped one winter to $-18.9°C$ ($-2°F$) and the next winter to $-17.8°C$ ($0°F$). Later an attempt to establish cold-hardened cyathurans from Massachusetts at the same site failed when beavers persisted in destroying the "cyathuran habitat" by damming up the runoff from the spring. These experiments, however, emphasize that these estuarine animals are truly adaptable.

VII. Conclusions

At present there are still many aspects of cyathuran ecology that are enigmas wrapped in mystery; among these is the question of the homogeneity of cyathuran species. For *C. polita,* we have demonstrated latitudinal differences in chromatophoral patterns, osmoregulation, reproductive cycles, percentages of amino acids, and size of individuals. For the first time, ecotypic status has been proposed for populations of *C. polita* on the Gulf of Mexico coast of Florida (Kruczynski and Subrahmanyam, 1978). With more latitudinal studies, particularly those dealing with possible chromosomal and electrophoretic differences, more ecotypes may emerge. Also, the question of how fertilization occurs in *Cyathura* should be investigated, and this information used to elucidate whether latitudinal differences are of environmental or genetic origin.

More information on cyathurans should reinforce their value as indicator species. Their fidelity (degree of restriction of a species to a particular situation) (Odum, 1971) to the estuarine situation is so great that the location of reproducing populations points to "healthy" environmental conditions. Reduction in size of a population, no evidence of reproduction within a population, or absence of *C. polita* from a typical estuarine habitat are evidences of disturbed conditions. Experimental transplantations of cyathurans, particularly if tagged with radioisotopes, might be a useful tool for assessment of the degree of naturalness of an estuary. Since latitudinal variations do exist for *C. polita,* care should be exercised in making transplantations or in applying data accumulated for Maine cyathurans, for example, to situations in Florida and *vice versa*.

Even in our present state of knowledge, the following conclusions may prove to be helpful:

1. *Cyathura polita* meets all of the criteria of an indicator species of the estuarine ecotone. When more information is available, *C. burbancki* may serve in the same capacity for more saline habitats.

2. *Cyathura polita* is the species most characteristic of estuaries from New Brunswick, Canada, to Louisiana, and the latitudinal differences that do occur do not lessen its value as an indicator species.

3. Densities of reproducing cyathuran populations, when once determined, are so predictable that abnormal fluctuations or disappearance of such populations are a key to environmental changes.

4. Because they are euryokous, populations of *Cyathura polita* can be transplanted to serve as monitors of possible environmental perturbations.

References

Amanieu, M. (1969). Variations saisonnières de la taille et cycle reproducteur à Arcachon de *Cyathura carinata* (Kröyer). *J. Exp. Mar. Biol. Ecol.* **4**, 79–89.

Boesch, D. F.)1973). Classification and community structure of macrobenthos in the Hampton Roads area, Virginia. *Mar. Biol.* **21**, 226–244.

Bowman, T. E. (1955). The isopod genus *Chiridotea* Harger, with a description of a new species from brackish waters. *Wash. Acad. Sci.* **45**, 224–229.

Bowman, T. E. (1972). *Cithadius cyathurae*, a new genus and species of Tachidiidae (Copepoda: Harpacticoida) associated with the estuarine isopod, *Cyathura polita*. *Proc. Biol. Soc. Wash.* **85**, 249–254.

Burbanck, M. P., and Burbanck, W. D. (1970). Recognition of stages in spermatogenesis in *Cyathura polita* (Crustacea, Isopoda). *ASB Bull.* **17**, 34.

Burbanck, M. P., and Burbanck, W. D. (1974). Sex reversal of female *Cyathura polita* (Stimpson, 1855) (Isopoda, Anthuridae). *Crustaceana (Leiden)* **26**, 110–112.

Burbanck, M. P., and Burbanck, W. D. (1975). Factors affecting survival of *Cyathura polita* (Curstacea, Isopoda) populations in estuaries of Cape Cod, Massachusetts. *ASB Bull.* **22**, 44.

Burbanck, M. P., Burbanck, W. D., Dadswell, M. J., and Gillis, G. F. (1979). Occurrence and biology of *Cyathura polita* (Stimpson) (Isopoda, Anthuridae) in Canada. *Crustaceana (Leiden)* (in press).

Burbanck, W. D. (1959). The distribution of the estuarine isopod *Cyathura* sp., along the eastern coast of the United States. *Ecology* **40**, 507–511.

Burbanck, W. D. (1961). The biology of *Cyathura* sp., an estuarine isopod of eastern North America. *Verh. Int. Ver. Theor. Angew. Limnol.* **14**, 968–971.

Burbanck, W. D. (1962a). An ecological study of the distribution of the Isopod *Cyathura polita* (Stimpson) from brackish waters of Cape Cod, Massachusetts. *Am. Midl. Nat.* **67**, 449–476.

Burbanck, W. D. (1962b). Further observations on the biotope of the estuarine isopod, *Cyathura polita*. *Ecology* **43**, 719–722.

Burbanck, W. D. (1963). Some observations on the Isopod, *Cyathura polita,* in Chesapeake Bay. *Chesapeake Sci.* **4,** 104–105.

Burbanck, W. D. (1967). Evolutionary and ecological implications of the zoogeography, physiology, and morphology of *Cyathura* (Isopoda). *Publ. Am. Assoc. Adv. Sci.* **83,** 564–573.

Burbanck, W. D., and Burbanck, G. P. (1967). Parameters of interstitial water collected by a new sampler from the biotopes of *Cyathura polita* (Isopoda) in six southeastern states. *Chesapeake Sci.* **8,** 14–27.

Burbanck, W. D., and Burbanck, M. P. (1961). Variations in the dorsal pattern of *Cyathura polita* (Stimpson) from estuaries along the coasts of eastern United States and the Gulf of Mexico. *Biol. Bull. (Woods Hole, Mass.)* **121,** 257–264.

Burbanck, W. D., and Burbanck, M. P. (1964). A comparison of dorsal patterns of the estuarine isopods *Cyathura polita* of North America and *C. carinata* of Europe. *Verh. Int. Ver. Theor. Angew. Limnol.* **15,** 865–870.

Burbanck, W. D., and Burbanck, M. P. (1967). Chromosomes of *Cyathura polita* from Pocasset River, Massachusetts. *Biol. Bull. (Woods Hole, Mass.)* **133,** 460.

Burbanck, W. D., and Burbanck, M. P. (1975). Growth of a transplanted population of the estuarine *Cyathura polita* (Crustacea: Isopoda), Cape Cod, Massachusetts. *Verh. Int. Ver. Theor. Angew. Limnol.* **19,** 3001–3006.

Burbanck, W. D., and Payne, R. G. (1970). Substratum preference and population distribution of *Cyathura polita* (Crustacea: Isopoda). *ASB Bull.* **17,** 34.

Burbanck, W. D., Pierce, M. E., and Whiteley, G. C., Jr. (1956). A study of the bottom fauna of Rand's Harbor, Massachusetts: An application of the ecotone concept. *Ecol. Monogr.* **26,** 213–243.

Burbanck, W. D., Grabske, R., and Comer, J. R. (1964). The use of the radio-isotope, zinc 65, in a preliminary study of population movements of the estuarine isopod, *Cyathura polita* (Stimpson, 1855). *Crustaceana (Leiden)* **7,** 17–20.

Burbanck, W. D., Goodchild, C. G., Dennis, E. S., Styron, C. E., and Burbanck, M. P. (1969). Chromatographic studies of three species of *Cyathura* (Isopoda, Anthuridea) and *Lirceus fontinalis* (Isopoda, Asellota). *Verh. Int. Ver. Theor. Angew. Limnol.* **17,** 848–854.

Cammen, L. M. (1976a). Macroinvertebrate colonization of *Spartina* marshes artificially established on dredge spoil. *Estuarine Coastal Mar. Sci.* **4,** 357–372.

Cammen, L. M. (1976b). Abundance and production of macroinvertebrates from natural and artificially established salt marshes in North Carolina. *Am. Midl. Nat.* **96,** 487–493.

Carriker, M. R. (1967). Ecology of estuarine benthic invertebrates: A perspective. *Publ. Assoc. Adv. Sci.* **83,** 442–487.

Cléret, J.-J. (1959). Polytypisme antennulaire et rapport numérique des sexes chez *Cyathura carinata* (Kröyer) (Isopode, Anthuridae). *C. R. Hebd. Seances Acad. Sci.* **248,** 2508–2510.

Cléret, J.-J. (1960). Etude de *Cyathura carinata* (Kröyer) (Isopode, Anthuridae). *Cah. Biol. Mar.* **1,** 433–452.

Cléret, J.-J. (1962). Sur un lot de "Cyathura carinata" (Kröyer) (Isopode, Anthuridae) provenant du Bassin d'Arcachon. Comparaison aux "C. carinata" vivant sur les côtes de la Manche. *Bull. Soc. Linn. Normandie* **3,** 142–144.

Crumb, S. E. (1977). Macrobenthos of the tidal Delaware River between Trenton and Burlington, New Jersey. *Chesapeake Sci.* **18,** 253–265.

Dadswell, M. J. (1976). Biology of the shortnose sturgeon (*Acipenser brevirostrum*) in the Saint John River estuary, New Brunswick, Canada. *Trans. Atl. Chap. Can. Soc. Environ. Biol., Annu. Meet., 1975* pp. 20–72.

Dean, D., and Haskin, H. H. (1964). Benthic repopulation of the Raritan River estuary following pollution abatement. *Limnol. Oceanogr.* **9,** 551–563.

Degens, E. T., Spencer, D. W., and Parker, R. H. (1967). Paleobiochemistry of molluscan shell proteins. *Comp. Biochem. Physiol.* **20,** 553–579.

Elton, C. (1927). "Animal Ecology." Macmillan, New York.

Frankenberg, D. (1962). A comparison of the physiology and ecology of the estuarine isopod *Cyathura polita* (Stimpson) in Massachusetts and Georgia. Ph.D. Dissertation, Woodruff Memorial Library, Emory University, Atlanta, Georgia.

Frankenberg, D. (1965). A new species of *Cyathura* (Isopoda, Anthuridae) from coastal waters off Georgia, U. S. A. *Crustaceana (Leiden)* **8**, 206-212.

Frankenberg, D., and Burbanck, W. D. (1963). A comparison of the physiology and ecology of the estuarine isopod *Cyathura polita* in Massachusetts and Georgia. *Biol. Bull. (Woods Hole, Mass.)* **125**, 81-95.

Haefner, P. A., Jr., Mazurkiewicz, M., and Burbanck, W. D. (1969). Range extension of the North American estuarine isopod crustacean *Cyathura polita* (Stimpson, 1855). *Crustaceana (Leiden)* **17**, 314-317.

Hamilton, D. H., Jr., and LaPlante, R. S. (1972). "Cove Point Benthic Study," 1st Annu. Rep., Ref. No. 72-36. Nat. Resour. Inst., Chesapeake Biol. Lab., Solomons, Maryland.

Hare, P. E., and Abelson, P. H. (1964). Proteins in mollusk shells. *Carnegie Inst. Washington, Yearb.* **63**, 267-270.

Harman, W. N. (1974). Snails (Mollusca: Gastropoda). *In* "Pollution Ecology of Freshwater Invertebrates" (C. W. Hart, Jr., and S. L. H. Fuller, eds.), pp. 275-312. Academic Press, New York.

Hogan, T. M., Williams, B. S., and Zo, Z. (1973). Ecology of the estuarine isopod, *Cyathura polita* (Stimpson), in the lower Hudson River, in Indian Point Benthic Studies 1972, Paper No. 19. *In* "Hudson River Ecology, Proceedings Third Symposium, Bear Mountain, N. Y." (Hudson River Environmental Society, Inc., ed.).

Jazdzewski, K. (1969). Biology of two hermaphroditic Crustacea, *Cyathura carinata* (Kröyer) (Isopoda) and *Heterotanais oerstedi* (Kröyer) (Tanaidacea) in waters of the Polish Baltic Sea. *Zool. Pol.* **19**, 5-25.

Jones, N. S., and Burbanck, W. D. (1959). *Almyracuma proximoculi* gen. et sp. nov. (Crustacea, Cumacea) from brackish water of Cape Cod, Massachusetts. *Biol. Bull. (Woods Hole, Mass.)* **116**, 115-124.

Juchault, P. (1966). Contribution à l'etude de la différenciation sexuelle mâle chez les Crustacés Isopodes. Thèses Fac. Sci., Université de Poitiers. No. 57, pp. 5-111.

Kelley, B. J., Jr., and Burbanck, W. D. (1972). Osmoregulation in juvenile and adult *Cyathura polita* (Stimpson) subjected to salinity changes and ionizing gamma irradiation (Isopoda, Anthuridea). *Chesapeake Sci.* **13**, 201-205.

Kelley, B. J., Jr., and Burbanck, W. D. (1976). Responses of embryonic *Cyathura polita* (Stimpson) (Isopoda: Anthuridea) to varying salinities. *Chesapeake Sci.* **17**, 159-167.

Kelley, B. J. McL., Jr. (1967). Effects of salinity and acute gamma irradiation on the osmoregulation of juvenile and adult estuarine isopods, *Cyathura polita* (Stimpson). M.S. Thesis, Woodruff Memorial Library, Emory University, Atlanta, Georgia.

Kelley, B. J. McL., Jr. (1969). Effect of salinity and ionizing gamma irradiation on the developmental stages of the anthurid isopod, *Cyathura polita* (Stimpson). Ph.D. Dissertation, Woodruff Memorial Library, Emory University, Atlanta, Georgia.

Kendall, D. R. (1974). The ecology of the macrobenthos of a tidal creek, St. Simon's Island, Georgia. M.S. Thesis, Woodruff Memorial Library, Emory University, Atlanta, Georgia.

Kruczynski, W. L., and Subrahmanyam, C. B. (1978). Distribution and breeding cycle of *Cyathura polita* (Isopoda: Anthuridae) in a *Juncus roemerianus* marsh of northern Florida. *Estuaries* **1**, 93-100.

Legrand, J.-J., and Juchault, P. (1961). Glande androgène, cycle spermatogénétique et caractères sexuels temporaires mâles chez *Cyathura carinata* (Kröyer) (Crustacé Isopode Anthuride). *C. R. Hebd. Seances Acad. Sci.* **252**, 2318-2320.

Legrand, J.-K., and Juchault, P. (1963). Mise en évidence d'un hermaphrodisme protogynique fonctionnel chez l'Isopode Anthuridé *Cyathura carinata* (Kröyer) et étude du mécanisme de l'inversion sexuelle. *C. R. Hebd. Seances Acad. Sci.* **256**, 2931-2933.

McLane, W. M. (1947). The seasonal food of the large-mouth black bass, *Micropterus salmoides floridanus* (Lacepede), in the St. Johns River, Welaka, Florida. *Q. J. Fla. Acad. Sci.* **10**, 103-138.

Menzies, J. J., and Frankenberg, D. (1966). "Handbook on the Common Marine Isopod Crustacea of Georgia." Univ. of Georgia Press, Athens.

Miller, M. A., and Burbanck, W. D. (1961). Systematics and distribution of an estuarine isopod crustacean, *Cyathura polita* (Stimpson, 1855), new comb., from the Gulf and Atlantic seaboard of the United States. *Biol. Bull. (Woods Hole, Mass.)* **120**, 62-84.

Odum, E. P. (1971). "Fundamentals of Ecology," 3rd ed. Saunders, Philadelphia, Pennsylvania.

Payne, R. G. (1969). A comparative study of population dynamics of the estuarine isopod *Cyathura polita* (Stimpson) from Florida and Georgia. M.S. Thesis, Woodruff Memorial Library, Emory University, Atlanta, Georgia.

Pettibone, M. H. (1977). The synonymy and distribution of the estuarine *Hypaniola florida* (Hartman) from the east coast of the United States (Polychaeta: Ampharetidae). *Proc. Biol. Soc. Wash.* **90**, 205-208.

Sanders, H. L., Mangelsdorf, P. C., Jr., and Hampson, G. R. (1965). Salinity and faunal distribution in the Pocasset River, Massachusetts. *Limnol. Oceanogr.* **10**, Suppl., R216-R229.

Schultz, G. A. (1969). "How to Know the Marine Isopod Crustaceans." W. C. Brown, Dubuque, Iowa.

Segal, E., and Burbanck, W. D. (1963). Effects of salinity and temperature on osmoregulation in two latitudinally separated populations of an estuarine isopod, *Cyathura polita* (Stimpson). *Physiol. Zool.* **36**, 250-263.

Segerstråle, S. G. (1957). Baltic Sea. *Mem., Geol. Soc. Am.* **67**, 751-800.

Shaw, U. (1972). Sexual maturity of *Cyathura polita* (Stimpson, 1855) (Isopoda: Anthuridae) as related to size. M.S. Thesis, Woodruff Memorial Library, Emory University, Atlanta, Georgia.

Smith, R. I. (1964). "Keys to Marine Invertebrates of the Woods Hole Region," Contrib. No. 11. Syst. Ecol. Program, Mar. Biol. Lab., Woods Hole, Massachusetts.

Strömberg, J.-O. (1967). Segmentation and organogenesis in *Limnoria lignorum* (Rathke) (Isopoda). *Ark. Zool.* [2], **20**, 91-139.

Strömberg, J.-O. (1972). *Cyathura polita* (Crustacea, Isopoda), some embryological notes. *Bull. Mar. Sci.* **22**, 463-482.

Subrahmanyam, C. B., Kruczynski, W. L., and Drake, S. H. (1976). Studies on the animal communities in two north Florida salt marshes. Part II. Macroinvertebrate communities. *Bull. Mar. Sci.* **26**, 172-195.

Tabb, D. C., and Manning, R. B. (1961). A check-list of the flora and fauna of northern Florida Bay and adjacent brackish waters of the Florida mainland collected during the period July, 1957 through September, 1960. *Bull. Mar. Sci. Gulf Caribb.* **11**, 552-649.

Verrill, A. E., and Smith, S. I. (1874). Report upon the invertebrate animals of Vineyard Sound and adjacent waters. *U. S. Comm. Fish Fish. Rep., 1871-1872* pp. 295-852.

CHAPTER 10

Isopods Other Than *Cyathura* (Arthropoda: Crustacea: Isopoda)

CHARLES E. POWELL, JR.

I. Introduction

The order Isopoda is an extremely successful group. Representatives of the more than 4000 species of isopods occur in the world's oceans from hadal depths (Birstein, 1963) to surface waters, in estuaries and fresh waters, and even on land (Kaestner, 1970). In their exploitation of terrestrial habitats, the isopods are the most successful of the many crustacean groups (Kaestner, 1970). In aquatic systems, the isopods are represented by swimmers, crawlers, borers, burrowers, and clinging forms. Most isopods are omnivorous, a few are predators, and a substantial number are parasites, either temporary parasites of fish, such as the flabelliferan families Cymothoidae and Cirolanidae (Richardson, 1905; Schultz, 1969) or obligatory parasites (the suborder Epicaridea) of crustaceans, including other isopods (Anderson, 1975; Bourdon, 1968; Bourdon and Bowman, 1970; Markham, 1973). Isopods are important constituents in the trophic dynamics of estuaries as they recycle plant and animal tissues and are themselves important food items in the diets of bottom-feeding fishes (Bright, 1970; Okata, 1975; Stickney *et al.,* 1975a,b).

In addition to their value as a source of food for fishes, isopods are of economic importance in the negative sense because of the destructive boring of submerged timbers, particularly by species of the genus *Limnoria* (Flabellifera), commonly called "gribbles" (Eltringham, 1965a; Eltringham and Hockley, 1961; Henderson, 1924; Menzies, 1957, 1958; Miller, 1926). In fact, some of the earliest references on the response of isopods to toxic compounds such as metallic salts, naphthols, diphenyl chlorides, and creosote and other oil derivatives, were tested on *Limnoria* species in order to find the best way to kill them (Shackell, 1924; White, 1929). Even in today's more ecologically oriented society, a motion to conserve the "gribble" would receive little support along the world's waterfronts. Thermal pollution, as we shall note elsewhere in this chapter, may actually enhance conditions for *Limnoria,* and bring about even more extensive damage (P. R. O. Barnett, 1971; Beckmann and Menzies, 1960; Naylor, 1965a).

One of the features that distinguish the superorder Peracarida from other Malacostraca (and from all other Crustacea) is the "epimorphic" development of the peracarid larvae. In all peracarids (Isopoda, Tanaidacea, Amphipoda, Mysidacea, and Cumacea), the eggs develop in a brood pouch (the marsupium) formed by overlapping plates (oostegites) on the venter of the thoracic somites of the female (Kaestner, 1970). The early stages of postembryonic development characteristic of most Crustacea (the nauplius, metanauplius, and zoea), which occur as planktonic larvae in other crustaceans, in the isopods take place in the marsupium. The young isopod leaves the brood pouch at the manca stage of development, i.e., as a juvenile that is a miniature version of the adult, with the exceptions that the last pair of peraeopods (thoracic appendages) and the sexual characters are absent (Kaestner, 1970). After several juvenile molts, the adult stage is attained. Metamorphosis between the manca and adult stages occurs only in the suborders Gnathiidea (Monod, 1926) and Epicaridea (Anderson, 1975; Bourdon and Pike, 1972; Kaestner, 1970).

In isopods, unlike most other crustaceans, and even most other peracarids, the respiratory function is not carried out by thoracic gills, but by some of the pleopods (abdominal appendages) (M. B. Jones, 1975; Kaestner, 1970; Wieser,

1963). During molt and the subsequent increase in size the isopods, like all other aquatic crustaceans, take in large amounts of water across the gills (Passano, 1960).

The molting process in isopods is biphasic (Bulnheim, 1974; Klapow, 1972). The isopod sheds the posterior portion of the exoskeleton first and, a day or so later, sheds the anterior portion. In his study of molting in *Excirolana chiltoni* (Richardson) (Flabellifera), Klapow (1972) noted that the split in the exoskeleton occurs between the fifth and sixth thoracic somites and that the anterior portion was shed about 25 hours after the posterior portion. *Excirolana chiltoni* molts at 2-week intervals (Klapow, 1972). Bulnheim (1974) examined the molting process in *Idotea balthica* (Pallas) (Valvifera), and observed an interval of 10 to 17 hours between the posterior and anterior molts. He found that small *I. balthica* molted more frequently than large specimens; males 10 mm in length molted at 2- to 3-week intervals, while males 20 mm long molted at 4-week intervals (Bulnheim, 1974).

Molting is an important consideration in the pollution ecology of estuarine isopods, not so much because it is the process by which the organism attains greater size (Bulnheim, 1974; Kaestner, 1970; Klapow, 1972; Passano, 1960) as it is because the female molt immediately precedes fertilization (Kaestner, 1970), and the animal is under great metabolic stress during molt (Bulnheim, 1974), regardless of the sex of the animal, and the organism is less tolerant of stresses due to pollutants or of natural stresses than other times (Swedmark *et al.,* 1973). The increased effects of specific pollutants during molting are discussed in this chapter.

II. Pollution Effects

A. ORGANIC POLLUTION

I know of no quantitative studies on isopod tolerance to organic pollution. Domestic sewage and sludge form such a variable amalgam (O'Sullivan, 1971) that their ecological impact defies component-specific analysis. We have only distributional data from which to infer tolerance to organic pollution and are thus forced to equate presence of a species with tolerance and absence with intolerance. Table I lists the distributions of some isopod species with reference to their tolerance or intolerance of organic pollution. Tulkki (1968) listed *Cyathura carinata* (Krøyer) (Anthuridea) and *Idotea chelipes* (Pallas) (Valvifera) as "transgressive species", i.e., species that may have extended the range of their distribution in the direction of organic pollution. Additional research may place still more isopod species within this category. Organic pollution may provide food for isopods in the form of particulate organic matter (Littler and Murray, 1975),

TABLE I

TOLERANCE OF ISOPOD SPECIES INFERRED FROM DISTRIBUTIONAL DATA

Species	Intolerant	Tolerant	Location	Reference
Angeliera phreaticola		+	Natal, South Africa	O'Sullivan (1971)
Cyathura carinata		+	Gothenberg, Sweden	Tulkki (1968)
Edotea triloba		+	Long Island Sound, New York	U.S. Dept. of Commerce (1972)
Gnorimosphaeroma oregonensis		+	San Francisco Bay, California	Filice (1959)
Idotea baltica		+	Kiel Bay, Germany	Anger (1975)
I. chelipes		+	Gothenberg, Sweden	Tulkki (1968)
I. neglecta		+	Loch Linnhe and Loch Eil, Scotland	Pearson (1972)
Jaera albifrons	+		Kiel Bay, Germany	Anger (1975)
Ligia occidentalis		+	San Clemente, Calif.	Littler and Murray (1975)
Mesidotea entomon		+	Kokkola Bay, Finland	Bagge (1969)

or isopods may find suitable habitat in such debris of organic pollution as wood pulp and fibers from paper mills (Pearson, 1972).

Of course, the effects of organic pollution are generally far from beneficial, and it is certain that some of the factors that enhance the habitat for some species would tend to exclude others; for example, a change in the texture of the bottom substrate. Furthermore, heavy metals associated with domestic wastes could eliminate species sensitive to these elements (O'Sullivan, 1971), and the high biochemical oxygen demand (BOD) that organic wastes impose on the medium could easily reduce levels of dissolved oxygen to concentrations below those required by many species, particularly more active ones. Furthermore, a large volume of fresh water used to flush the organics into the estuary would, by the induction of salinity stress, increase the oxygen demand of the organism struggling to maintain osmotic equilibrium while, simultaneously, the organic matter present would reduce the amount of available oxygen.

B. PESTICIDES

In addition to direct application, as in the spraying of marshes for mosquito control, estuaries commonly receive pesticides from their tributaries (Butler, 1971; Cox, 1971; Manigold and Schultze, 1969). Although the toxic effects of pesticides on crustaceans has been known for quite some time (Cottam and Higgins, 1946), I know of no relevant studies on the effects of pesticides on estuarine isopods. We can, however, infer that because of the close relation of Crustacea to insects there will be a deleterious effect on the isopod fauna as a direct result of pesticide application. Studies on other Crustacea demonstrate their susceptibility to pesticides as well as certain nonlethal effects that may be anticipated.

Organochlorine pesticides, generally insoluble in aqueous media, are readily adsorbed on particulate matter (Portmann, 1975). Uptake by estuarine organisms, including isopods, may take place by the ingestion of food material tainted with the pesticide, from water passing through the gills, or by diffusion across the cuticle (R. Barnett, 1971; Livingston, 1976). Odum *et al.* (1969) studied the sublethal effects of DDT residues on the fiddler crab *Uca pugnax*, and Klein and Lincer (1974) conducted a similar experiment using dieldrin on *U. pugilator*. In both cases, the insecticide, at sublethal concentrations, interfered with coordination to the point where the organism was no longer able to escape predators. Farr (1977) exposed grass shrimps, *Palaemonetes pugio*, to sublethal concentrations of the insecticide parathion and noted an increase in spontaneous activity; the shrimp did not seek concealment, as they ordinarily would have, and as the controls did, but rather swam actively about the tank. Farr (1977) also noted that the insecticide affected endurance such that the shrimp treated with parathion tired more easily than did the controls, and states that parathion is an anticholinesterase, hence affecting synapses. The phenomenon of "hyperactivity'' followed by increasing restriction of motion and paralysis was observed by Caldwell (1974) when he exposed the crabs *Cancer magister* and *Hemigrapsus nudus* to methoxyclor. Caldwell (1974) stated that although survival of these two species was impaired by the presence of methoxychlor at reduced salinities, osmotic and ionic regulation were not so strongly affected as was the nervous system; nerve adenosinetriphosphatase (ATPase) was inhibited, affecting the control of sodium and potassium at the gill membrane. It is likely that studies on isopods will demonstrate that pesticides also affect these organisms behaviorally, in terms of predator response, and neurologically, in terms of ionic maintenance.

C. Detergents

Detergents, including oil dispersant compounds, may be divided into two categories (Nagell *et al.,* 1974): water-base compounds which although highly toxic to fish and bivalves are not strongly toxic to Crustacea, and solvent-base compounds for which the reverse is true—a much stronger effect on Crustacea than upon fish or bivalve mollusks. This phenomenon is explained by the fact that fish and bivalves possess hydrophylic gill surfaces while Crustacea have a waxy, hydrophobic outer layer on their gills (Nagell *et al.,* 1974).

Kaim-Malka (1972a,b,c) tested the toxicities on *Idotea baltica basteri* Audouin (Valvifera) and *Sphaeroma serratum* (Fabricius) (Flabellifera) of various nonionic detergents in concentrations ranging from 0.1 to 800 mg/liter. The individual detergents affected the two isopod species differently: (a) acid- and ester-base detergents, inactive against *Idotea baltica,* were lethal to *Sphaeroma serratum* at concentrations of 10 to 25 mg/liter; (b) ether-base detergents were four times more toxic to *I. baltica* than to *S. serratum;* (c) alcohol-function detergents were twice as toxic to *S. serratum* than to *I. baltica;* and (d) alkyl-aryl polyoxythelenes were ten times as toxic to *I. baltica* as to *S. serratum.* It is obvious that the different isopod suborders differ greatly in their susceptibility to individual detergent formulas and that much additional work must be done before firm conclusions as to the reasons for this may be ascertained. It is quite possible that the structure of the gills and the presence (Valvifera) or absence (Flabellifera) of operculate uropods to protect the gills may be directly involved.

As alluded to in the Introduction, Swedmark *et al.* (1973) found that, in general, crustaceans are most sensitive to surfactants during molting. Surfactants would demonstrate a strong affinity for the waxy outer layer of the gills, interfering with the uptake of required water (Passano, 1960) and with respiration. The great metabolic stress of molting has, likewise, already been noted.

D. Ammonium Nitrate

Brown (1974) noted that ammonium nitrate was a common element in industrial effluents at Table Bay, South Africa, and tested the lethality of various concentrations of this compound on adult specimens of *Eurydice longicornis, Exosphaeroma truncatitelson,* and *Pontogeloides latipes.* At concentrations of 10 ppm, no effects were observed for any of the species. At 100 ppm *Eurydice longicornis* began to swim erratically and its burrowing activity ceased, but the other species showed no ill effects. At 300 ppm the swimming activity of *Eurydice longicornis* became "more frenzied," and *Exosphaeroma truncatitelson* began swimming upside down at the water's surface. Brown had observed the latter response in the presence of other pollutants and under conditions of low dissolved oxygen. A 300-ppm concentration had no apparent effects upon *Pontogeloides latipes.* When the concentration of ammonium nitrate was increased to 500 ppm, 60% of the *E. longicornis* died, and the remaining 40% continued swimming erratically, but the responses of the other two species remained the same as at a concentration of 300 ppm. At a concentration of 750 ppm, the LD_{50} for *Eurydice longicornis* was 58 min, and for *Exosphaeroma truncatitelson,* 5.5 hours. *Pontogeloides latipes,* although beginning to demonstrate stress after 90 minutes, had no mortalities even after 12 hours. Considering the plethora of compounds contained in industrial effluents, it is indeed unfortunate that more studies like this have not been conducted on isopods.

E. SALINITY

Most of the isopods, and, indeed, most of the organisms, found in estuaries are either of marine origin or of close affinity to marine species (Dorgelo, 1976; Gunter, 1956, 1961; Gunter *et al.,* 1974; Pearse and Gunter, 1957). By definition, salinity fluctuations are characteristic of estuaries, but human activities can alter salinity and salinity distributions within an estuary on a long-term basis, far longer than the tidal or seasonal variations to which estuarine animals must adapt (Gunter *et al.,* 1974; Harrell *et al.,* 1976). Human influences on salinity which immediately come to mind are diversion of water and damming. Diversion will, of course, raise salinity by reducing freshwater flow and, if the water is not returned to the same river or stream from which it has been taken, the salinity will increase at all points below the diversion. If, on the other hand, water diverted from one waterway is discharged into another, the salinity of the receiving body of water may well be reduced. Dams, depending upon their placement, will likewise either raise or lower salinity, and dams can further degrade water quality by changing water circulation patterns, increasing siltation, and by concentrating pollutants (Gunter *et al.,* 1974; Harrell *et al.,* 1976).

Because of their euryhaline marine nature, most estuarine animals, including isopods, can tolerate even full-strength seawater, so the raising of salinity to this level poses no physiological problem to them, but many species in estuaries are at or near the lower limits of their salinity tolerance range such that a further reduction in salinity is lethal (Gunter *et al.,* 1974). This lethality of reduced salinity may be due not only to hypotonicity of the aqueous medium relative to body fluids of the organism, but, perhaps, also to the changes in the relative concentrations of calcium and potassium to sodium and chlorine which occur in the salinity range of 5-8‰ (Dorgelo, 1976), ions physiologically important in maintaining osmotic balance.

There have been many studies listing the distributions of isopods in estuaries relative to salinity (e.g., Conover and Reid, 1975; Eriksen, 1968; Gillespie, 1969; Watling *et al.*, 1974), but we shall concentrate on laboratory studies of salinity tolerance for, as Dorgelo points out (1976, p. 261), "distribution patterns essentially cannot exceed the salt range tolerances." While exceptionally salinity ranges for isopods may be extremely broad, as with the Australian brine pool isopod *Haloniscus searli* Chilton (Oniscoidea), reported to tolerate salinities from 4.2 to 267‰ (Dorgelo, 1976; Ellis and Williams, 1970), this is highly unusual.

In his study of *Idotea baltica* (Pallas), Bulnheim (1974) acclimated his test animals to either 10 or 30‰, and then rapidly immersed them in the other medium. [Dorgelo (1976) considers this approach to be the best method of testing the osmoregulatory ability of an organism.] Bulnheim found (1974), in all cases, an increase in oxygen consumption due to compensation for the change in salinity, but found significant differences in the lengths of time required for compensation. When the animals were transferred from 10 to 30‰, a metabolic steady state was attained in 4–6 hours, but for those animals transferred from 30 to 10‰, 30 hours were required for oxygen consumption to return to the normal level. Studies on other species of *Idotea* (Hørlyck, 1973; M. B. Jones, 1974; Naylor, 1955) indicate that salinity is of primary importance in limiting the distributions of the following species: *I. viridis* [= *I. chelipes* (Pallas)] can tolerate all salinities from limnetic to marine, *I. baltica* is limited to brackish and marine waters, and *I. granulosa* Rathke is limited to the marine salinities at the mouths of estuaries.

In the distributions of the asellote species *Jaera albifrons* Leach and *J. nordmanni* (Rathke), however, competition, rather than salinity, determines distribution (Naylor *et al.*, 1961). Survival of the two species in salinities from 3.4 to 34‰ is about equal (Harvey *et al.*, 1973; M. B. Jones, 1972; Naylor and Haahtela, 1966; Naylor and Slinn, 1958; Naylor *et al.*, 1961; Sjöberg, 1967), but *Jaera albifrons* "outcompetes" *J. nordmanni* in brackish waters and hence limits its distribution (Naylor *et al.*, 1961).

The role of salinity in determining the distribution of *Sphaeroma* species has been studied by Jansen (1970) and Kerambrun (1970). After the Berre estuary was dammed, salinities dropped from about 30‰ to between 9 and 13‰, and the more brackish-water *Sphaeroma serratum* (Fabricius) was replaced by *S. hookeri* Leach as, in this instance, the division between the ecological niches of the two species was solely salinity (Kerambrun, 1970). In a study comparing the effects of salinity on *Sphaeroma hookeri* Leach and *S. rugicauda* Leach, Jansen

1970) found that not only did adult *S. hookeri* survive better at reduced salinities, but *S. hookeri* produced greater numbers of young and more viable young than did *S. rugicauda*. *Sphaeroma rugicauda* tends to lose fluid from the blood to the cells in dilute media (Harris, 1970), and the permeability of the integuement doubles at molt (Lockwood and Inman, 1973). Further data on the tolerance of *S. hookeri* to fluctuations in ecological variables may be found in de Casabianca and Kerambrun (1972).

. TEMPERATURE

The most common source of human influence on temperature in estuarine waters is in the form of heated effluents from industries and power plants (P. R. O. Barnett, 1971; Markowski, 1959; Naylor, 1955). *Eurydice* sp. and *Gnathia* spp. are among those estuarine isopods apparently unaffected by heated effluents in field studies (Markowski, 1957; Naylor, 1965a).

Bulnheim (1974) acclimated *Idotea baltica* (Pallas) to 5° or 15°C, then transferred the test animals from one temperature to the other. In the transfer from 15° to 5°C, oxygen consumption increased and then returned to normal after about 3 hours. In the animals transferred from 5° to 15°C, however, a metabolic steady state was not reached for 15 hours (Bulnheim, 1974).

Markowski (1959) found *Sphaeroma serratum* (Fabricius) and *S. hookeri* (Leach) near power stations. Harvey *et al.* (1973) observed that the rate of respiration, as measured by the rate of beating of the pleopods, increased with increasing temperature in these species, and Charmantier (1972) found that temperatures at or less than 2°C, or higher than or equal to 30°C partially inhibit osmoregulation in *S. serratum* and *S. hookeri*. He found (1971, 1972) that 10°C was the optimal temperature for osmoregulation in *S. serratum*. Marsden (1973) notes that oxygen consumption by *S. rugicauda* is strictly dependent upon temperature; he could find no compensation for temperature variation in this animal's metabolism. He also observed that physical activity and oxygen consumption increased with increasing temperature; this response had been noted for *Cyathura polita* (Stimpson) (Anthuridea) also (Frankenberg and Burbanck, 1963).

Limnoria tripunctata Menzies has been reported as "thriving" near power plant effluents (P. R. O. Barnett, 1971; see also Naylor, 1965a). The optimal survival temperature for *L. lignorum* (Rathke), *L. tripunctata* Menzies, and *L. quadripunctata* Holthuis is from 5 to 10°C (Eltringham, 1965a), and threshold reproductive temperatures are 10°C for *L. lignorum* and *L. quadripunctata* and 2°C for *L. tripunctata* (Eltringham and Hockley, 1961). As water temperature increases from 5° to 20°C, boring activity by the three *Limnoria* species increases, declining somewhat from 20° to 25°C and declining sharply as temperature increases from 25° to 35°C (Eltringham, 1965b). There is no boring activity at temperatures lower than 5°C or higher than 35°C. In examining the development of *Limnoria* eggs in relation to temperature, Eltringham (1967) found that, whereas egg viability decreased with increasing temperature, the time required for the egg to develop to the manca stage was abbreviated. Hence, in addition to increasing the boring activity of *Limnoria* species (Eltringham, 1965b, 1967), moderate heating of estuarine waters may also accelerate damage to submerged timbers by shortening the developmental period and perhaps increasing the number of broods per year (Eltringham, 1967).

G. RADIATION

The major sources of radioactive pollution in the marine ecosystem are the testing of nuclear weapons, disposal of radioactive waste materials, and effluents from the production of nuclear power (Broom *et al.*, 1975; Fowler *et al.*, 1975; Mauchline and Templeton, 1964; Preston, 1972). In estuaries, however, the greatest of these contributors of radionuclides is nuclear power production (Broom *et al.*, 1975; Mauchline and Templeton, 1964; Preston, 1972).

Before considering the effects of radioactive pollution, however, it is well to point out that there are areas in which levels of radiation are high owing to purely natural processes. De Loyola y Silva (1965) mentioned Guarapari, Espirito Santo, Brazil, as such a place and noted that certain isopods were found there where levels of natural radiation were 0.4 to 2.0 mR/hour and where even the algae upon which some of the isopods fed was radioactive. Although he did not, unfortunately, define the source and nature of the radiation or the physiological state, or even the health, of the isopods he collected, de Loyola y Silva did remark that *Cymodoce barrerae* (Boone) and *Cymodocella guarapariensis* de Loyola y Silva (Flabellifera) do ingest radioactive particles along with their food and that *Microcerberus delamarei* Remane and Siewing (Microcerberidea), an infaunal species, was in continuous and "intimate contact with radioactive materials."

Estuaries may receive any of a fairly wide variety of radionuclides (Mauchline and Templeton, 1964). Broom *et al.* (1975) noted that cesium-137 was a "major constituent of the low-level radioactive liquid effluent" of a nuclear power plant at Windscale, on the Irish Sea in England, and Preston (1972) found that in the Blackwater estuary, also in England, that zinc-65, manganese-54, cesium-137, phosphorus-32, chromium-51, cobalt-60, silver-110, and antimony-124 were retained and concentrated by interactions with estuarine silts. Although few studies have been made of the effects of radionuclides on estuarine isopods, it is apparent from these studies, and from studies on other Crustacea, that two factors are of primary importance in assessing the effects of radioactive pollution on isopods: accessibility of the radionuclide to the organism and, if the radionuclide is concentrated by the organism, the site of the bioconcentration.

H. MANGANESE

Preston (1972) found that manganese, zinc, and like elements, were "readily removed to the sediments" by interactions with estuarine silts and stated that this increased the accessibility of these elements to the estuarine biota. Whereas no similar studies, to my knowledge, have been applied to isopods, Bryan and Ward (1965) exposed the lobster *Homarus vulgaris* to both radioactive and nonradioactive manganese. They found that 80% of the radioactive manganese and 98% of the nonradioactive manganese was concentrated in the lobster's calcified exoskeleton, which was, of course, shed at molt. This correlates with the findings of Bertine and Goldberg (1972) that the heavy metals iron, cobalt, zinc, selenium, antimony, silver, and mercury are concentrated in the shells and exoskeletons of mollusks and crustaceans. This being true for such crustaceans as lobsters (Bryan and Ward, 1965) and shrimp (Bertine and Goldberg, 1972), it seems likely that the concentration of manganese and similar elements in the exoskeleton, whether the element be radioactive or not, will be demonstrated for isopods also.

I. CESIUM

Preston (1972) noted that cesium-137 was not so readily adsorbed by estuarine silts as were manganese and zinc and felt that, considering the flushing time of the Blackwater estuary, cesium would be available to the fauna of that estuary for less time than either manganese or zinc would be. This lack of adsorption, however, might well be a large part of the problem that cesium poses for estuarine organisms in general and estuarine isopods in particular. Broom *et al.* (1975) noted that, in seawater, almost all of the cesium present was in the ionic state and mentioned that "ionic species can pass easily through semipermeable membranes." Cesium, furthermore, is similar to potassium both in its chemistry and its uptake by organisms (Bryan, 1961, 1963a,b; Bryan and Ward, 1962; Broom *et al.*, 1975). Bryan (1963a) exposed *Sphaeroma serratum* (Fabricius) (Flabellifera), a euryhaline marine species commonly found in estuaries, to cesium-137 and found that this radionuclide was accumulated "relatively rapidly to a concentration which exceeds that for" potassium. Bryan (1963b) exposed *S. hookeri* Leach also to cesium-137 and found that this radionuclide was assimilated such that its concentration was almost twice as high as that for potassium. It should be noted also that *S. hookeri* is a characteristically brackish-water species, and Bryan (1963b) states that, in dilute media, such estuarine species accumulate cesium-137 to much higher levels than do marine species at their normal marine salinities. As previously stated, most estuarine species are existing at, or near to, their lower limit of tolerance in terms of salinity and their body fluids are most likely hyperosmotic compared to the medium.

The concentration of cesium-137 is especially critical should the isopod be exposed to this element during the period of molt. Immediately after the old exoskeleton is cast off, the isopod takes in a large volume of water and, because the new exoskeleton is not yet rigid, thereby attains an increase in size (Kaestner, 1970; Lockwood and Inman, 1973; Passano, 1960). At this time the ionic cesium also would be taken in across the gill membranes. Additionally, adult female isopods molt immediately before reproduction (Bulnheim, 1974); thus the high concentrations of cesium, given its similarity to potassium, could reasonably be expected to affect the viability and development of the eggs.

J. SYNERGISTIC EFFECTS OF SALINITY AND TEMPERATURE

In all studies on the effects of varying salinity and temperature (e.g., Bulnheim, 1974; Harvey *et al.,* 1973; Jansen, 1970; M. B. Jones, 1972; Wilson, 1970), estuarine isopods survived best in conditions of high salinity (30–34‰) and low temperature (4°–10°C). Salinity stress and temperature stress each cause an organism to consume greater amounts of oxygen than normally required because the organism is metabolically coping with the stress (Bulnheim, 1974), and the hypotonicity of the medium may well cause a loss of potassium and sodium from the hemolymph (Harris, 1970; Lockwood and Inman, 1973). The reduced capacity of warm water to retain dissolved oxygen in comparison to that of cooler water means that less oxygen is available to the organism at the precise time that it requires it. Adding still further to the effects of hypotonicity of the medium is the fact that at salinities of 5–8‰, ionic concentrations of calcium and potassium to sodium and chlorine change in their relative proportions, perhaps depriving organisms of the very ions which they need to maintain osmotic equilibrium (Dorgelo, 1976; Harris, 1970; Lockwood and Inman, 1973). Table II lists the isopod species for which data on the synergistic effects of salinity and temperature are known.

K. SYNERGISTIC EFFECTS OF SALINITY, TEMPERATURE, AND HEAVY METALS

M. B. Jones studied the effects of salinity and temperature on the toxicities of mercury (1973) and of cadmium, zinc, and lead (1975) to some species of marine and estuarine isopods. Mercury was tested at various salinities and temperatures on the marine isopods *Idotea emarginata* and *I. neglecta* (Pallas) and on the estuarine species *Jaera nordmanni* (Rathke) and *J. albifrons* Leach. Jones (1973) found that the toxicity of mercury was greatly increased in the presence of decreased salinity and increased temperature and that the estuarine species were more strongly affected than the marine species.

In assessing the effects of salinity and temperature on the toxicities of cadmium, zinc, and lead, M. B. Jones (1975) tested the marine species *Idotea*

TABLE II

Isopods Tested for the Synergistic Effects of
Temperature and Salinity

Species	Reference
Ligia occidentalis	Wilson (1970)
L. pallasii	Wilson (1970)
Idotea baltica	Bulnheim (1974)
I. baltica	Harvey *et al.* (1973)
I. granulosa	Harvey *et al.* (1973)
I. chelipes	Harvey *et al.* (1973)
Dynamene bidentata	Harvey *et al.* (1973)
Campecopea hirsuta	Harvey *et al.* (1973)
Gnathia maxillaris	Harvey *et al.* (1973)
Paragnathia formica	Harvey *et al.* (1973)
Jaera nordmanni	Harvey *et al.* (1973)
Sphaeroma serratum	Harvey *et al.* (1973)
S. monodi	Harvey *et al.* (1973)
S. rugicauda	Harvey *et al.* (1973)
S. rugicauda	Jansen (1970)
S. hookeri	Jansen (1970)
Jaera ischiosetosa	M. B. Jones (1972)
J. albifrons	M. B. Jones (1972)
J. praehirsuta	M. B. Jones (1972)
J. forsmani	M. B. Jones (1972)

baltica, I. neglecta, I. emarginata, and *Eurydice pulchra* and the estuarine species *Jaera albifrons* and *J. nordmanni.* All three of these metals were quite toxic to the marine and the estuarine species, even in waters of marine salinity and at a metal concentration of 10 ppm. Again, a decrease in salinity accompanied by an increase in temperature increased the toxicity of each of these metals. Jones (1975) found also that estuarine species tested near the lower limits of their respective salinity tolerances were much more strongly affected by heavy metal pollution than were marine species tested at their normal salinities. Biochemical tests showed that in all of the species tested, whether marine or estuarine, cadmium, zinc, and mercury altered the osmotic concentration of the hemolymph (M. B. Jones, 1975). Jones (1975) felt that destruction of gill tissues of the pleopods is caused by these heavy metals and that that would account for the effects noted on the osmoregulatory capacities of the animals. Nimmo *et al.* (1977) exposed the pink shrimp *Penaeus duorarum* (Decapoda: Penaeidea) and the grass shrimps *Palaemonetes pugio* and *P. vulgaris* (Decapoda: Caridea) and reported that in these species gill lesions formed after the animals had been exposed to cadmium.

III. Summary

The net result of pollution is an alteration of the environment, and, because many estuarine species are living at or near their physiological limits for such parameters as salinity or dissolved oxygen, any added stress may prove lethal. I feel that, all too often, we tend to overlook the fact that an organism is, at any point in time, reacting to its total environment. The organisms are never reacting to any one given stress—although for purposes of analysis it is convenient to isolate one parameter in order to gain legible data—but are simultaneously reacting to all that affects them: temperature, salinity, pH, dissolved oxygen, ionic concentrations, diet, etc. Each factor that physiologically affects an organism, then, is potentially synergistic with every other factor. The survival limits where noted in this chapter, then, should be taken as those under optimal conditions.

The isopods are a rather convenient group for study in assessing the effects of estuarine pollution; they are common, their taxonomy is in good order, they fill a variety of ecological niches, and isopod life cycles are well-documented. There have been, as evidenced by the contents of this chapter, a good number of well-run and well-written experiments testing the effects of pollutants on isopods and an even larger number of good field studies. Yet and still, admitting that I may have missed some things (for which I accept responsibility), I feel that we are left with more questions than answers. For each of the heavy metals or radionuclides studied, for example, there is at least a half-dozen remaining to be tested. We still have no concept of the effects of heavy metal accumulation in adult female isopods on egg viability and development, or of the effects of cesium concentration on the eggs and young. I have attempted to portray the state of the art as accurately as possibly so as to spur further research, for there is an inestimable amount of research that may still be accomplished.

References

Anderson, G. (1975). Larval metabolism of the epicaridean isopod parasite *Probopyrus pandalicola* and metabolic effects of *P. pandalicola* on its copepod intermediate host *Acartia tonsa*. *Comp. Biochem. Physiol. A* **50**, 747–751.

Anger, K. (1975). On the influence of sewage pollution on inshore benthic communities in the South of Kiel Bay. Part 2, quantitative studies on community structure. *Helgol. Wiss. Meeresunters.* **27**, 408–438.

Bagge, P. (1969). The succession of the bottom fauna communities in polluted estuarine habitats. *Limnologica* **7**, 87–94.

Barnett, P. R. O. (1971). Some changes in intertidal sand communities due to thermal pollution. *Proc. R. Soc. London, Ser. B* **177**, 353–364.

Barnett, R. (1971). DDT residues: Distribution of concentrations in *Emerita analoga* (Stimpson) along coastal California. *Science* **174**, 606-608.

Beckman, C., and Menzies, R. J. (1960). The relationship of reproductive temperature and the geographical range of the marine woodborer *Limnoria tripunctata*. *Biol. Bull. (Woods Hole, Mass.)* **118**, 9-16.

Bertine, K. K., and Goldberg, E. D. (1972). Trace elements in clams, mussels, and shrimp. *Limnol. Oceanogr.* **17**, 877-884.

Birstein, J. A. (1963). Isopods (Crustacea, Isopoda) from the ultraabyssal zone of the Bougainville Trench. *Zool. Zh.* **42**, 814-834.

Bourdon, R. (1968). Les Bopyridae des mers Européenes. *Mem. Mus. Natl. Hist. Nat., Paris, Ser. A* [N.S.] **50**, 77-424.

Bourdon, R., and Bowman, T. E. (1970). Western Atlantic species of the parasitic genus *Leidya* (Epicaridea: Bopyridae). *Proc. Biol. Soc. Wash.* **83**, 409-424.

Bourdon, R., and Pike, R. B. (1972). Description des larves et du développement post-larvaire de *Pseudione affinis* (G. O. Sars). *Crustaceana (Leiden), Suppl.* **3**, 148-154.

Bright, T. J. (1970). Food of deep-sea bottom fishes. *Contrib. Biol. Gulf Mex.* **1**, 245-252.

Broom, M. J., Grimwold, P. D., and Bellinger, E. G. (1975). Caesium-137: Its accumulation in a littoral community. *Mar. Pollut. Bull.* **6**, 24-26.

Brown, A. C. (1974). Observations on the effect of ammonium nitrate solutions on some common marine animals from Table Bay. *Trans. R. Soc. S. Afr.* **41**, 217-223.

Bryan, G. W. (1961). The accumulation of radioactive caesium by crabs. *J. Mar. Biol. Assoc. U.K.* **41**, 55-75.

Bryan, G. W. (1963a). The accumulation of radioactive caesium by marine invertebrates. *J. Mar. Biol. Assoc. U.K.* **43**, 519-539.

Bryan, G. W. (1963b). The accumulation of ^{137}Cs by brackish water invertebrates and its relation to the regulation of potassium and sodium. *J. Mar. Biol. Assoc. U.K.* **43**, 541-565.

Bryan, G. W., and Ward, E. (1962). Potassium metabolism and the accumulation of [137]Caesium by decapod Crustacea. *J. Mar. Biol. Assoc. U.K.* **42**, 199-241.

Bryan, G. W., and Ward, E. (1965). The absorption and loss of radioactive manganese by the lobster *Homarus vulgaris*. *J. Mar. Biol. Assoc. U.K.* **45**, 65-95.

Bulnheim, H.-P. (1974). Respiratory metabolism of *Idotea baltica* (Crustacea, Isopoda) in relation to environmental variables, acclimation processes and moulting. *Helgol. Wiss. Meeresunters.* **26**, 464-480.

Butler, P. A. (1971). Influence of pesticides on marine ecosystems. *Proc. R. Soc. London, Ser. B* **177**, 321-329.

Caldwell, R. S. (1974). Osmotic and ionic regulation in decapod Crustacea exposed to methoxychlor. *In* "Pollution and Physiology of Marine Organisms" (F. J. Vernberg and W. B. Vernberg, eds.), pp. 197-223. Academic Press, New York.

de Casabianca, M.-L., and Kerambrun, P. P. (1972). Ecologie comparee de *Sphaeroma ghigii* et *Sphaeroma hookeri* (Crustacea: Isopoda Flabellifera) dans les étangs Corses. *Tethys* **4**, 935-946.

Charmantier, G. (1971). Recherches physiologiques chez *Sphaeroma serratum* Fabricius (Isopode, Flabellifere). Influence de la température et de la concentration du milieu externe sur la régulation ionique. *C.R. Hebd. Seances Acad. Sci., Ser. D* **272**, 444-446.

Charmantier, G. (1972). Recherches écophysiologiques chez *Sphaeroma serratum* (Fabricius) *Bull. Soc. Zool. Fr.* **97**, 35-44.

Conover, D. O., and Reid, G. K. (1975). Distribution of the boring isopod *Sphaeroma terebrans* in Florida. *Fla. Sci.* **38**, 65-72.

Cottam, C., and Higgins, E. (1946). DDT: Its effect on fish and wildlife. *U. S., Fish Wildl. Serv., Circ.* **11**, 1-14.

Cox, J. L. (1971). DDT residues in sea water and particulate matter in the California current system. *Fish. Bull.* **69**, 443–450.

de Loyola y Silva, J. (1965). Marine isopods from highly radioactive places (Guarapari, Espirito Santo). *An. Acad. Bras. Cienc.* **37**, Suppl., 259.

Dorgelo, J. (1976). Salt tolerance and the influence of temperature upon it. *Biol. Rev. Cambridge Philos. Soc.* **51**, 255–290.

Ellis, P., and Williams, W. D. (1970). The biology of *Haloniscus searli* Chilton, an oniscoid isopod living in Australian salt lakes *Aust. J. Mar. Freshwater Res.* **21**, 51–69.

Eltringham, S. K. (1965a). The effect of temperature upon the boring activity and survival of *Limnoria* (Isopoda). *J. Appl. Ecol.* **2**, 149–157.

Eltringham, S. K. (1965b). The respiration of *Limnoria* (Isopoda) in relation to salinity. *J. Mar. Biol. Assoc. U. K.* **45**, 145–152.

Eltringham, S. K. (1967). The effects of temperature on the development of *Limnoria* eggs (Isopoda: Crustacea). *J. Appl. Ecol.* **4**, 521–529.

Eltringham, S. K., and Hockley, A. R. (1958). Coexistence of three species of the wood-boring isopod genus *Limnoria* in Southampton water. *Nature (London)* **181**, 1959–1660.

Eltringham, S. K., and Hockley, A. R. (1961). Migration and reproduction in the wood-boring isopod, *Limnoria,* in Southampton water. *Limnol. Oceanogr.* **6**, 467–482.

Eriksen, C. H. (1968). Aspects of the limnoecology of *Corophium spinicorne* Stimpson (Amphipoda) and *Gnorimosphaeroma oregonensis* (Dana) (Isopoda). *Crustaceana (Leiden)* **14**, 1–12.

Farr, J. A. (1977). Impairment of antipredator behavior in *Palaemonetes pugio* by exposure to sublethal doses of parathion. *Trans. Am. Fish. Soc.* **160**, 287–290.

Filice, F. P. (1959). The effect of wastes on the distribution of bottom invertebrates in the San Francisco Bay estuary. *Wasmann J. Biol.* **17**, 1–17.

Fowler, S. W., La Rosa, J., Heyraud, M., and Renfro, W. C. (1975). Effect of different radiotracer labelling techniques on radionuclide excretion from marine organisms. *Mar. Biol.* **30**, 297–304.

Frankenberg, D., and Burbanck, W. D. (1963). A comparison of the physiology and ecology of the estuarine isopod *Cyathura polita* in Massachusetts and Georgia. *Biol. Bull. (Woods Hole, Mass.)* **125**, 81–95.

Gillespie, M. C. (1969). The polyhaline Isopoda of the western Mississippi River delta. *Proc. La. Acad. Sci.* **32**, 53–56.

Gunter, G. (1956). Some relations of faunal distribution to salinity in estuarine waters. *Ecology* **37**, 616–619.

Gunter, G. (1961). Some relations of estuarine organisms to salinity. *Limnol. Oceanogr.* **6**, 182–190.

Gunter, G., Ballard, B. S., and Venkataramiah, A. (1974). A review of salinity problems of organisms in United States coastal areas subject to the effects of engineering works. *Gulf Res. Rep.* **4**, 380–475.

Harrel, R. C., Ashcroft, J., Howard, R., and Patterson, L. (1976). Stress and community structure of macrobenthos in a Gulf Coast riverine estuary. *Contrib. Mar. Sci.* **20**, 69–82.

Harris, R. R. (1970). Sodium uptake in the isopod *Sphaeroma rugicauda* Leach. (sic) during acclimatization to high and low salinities. *Comp. Biochem. Physiol. A* **32**, 763–773.

Harvey, C. E., Jones, M. B., and Naylor, E. (1973). Some factors affecting the distribution of estuarine isopods (Crustacea). *Estuarine Coastal Mar. Sci.* **1**, 113–124.

Henderson, J. T. (1924). The Gribble: A study of the distribution factors and life-history of *Limnoria lignorum* at St. Andrews, N. B. *Contrib. Can. Biol.* [N.S.] **2**, 309–325.

Hørlyck, V. (1973). The osmoregulatory ability in three species of the genus *Idotea* (Isopoda, Crustacea). *Ophelia* **12**, 129–140.

Jansen, K. P. (1970). Effect of temperature and salinity on survival and reproduction in Baltic populations of *Sphaeroma hookeri* Leach, 1814 and *S. rugicauda* Leach, 1814 (Isopoda). *Ophelia* **7**, 177–184.

Jones, M. B. (1972). Effects of salinity on the survival of the *Jaera albifrons* Leach group of species (Crustacea: Isopoda). *J. Exp. Mar. Biol. Ecol.* **9**, 231–237.

Jones, M. B. (1973). Influence of salinity and temperature on the toxicity of mercury to marine and estuarine isopods (crustacea). *Estuarine Coastal Mar. Sci.* **1**, 425–431.

Jones, M. B. (1974). Survival and oxygen consumption in various salinities of three species of *Idotea* (Crustacea, Isopoda) from different habitats. *Comp. Biochem. Physiol. A* **48**, 501–506.

Jones, M. B. (1975). Synergistic effects of salinity, temperature and heavy metals on mortality and osmoregulation in marine and estuarine isopods (Crustacea). *Mar. Biol.* **30**, 13–20.

Kaestner, A. (1970). "Invertebrate Zoology," Vol. III. Wiley (Interscience), New York.

Kaim-Malka, R. A. (1972a). Action des détergents sur deux espèces de Crustaces. *Rapp. P.-V. Reun., Comm. Int. Explor. Mer Medit., Monaco* **21**, 255-258.

Kaim-Malka, R. A. (1972b). Action in vitro des détergents non ioniques sur l'isopode valvifère *Idotea balthica basteri* Audouin 1827. *Tethys* **4**, 51–61.

Kaim-Malka, R. A. (1972c). Action in vitro des détergents non ioniques sur l'isopode *Sphaeroma serratum* (Fabricius). *Tethys* **4**, 587–596.

Kerambrun, P. (1970). Remplacement de *Sphaeroma serratum* par *S. hookeri* dans l'Etang de Berre par suite de sa dessalure. *Mar. Biol.* **6**, 128–340.

Klapow, L. A. (1972). Fortnightly molting and reproductive cycles in the sand-beach isopod, *Excirolana chiltoni*. *Biol. Bull. (Woods Hole, Mass.)* **143**, 568–591.

Klein, M. L., and Lincer, J. L. (1974). Behavioral effects of dieldrin upon the fiddler crab, *Uca pugilator*. *In* "Pollution and Physiology of Marine Organisms" (F. J. Vernberg and W. B. Vernberg, eds.), pp. 181–196. Academic Press, New York.

Littler, M. M., and Murray, S. N. (1975). Impact of sewage on the distribution, abundance, and community structure of rocky intertidal macro-organisms. *Mar. Biol.* **30**, 277–291.

Livingston, R. J. (1976). Dynamics of organochlorine pesticides in estuarine systems: Effects on estuarine biota. *In* "Estuarine Processes" (M. Wiley, ed.), Vol. 1, pp. 507–522. Academic Press, New York.

Lockwood, A. P. M., and Inman, C. B. E. (1973). Changes in the apparent permeability to water at moult in the amphipod *Gammarus duebeni* and the isopod *Idotea linearis*. *Comp. Biochem. Physiol. A* **44**, 943–952.

Manigold, D. B., and Schultze, J. A. (1969). Pesticides in selected western streams. *Pestic. Monit. J.* **3**, 124–135.

Markham, J. C. (1973). Six new species of bopyrid isopods parasitic on galatheid crabs of the genus *Munida* in the western Atlantic. *Bull. Mar. Sci.* **23**, 613–648.

Markowski, S. (1959). The cooling water of power stations: A new factor in the environment of marine and freshwater invertebrates. *J. Anim. Ecol.* **28**, 243–258.

Marsden, I. D. (1973). The influence of salinity and temperature on the survival and behaviour of the isopod *Sphaeroma rugicauda* from a salt-marsh habitat. *Mar. Biol.* **21**, 75–85.

Mauchline, J., and Templeton, W. L. (1964). Artificial and natural radioisotopes in the marine environment. *Oceanogr. Mar. Biol.* **2**, 229–279.

Menzies, R. J. (1957). The marine borer famils Limnoriidae (Crustacea, Isopoda). *Bull. Mar. Sci. Gulf Caribb.* **7**, 101–200.

Menzies, R. J. (1958). The distribution of woodboring *Limnoria* in California. *Proc. Calif. Acad. Sci.* **29**, 267–272.

Miller, R. C. (1926). Ecological relationships of marine wood-boring organisms in San Francisco Bay. *Ecology* **7**, 247–254.

Monod, T. (1926). Les Gnathiidae, essai monographique (morphologie, biologie, systematique). *Mem. Soc. Sci. Nat. Maroc* **13**, 1–667.

Nagell, B., Notini, M., and Grahn, O. (1974). Toxicity of four oil dispersants to some animals from the Baltic Sea. *Mar. Biol.* **28**, 237–243.

Naylor, E. (1955). The ecological distribution of British species of *Idothea* (Isopoda). *J. Anim. Ecol.* **24**, 255–269.

Naylor, E. (1965a). Biological effects of a heated effluent in docks at Swansea, S. Wales. *Proc. R. Soc. London, Ser. B* **144**, 253–268.

Naylor, E. (1965b). Effects of heated effluents upon marine and estuarine organisms. *Adv. Mar. Biol.* **3**, 83–109.

Naylor, E., and Haahtela, I. (1966). Habitat preferences and interspersion of species within the species *Jaera albifrons* Leach (Crustacea, Isopoda). *J. Anim. Ecol.* **35**, 209–216.

Naylor, E., and Slinn, D. J. (1958). Observations on the ecology of some brackish water organisms in pools at Scarlett Point, Isle of Man. *J. Anim. Ecol.* **27**, 15–25.

Naylor, E., Slinn, D. J., and Spooner, G. M. (1961). Observations on the British species of *Jaera* (Isopoda: Asellota). *J. Mar. Biol. Assoc. U. K.* **41**, 817–828.

Nimmo, D. R., Lightner, D. V., and Bahner, L. H. (1977). Effects of cadmium on the shrimps *Penaeus duorarum, Palaemonetes pugio,* and *Palaemonetes vulgaris. In* "Physiological Responses of Marine Biota to Pollutants" (F. J. Vernberg *et al.,* eds.), pp. 131–183. Academic Press, New York.

Odum, W. E., Woodwell, G. M., and Wurster, C. F. (1969). DDT residues absorbed from organic detritus by fiddler crabs. *Science* **164**, 576–577.

Okata, A. (1975). Ecological studies on the biological production of young amberfish community in the Sendai Bay. I. Food chains in the amberfish community. *Bull. Jpn. Soc. Sci. Fish.* **41**, 1247–1262.

O'Sullivan, A. J. (1971). Ecological effects of sewage discharge in the marine environment. *Proc. R. Soc. London, Ser. B* **177**, 331–351.

Passano, L. M. (1960). Molting and its control. *In* "The Physiology of Crustacea" (T. H. Waterman, ed.), pp. 473–536. Academic Press, New York.

Pearse, A. S., and Gunter, G. (1957). Salinity. *Mem., Geol. Soc. Am.* **67**, 129–157.

Pearson, T. H. (1972). The effect of industrial effluent from pulp and paper mills on the marine benthic environment. *Proc. R. Soc. London, Ser. B* **180**, 469–495.

Portmann, J. E. (1975). The bioaccumulation and effects of organochlorine pesticides in marine animals. *Proc. R. Soc. London, Ser. B* **189**, 291–304.

Preston, A. (1972). Artificial radioactivity in freshwater and estuarine systems. *Proc. R. Soc. London, Ser. B* **180**, 421–436.

Richardson, H. (1905). A monograph on the Isopoda of North America. *U. S., Natl. Mus., Bull.* **54**, 1–272.

Schultz, G. A. (1969). "How to Know the Marine Isopod Crustacea." W. C. Brown, Dubuque, Iowa.

Shackell, L. F. (1924). Toxicities of coal tar creosote, creostoe distillates, and individual constituents for the marine wood borer *Limnoria lignorum. Bull. U. S. Bur. Fish.* **39**, 221–230.

Sjöberg, B. (1967). On the ecology of the *Jaera albifrons* group (Isopoda). *Sarsia* **29**, 321–347.

Stickney, R. J., Taylor, G. L., and Heard, R. W., III (1975a). Food habits of Georgia estuarine fishes. I. Four species of flounder (*Pleuronectes:* Bothidae). *Fish. Bull.* **72**, 515–525.

Stickney, R. J., Taylor, G. L., and White, D. B. (1975b). Food habits of five species of young Southeastern United States estuarine Sciaenidae. *Chesapeake Sci.* **16**, 104–114.

Swedmark, M., Granmo, A., and Kollberg, S. (1973). Effects of oil dispersants and oil emulsions on marine animals. *Water Res.* **7**, 1649–1672.

Tulkki, P. (1968). Effect of pollution on the benthos off Gothenburg. *Helgol. Wiss. Meeresunters.* **17,** 209-215.

U. S. Dept. of Commerce (1972). The effects of waste disposal in the New York Bight. Summary Final Report. *U. S. Natl. Mar. Fish. Serv., Inform. Rep.* **2,** 1-70.

Watling, L., Lindsay, J., Smith, R., and Maurer, D. (1974). The distribution of Isopoda in the Delaware Bay region. *Int. Rev. Gesamten Hydrobiol.* **59,** 343-351.

White, F. D. (1929). Studies on marine borers. I. The toxicity of various substances on *Limnoria lignorum. Contrib. Can. Biol. Fish.* [N.S.] **4,** 1-7.

Wieser, W. (1963). Adaptations of two intertidal isopods. II. Comparison between *Campecopea hirsuta* and *Naesa bidentata* (Sphaeromatidae). *J. Mar. Biol. Assoc. U. K.* **43,** 97-112.

Wilson, W. J. (1970). Osmoregulatory capabilities in isopods: *Ligia occidentalis* and *Ligia pallasii. Biol. Bull. (Woods Hole, Mass.)* **138,** 96-108.

CHAPTER 11

Amphipods (Arthropoda: Crustacea: Amphipoda)

DONALD J. REISH and J. LAURENS BARNARD

I. Introduction

The order Amphipoda is a large marine, freshwater, terrestrial group of crustaceans which are divisible into four suborders. The Gammaridea, which the main subject of this chapter, are the ordinary amphipods of greatest diversity and abundance, occurring in almost all aquatic habitats, as well as some damp terrestrial ones. Hyperiidea are entirely marine and pelagic mainly in the bathyal zone; they have various pelagic adaptations and lack a maxillipedal palp. Caprellidea, the skeleton shrimps, are sedentary marine species with thin bodies and vestigial

345

abdomens (one exception, the Caprogrammaridae). The Ingolfiellidea are a tiny interstitial group with predominance of gnathopodal dactyls and widespread distribution through the seas in anchialine habitats.

Gammaridean amphipods are generally flattened laterally and usually have a curved dorsal surface. The body is divided into a head composed of five fused segments, a thorax of seven freely articulating somites, and an abdominal region with an anterior pleon of three articulating segments and a posterior urosome consisting of three articulating segments and a terminal telson. Each segment bears a pair of jointed appendages which are highly variable according to the family or genus. Appendages are arranged accordingly: head—two pairs of antennae, a heavily chitinized mandible, and two pairs of maxillae; thorax—maxillipeds (as the last pair of mouth parts, two pairs of gnathopods that terminate in a glovelike, more or less chelated structure, and five pairs of pereopods; abdomen—three pairs of pleopods and three pairs of uropods. Plates, or coxae, are lateral extensions of the thoracic pereon. Branchiae or gills are attached medial to coxae 2–7 on each side. They are usually simple fleshy plates supplied with a rich capillary network for gaseous exchange.

A. Ecological Occurrence

Marine gammaridean amphipods are present in all oceanic regions of the world, in all habitats and all depths. Semiterrestrial forms, primarily members of the family Talitridae, are found burrowing into high intertidal sandy beaches. Many intertidal species occur along rocky shores where they can be found on fronds or within algal holdfasts, under rocks, or wherever some type of protection can be afforded. Subtidally they are abundant either on or within the sediment and specimens are generally encountered within each quantitative benthic sample taken. Gammarideans are outnumbered by hyperiideans within the pelagic environment; however, they are capable of swimming to another locality, especially when disturbed.

A faunule, in this case, is an aggregate of species of Gammaridea in one geographic province, region, or habitat. Five well-known megafaunules are known in the world: cold temperate of northeast Atlantic Ocean, cold temperate of northwest Pacific Ocean including the Okhotsk Sea, circumsubarctic encompassing the Norwegian basin and the Siberian coast, the warm temperate of northeast Atlantic Ocean including the western Mediterranean Sea, and the circumsubantarctic. Secondarily well-known faunules include a northeastern Pacific warm temperate and a South African warm temperate. Intertidal amphipods have been more extensively studied because of the relative ease in collection. Table I summarizes the number of species and genera of intertidal gammarideans in some selected known faunules. At first view, apparently a greater number of intertidal amphipods are known from California and the

TABLE I

THE NUMBER OF GENERA AND SPECIES OF INTERTIDAL
GAMMARIDEA FROM KNOWN FAUNULES[a]

Region	Genera	Species
California	66	138
Plymouth, England	63	92
Isle of Man	40	66
French Mediterranean Sea	41	72
French Atlantic Ocean	69	109
South Georgia Island	45	62
Falkland Islands	48	63
Magellan, continental	57	77

[a] Modified from Barnard (1969a).

French Atlantic coast, but, undoubtedly, this is more the result of a larger geographical collecting unit than the other examples given. This table does, however, serve to give some indication of the number of species at selected localities. More precise definitions of the number of species in a particular faunule depend upon more data.

1. Habitats

Marine gammaridean amphipods are divisible into five major types according to their ecological position: nestler, domicolous, inquiline, fossorial, and nektonic. Wood and algal borers are minor. Nestlers scarcely dominate the group and they generally hide among algal fronds, under debris, or in crevices. There are many domicolous species which spin tubes or cradles from a web emitted from the apices of four of the 14 main legs. The sticky web often agglutinates silt particles, as in *Corophium acherusicum,* which can become very thick when animals attach tube over tube. The tubes are rarely obligatory homes as species will often use abodes of other phyla, such as *C. acherusicum* using the empty calcareous tubes of the serpulid polychaete *Hydroides pacificus.* Under laboratory conditions specimens will occupy cradlelike nests for months, but this longer period may be the result of absence of competition. Tubes or nests are attached to any hard substrate and thus domicolous amphipods are frequently common on pilings in harbors.

Fossorial species are not too numerous and are confined especially to species in families Haustoriidae, Oedicerotidae, and Phoxcephalidae. Burrowing species are prominent members of the deep-sea fauna because of the loss of other types of habitat. Setae of the distal articles of the posterior pereopods, which are used for burrowing, are more excessive and longer than in species from other families.

Little is known about the duration of the burrow; presumably the activity of many species resembles that of mole or sand crabs.

Inquilinous species are dominant in the tropics, where numerous sedentary invertebrates probably act as hosts for many kinds of commensalistic or sub-parasitic activities of these amphipods. Amphilochids are presumed to be slime-lappers on corals; leucothoids and anamixids live within sponges and tunicates. Gammarideans have been found within sea anemones, and liljeborgiids are associates of the tube-building maldanid polychaetes. Virtually nothing is known of the relationship between the inquilinous species and their hosts.

Nektonic species include all the hyperiidean amphipods that dominate bathypelagic depths. Species either have well-developed swimming or antisinking devices or are associated within medusae or salps. Neritic waters contain a few species of gammarideans of purely nektonic existence which may be predators or be species which are normally benthic but which in various life phases swim upward into the shallow seas for short periods of time for either mating or dispersal.

2. Feeding

Amphipods have a wide variety of feeding modes and thus potentially a variety of responses to various kinds of macromolecular toxicants. The basic or primitive mouthparts are adaptable themselves to many feeding modes, depending on other morphological features. For example, the mouthparts of a nestler such as *Elasmopus* are generally indistinct from those of a domicolous species such as *Gammaropsis* or from those of a fossorial genus *Phoxocephalus*. Fixation to a category of habitat on the basis of mouthparts therefore appears to have little to do with the evolution of these appendages. In inquilinous amphipods the mouthparts become increasingly modified for piercing, sucking, or slime lapping. However, no strong discontinuity exists between the basic and specialized mouthparts as fine morphological transitions occur between the two types. Several families such as Leucothiodae are entirely dominated by inquilinous mouthparts, whereas, certain genera or generic groups in families such as Acanthonotozomatidae and Dexaminidae are partially to fully inquilinous without strong evolutionary discontinuity from their congenera bearing basic mouthparts.

Hundreds of species may appear grossly to have identical mouthparts, but fine specific and functional distinctions have been found by Croker (1967a,b) in burrowing haustoriids. He demonstrated a consistent difference in the pore space between the setae originating from the maxellae, which appears to be correlated with the substrate particle size preferred by the different species. One could assume that the pore space, in analogy to baleen in whales, permits the escape of small substrate silt particles but retains the larger ones, which presumably have food value. Pore space may change with an increase in body size during growth, which would result in larger specimens requiring coarser sediments. In this instance we may have one of the few cases where amphipods might have different habitats according to the stage in their life cycle. If such is the case, then pollution could become a crucial problem to those adults living within the narrow bands of coarser sediments constantly exposed to sludge or dredging activity.

Omnivory is well known in amphipods. One lysianssid, *Orchomene* sp., has been found to feed on diatoms, whereas another, *Lysianassa holmesi,* with almost identical mouthparts, commonly feeds by ingesting mud. Other species of *Orchomene* or *Pseudalibrotus* have been collected primarily in traps baited with meat and vegetable material (Shulenberger and Hessler, 1974; Barnard, 1959), which suggests that many lysianassids are opportunistic feeders on carcass and vegetational windfalls, but return to ordinary feeding methods once these windfalls have been consumed. The presence of large numbers of ecdysical casts in traps suggests deep-sea amphipods exist vegetatively, then accelerate maturation suddenly on appearance of a windfall meal. Hundreds of specimens of *Paralicella* sp. were found to engorge themselves almost to bursting on a windfall feed. Amphipods therefore represent a positive factor in the kind of natural or man-induced pollution that occurs following mass mortalities of larger animals. As facultative scavengers able to clean up fine debris and detritus, they remain available in the environment ready to descend on much larger matter. Biting mandibles, perhaps superfluous to a detritus feeder, suddenly find great usefulness on large carcasses. This is not to say that biting mandibles are not also highly useful to detritus feeders, but the functional morphology of this operation has not been worked out. Amphipods, as well as other crustaceans, undoubtedly play an important role in cleaning up the environment.

Fenchel *et al.* (1975) showed that the domicolous species *Corophium volutator* could only feed effectively when the particle size ranged between 4 and 63 μm in diameter. Furthermore, they were only able to feed upon bacteria which were adsorbed to particles within this size range. *Corophium* would also select natural sediments to autoclaved ones, which is similar to the earlier findings of Wilson (1955) with the polychaete *Ophelia*. The distribution of *C. volutator* is highly dependent upon the presence of at least some sediments within the 4–63 μm size range.

Food habits of four species of Black Sea amphipods were found to vary as to type of food, amount of food, and time of feeding (Greze, 1968). *Dexamine spinosa* ate largely diatoms, brown and green algae, whereas detritus made up nearly 50% of the diet of the other three species. *Gammarus locusta* fed during the night, whereas the other three species fed during the daylight hours. Under laboratory conditions the contents of the intestine of young of *Gammarellus carinatus* changed 13–17 times per day, adult males 5–6 times per day, and adult females 2–4 times per day. These variables are of particular importance in the possible selection and use of amphipods as bioassay organisms. The effect of a particular toxicant may vary considerably because of the feeding state as well as stage in the life cycle.

3. Life Cycle

Amphipods lay yolky eggs into a brood pouch of eight interlacing setose lamellae located ventrally on the thoracic region and medially between some of the pereopods. Egg number varies between 1 and 100+, depending upon the species and age. Thin-bodied, tube-dwelling species appear to lay the fewest number of eggs, but these species generally lay the largest or yolkiest eggs. Developing eggs hatch into juvenile stages which resemble miniature adults; there is no larval stage as in higher crustaceans. The juveniles are retained in the brood pouch for a few hours and are then released. Sheader and Chia (1970) have shown that juveniles of two species of *Marinogammus* may reenter the brood chamber after release, especially in females carrying embryos in advanced stages of development. Under laboratory conditions they observed the juveniles of *M. finmarchius* entering the brood pouch of *M. obtusatus,* but not the reverse.

Eggs are laid by females through two ventral pores located on the sixth thoracic sternite. Expulsion presumably occurs just following ecdysis, when the exoskeleton is still soft and the pores pliable. Females engage in amplexus with males hours to days prior to egg laying. The pair swim about either on their sides or with the male on top. Minute penial spouts are located on the thoracic sternite seven, but how sperm are transferred and fertilization effected is unknown.

Seasonal cycles within a population have largely been studied in the colder temperate waters of the Northern Hemisphere. For example, Dexter (1971) found two generations per year in the intertidal species *Neohaustoruis schmitzi.* The reproductive season in North Carolina extends from February through October, with the winter generation living 8 months and the summer one living 4 months. In subarctic waters Dunbar (1954) found that species reproduced once in the spring and rarely a second time, resulting in a 1-year life cycle. Very little data are available on deep-sea species, but reproductive activity is thought to be triggered by "windfall" food availability rather than related to any season (Shulenberger and Barnard, 1976).

The biological cycle of six species of amphipods was followed by Greze (1968) in the Black Sea, where water temperatures ranged from 7° to above 20°C. The population dynamics were similar for all species. Up to early spring the population is composed entirely of adults in equal sex ratio. Juveniles appear from April to May. Females predominate during the warmer months, producing spring and autumn breeding maxima. Growth rate was faster above 20°C, with a molting frequency of 3–6 days as compared to 24 days at 7° to 9°C.

The relationship between populations of amphipods and water temperature can be shown by monthly analyses of fouling organisms or by suspension of test panels. Populations on boat floats in southern California of four species of gammarideans and one caprellidean reached peaks of 1690, 17,450, and 5121 specimens per 325 cm² surface area in successive summers in contrast to 3, 6, and 71 specimens collected during the winter months (Reish, 1964). Monthly suspension of test panels in Los Angeles–Long Beach Harbors over a 7-year period has yielded a considerable amount of data on settling of amphipods as well as other organisms (Harbor Environmental Projects, 1976). In terms of occurrence by month and station, *Corophium acherusicum* was the most commonly encountered species followed by *Jassa falcata, Podocerus brasiliensis,* and *Stenothoe valida,* whereas in terms of numbers of specimens the positions of *Jassa* and *Corophium* were reversed. Analyses of the data at one station in outer Los Angeles Harbor during the 1971–1978 (D. J. Reish, unpublished data) period indicate that population peaks in *Jassa* and *Corophium* of 1000 or more individuals each occur during the months of July and August each year, with lows of less than 50 specimens each during the December–January period. The significance of these data, in terms of environmental studies, points up the importance of considering the seasonal variability of amphipod populations. One-time surveys, as so often is the case in environmental impact studies when data are actually taken, can yield a different sort of result depending upon the season when collected. A minimum of 1 year's data is required in order to obtain an indication of the dynamics of an amphipod population.

B. IDENTIFICATION OF GAMMARIDEAN AMPHIPODS

The identification of amphipods is not an easy task. The problem is compounded by the absence of unanimity of familial classification and the general absence of faunal handbooks. A handbook to the world genera was compiled by Barnard (1969b), and is now out of print. This reference is essential for anyone embarking on a career of identifying these organisms. In addition to keys to families and genera of amphipods, the morphological features of the various appendages are described in detail. Especially useful is a detailed description of the equipment and techniques of dissecting amphipods. This section (pp. 507–511) should be read frequently in order to achieve greater efficiency in preparing the material for identification. Since amphipod taxonomy has its own specialized vocabulary, a glossary is included with references to figures. It is hoped a new edition of this work will appear in the near future.

Table II is a summary, by geographical region, of the references that will be helpful in identifying amphipods to species. It is readily apparent that vast geographical regions lack any convenient reference for identifying amphipods. This deplorable condition is the probable result of the deemphasis of systematics. All too many people, professional as well as lay public, believe that this type of work is not even science. As a result, with only one career taxonomist in gam-

TABLE II

<small>SUMMARY OF THE REFERENCES TO GAMMARIDEAN AMPHIPODS BY GEOGRAPHICAL REGION</small>

Geographical region	Notes	Key	Reference
World	Keys to families and genera of the world	+	Barnard (1969b)
Oregon		+	Barnard (1954)
California	Rocky intertidal only	−	Barnard (1969a)
Hawaiian Islands	Amphipods from 0–30 m	+	Barnard (1971)
New England		+	Bousfield (1973)
Norway	The classic of the field	+	Sars (1895)
France	In French	+	Chevreux and Fage (1925)
USSR	In Russian	+	Gurjanova (1951)
South Africa		+	K. H. Barnard (1940)
South Africa		+	Griffiths (1975)
Micronesia		+	Barnard (1965)
New Zealand	Families only	+	Hurley (1958)
New Zealand		+	Barnard (1972a)
New Zealand		+	Lowry (1974)
Australia		+	Barnard (1972b, 1974)

maridean amphipods in the United States and one in Canada, progress will be slow and dependent upon the aggregate efforts of small studies by part-time specialists. The problem is further compounded by the ever-increasing demand from the taxonomist by scientists in experimental science. Assistance from academia in the way of support for faculty positions or training for students is virtually absent. World monographers are needed to realign taxonomic concepts and find the best characters to use for identification. Even so, many ticklish problems with amphipod species will require the approach of population dynamicists to solve the interplay of genetic variability. Gammaridean amphipods are second only to polychaetes as the most important group in the subtidal benthos.

Identification to species is imperative if a study is to have merit. Gammaridean amphipods comprise about 900 genera and 4800 known species. New species are encountered frequently, even within intertidal or subtidal environments. Perfection of the means to identify specimens to species requires the ability to recognize not only known species but also undescribed ones as well. An additional problem is that juveniles are almost impossible to identify to species unless they belong to a genus with only one species in the area. They can be apportioned to the several species of the genus according to the proportion of the adults, but we realize this holds a great margin of error since juveniles of one species may be more abundant than those of another species at any one time, owing to the partitioning of their reproductive cycles.

A neophyte in gammaridean taxonomy should begin cautiously and with intertidal species. More generalized references, such as Barnard (1975), should be employed before embarking upon major works and terminology and dissection techniques should be learned and worked out. As experience is gained, the primary literature (e.g., Barnard, 1969b), can be consulted. Faunistic accounts, if they exist, will be useful in working out local species. Voucher specimens should be retained as a check for oneself as well as by others. Consultation with an experienced worker is especially valuable. Frequently a novice works with an experienced individual in a consulting firm or governmental agency concerned with surveys or monitoring programs. Such a relationship greatly shortens the learning period, and this procedure is recommended whenever possible.

While gammaridean amphipods are an important element in the benthic fauna, they are often lost during the processing of samples. Since they are small-bodied, many are lost unless the sample is washed through a screen with 250-μm openings. Such a screen size should be considered necessary for all studies where a knowledge of what is present at what population level is essential.

II. The Effects of Natural and Anthropogenic Environmental Changes on Gammaridean Amphipods

Our knowledge on the ecological requirements of gammaridean amphipods is fragmentary, and what is known has not been synthesized. Earlier ecological studies were hampered by the lack of systematic knowledge of the group. While we are still limited by systematics, ecological and experimental studies with single species have been progressing at a rate faster than basic systematics. In assessing what has been the effect of man upon the populations of amphipods, it is essential that we look also to natural changes to help us make this assessment. The forthcoming discussion will be based, in part, upon the subjective analysis by the authors. In this way we hope to indicate what is known, what it means, what to do with the data, and where to go from here.

A. Habitat Specificity

The niche limits have not been defined precisely for any amphipod species. A few species are believed to be very limited in their ecological space and activity, but their substrate preference, on the other hand, is very widespread. For example, certain species of burrowers have close tolerances on particle size, as discussed in section I,A,2, and inquilinous species, such as leucothoids, have a limited number of tunicate hosts. Table III demonstrates the relationship of particle size and amphipods. This group of 14 selected species from Barnard

TABLE III

RELATIONSHIP BETWEEN ABUNDANT AMPHIPODS AND MEDIAN DIAMETER OF SUBSTRATE IN BAHIA DE SAN QUENTÍN, BAJA CALIFORNIA[a,b]

	Median diameter (μm)					
	Silts		Very fine sands		Fine sands	
Species	8–30	31–50	51–80	81–110	111–150	151–220
Acuminodeutopus heteraropus	21	141	586	281	797	9
Ampelisca compressa	12	2808	292	14527	846	655
Amphideutopus oculatus	0	7	233	547	3	0
Ampithoe plumulosa	0	11	25	19	66	7
Aruga holmesi	0	0	6	1480	43	20
Ericthonius brasiliensis	0	878	4035	197	91	16
Heterophoxus oculatus	79	13	4	4	4	2
Hyale nigra	0	168	1315	216	91	114
Lembos macromanus	0	131	452	121	38	27
Orchomene magdalenensis	23	22	98	167	2	27
Paraphoxus obtusidens	1	91	132	276	311	66
Pontogeneia quinsana	0	7	181	27	4	36
Rudilemboides stenopropodus	3	51	777	1192	411	61
Uristes entalladurus	0	0	3	232	36	5
Totals	139	4328	8139	19285	2743	1045

[a]The data are expressed as numbers of amphipods per square meter.
[b]Modified from Barnard (1964).

(1964) can be divided into five groups depending upon the particle size and the maximum number of specimens per square meter. It should be noted that the sediments have been further divided within a geological category, i.e., two size groups and a part of a third fall within "silts." These five groups are (1) maximum population (*Heterophoxus oculatus*) at mean sediment size of 8–30 μm; (2) maximum population (*Ampelisca compressa*) at 31–50 μm; (3) maximum population (*Ericthonius brasiliensis* and three other species) at 51–80 μm; (4) maximum population (*Amphideutopus oculatus* and three other species) at 81–110 μm; and (5) maximum population (*Acuminodeutopus heterouropus* and two other species) at 111–150 μm. While many specimens were collected from sediments showing a median diameter of 151–200 μm, no species preferred this particular size class. From the totals given in Table III, these particular amphipods appear to prefer very fine sands in the 81–110 μm size class. How much of a role microflora play in determining population size is unknown. Most amphipod species will be dispersed through several communities based on other ecological definitions than sediment size. An experienced amphipod taxonomist can appraise a sample of amphipods with 15 or more species and draw conclusions about the habitat based on species composition and abundance of individuals. Estimates of depth, type of sediment, and associated macrofauna can be made with reasonable accuracy. It is not necessary for such a person to subject his data to statistical analyses or some computer program to draw these conclusions; it is simply based on experience with the group. We hasten to add that we do not mean to imply that intuitive reasoning is the only means to reach an answer, we wholeheartedly endorse statistics and computers as tools.

B. HARBOR FAUNAS

Large harbors in north temperate climes, especially in the United States, have specific compositions of amphipods. Their main substrate is pilings. Until recent improvements of polluted conditions in various harbors (Reish, 1971), the 1930–1960 picture showed few to no amphipods living on the muddy benthos nearest the docks (Reish, 1959) of major harbors such as Los Angeles–Long Beach, San Francisco, and San Diego. Introduction of pilings into natural or man-made harbors increases substrate surface substantially and amphipods are often a conspicuous component of piling faunas (Barnard, 1961, 1970). Although mussels, barnacles, tunicates, algae, and polychaetes generally form the dominant biomass, amphipods frequently are dominant in terms of number of specimens (Reish, 1964). Many if not all the amphipods in active harbors are introduced species of possible cosmopolitan distribution; among these are *Corophium acherusicum, Jassa falcata, Ericthonius brasiliensis, Corophium insidiosum, Podocerus brasiliensis, Elasmopus "rapax,"* and *Stenothoe valida.* The first four species form tubicolous domiciles on pilings or live within abandoned polychaete tubes, whereas *E. rapax* is a true nestler; *Podocerus brasiliensis* is generally associated with fine anastomoses made by hydroids, and *Stenothoe valida* is a presumed iniquiline on sedentary invertebrates. These species were apparently distributed by fouling matter on ships. The pilings represent local pollution and the introduced species represent biogeographic pollutants to native faunas, although native faunas presumably have few species adaptable to the piling habitat if the faunistics in Barnard (1961) are correct.

In California and Pacific Mexico little damage to native amphipods appears to have occurred, as all potential estuarine or lagoonal species are able to survive in other areas than harbors. No bay-restricted native amphipods in California are thought to have existed before harbors were developed. Outward excursions of imported amphipods to the open sea appear to be minimal. For example, *Corophium acherusicum, C. insidiosum,* and *Stenothoe valida* remain mostly within harbor confines, whereas *Ericthonius brasiliensis* is so widely spread in world seas that one concludes it was naturally distributed prehistorically. A slight excursion of *Jassa falcata* was noted by Barnard (1969a) but the taxonomy of this situation is unclear—whether or not the open-sea populations are specifically distinct from the harbor populations. *Jassa falcata* represents a famous case of polymorphy and worldwide harbor distribution (Sexton and Reid, 1951) but the case has been doubted and debated ever since. The likelihood that Sexton and Reid are correct in their acceptance of polymorphy in a single species is good, as one would expect a polymorphic species to be a good candidate for worldwide dispersal by shipping.

Not all imported amphipods are of cosmopolitan character or of epidemic success. *Corophium uenoi* apparently was brought to California from Japan and resides in confined populations near oyster beds, especially in Morro Bay (Barnard, 1952). A more recent introduction (*Grandidierella japonica*) is yet confined to San Francisco Bay (Chapman and Dorman, 1975). Natural transport of species such as *Corophium insidiosum* in Hilo Harbor by rafted logs is possible (Barnard, 1970) although towing of log rafts from the continent to Hawaii is a good alternative explanation.

The potential impact of introduced species can be predicted by thorough knowledge of the ecological background of the species being transported and the faunistics of the region of arrival. Clearly, the well-known harbor components noted above should succeed in most embayments, but their excursion into the open sea will be dictated by the biogeographic strength of the native fauna. If, as on islands, the native fauna is impoverished, the imported species may be highly successful, but an expert zoogeographer and phyleticist may also be able to recognize faunistic weaknesses on continental shores. For example, the western North Atlantic clearly is impoverished of several phyletic groups. Importation of fossorial phoxocephalids from the faunistically strong Australia to the west Atlantic might be very dangerous and result in pestiferous displacement of native species, not only in Amphipoda but also in other orders or phyla as well.

While foulers are pests and amphipods can occasionally contribute much web and silt content to fouling masses, the total economic loss so far from amphipodan pests is negligible. They do not appear as yet to have caused any local specific extinctions, although they have clearly caused minor habitat exclusions but these are directly the cause of human interference. Unless transported or further stimulated, the importations remain confined to areas being altered by humans, mainly because the species so far transported originated in estuaries, lagoons, harbors, or in oyster farm materials.

In the more severely polluted parts of harbors, such as in Consolidated Slip in Los Angeles Harbor, or in the upper West Basin, during the 1950s amphipods were absent, both on the benthos and on pilings (Barnard, 1958; Reish, 1959). The only macroscopic crustacean found in pilings in the West Basin was a species of *Nebaliacea,* but even that crustacean was absent in Dominguez Channel, where the pollution-indicating polychaete species *Capitella capitata* thrived.

Further evidence of the sensitivity of the amphipods may be seen in the reestablishment of fouling organisms and benthic species during the early stages of the pollution abatement program in the Consolidated Slip–East Basin area of Los Angeles Harbor (Reish, 1971). Many phyla and groups appeared rapidly, such as coelenterates, polychaetes, gastropods, pelecypods, barnacles, ectoprocts, and tunicates, but amphipods were limited in species and specimens. *Jassa falcata* and *Corophium acherusicum* were early amphipod inhabitants. Generally, amphipods and nebalians, and amphipods and *Capitella,* are mutually abhorrent. We have observed this in natural and seminatural situations also. For example, in deep-sea bottom samples off California dominated by *Capitella,* where presumably the bottom was influenced by emergent freshwater springs, amphipods were absent. In clear-ponded and confined waters of oyster farms in Baja California, filled with the alga *Caulerpa,* the only macroscopic crustacean was a nebaliacean in great abundance. Clearly, amphipods cannot withstand abnormalities of environment such as lowered oxygen values, dilutions, abnormal pH, or a host of other anomalies. Harbors easily provide conditions that no amphipod can withstand. On the other hand, long-term exposure to unusual conditions can result in an evolutionary incursion of amphipods. For example, the peculiar, diluted habitats of old Hawaiian lava fields, into which seawater percolates but is diluted by rainwater, have been invaded by 10 species of amphipods with marine affinities, seven of them endemic just to this anchialine habitat. Some of these species are so peculiar as to represent relics from a cooler age in Hawaii (Barnard, 1977).

The two best known amphipod pests are *Chelura terebrans,* the woodborer, and *Gammarus tigrinus,* the tiger scud of America, recently exported to European shores. *Chelura terebrans* is the oldest known pest occupying and enlarging tunnels in marine pilings opened by the gribble, *Limnoria* sp. (isopod). *Chelura* has been shown to be almost benign. The major crustacean damage to pilings occurs when *Limnoria* opens the burrow. *Chelura* arrives later, but apparently is such a poor borer that it simply enlarges the holes very slowly over many months. Without *Limnoria, Chelura* can only dig surface trenches (Barnard, 1955). Perhaps *Chelura* simply scrapes wood to obtain superficial microflora, whereas *Limnoria* burrows for emergency protection and can digest cellulose for its nutriment.

Gammarus tigrinus, a shoreline scud that nestles in plant material, is believed to have been transported from Atlantic America to Europe in 1921. Since that time it has caused severe displacement of several native European species of *Gammarus* not as widely adapted and reproductively prodigious as *G. tigrinus.* No extinctions have occurred as yet, although two species of *Gammarus* in Europe have been constrained narrowly to specific habitats not preadapted by *G. tigrinus* (Pinkster, 1975). *Gammarus tigrinus* has also been able to penetrate stream systems at great distances from the sea, and has been introduced by fishery biologists, for the purpose of providing food sources to game fish, into saline streams formerly deprived of amphipods.

C. PLATFORMS AT SEA

Oil platforms on the high seas are known to be infested with amphipods similar to those of harbors (Gunter and Geyer, 1955; Pequegnat and Pequegnat, 1968). Fixed pilings supporting these platforms in the open sea are even more foreign to the environment than they are within harbors. They become infested with *Jassa, Corophium,* and *Stenothoe,* which are possibly brought by fouling matter on boats that commute between a mainland harbor and the platform. This seems to be the logical route, as there are very few epipelagic amphipods—few of these, if any, are adapted to a nonpelagic habitat. Since most of the harbor amphipod fauna are only adapted to the upper 10–15 m, it would be of interest to investigate the piling surfaces for amphipods below this depth. Since open-sea oil platforms frequently exceed these depths, do such species as *Jassa falcata* and *Corophium acherusicum* extend throughout the length of the pilings or are they replaced by other cosmopolitan species or species unique to the area?

D. PASSIVE TRANSPORT

The opening of pathways between seas, such as the Suez Canal and the Panama Canal, may result in passive transport of amphipods between ocean systems, which can be considered as a form of pollution or biotic contamination. As yet, no amphipod has been proved to have transgressed the Panama Canal, mainly because the waterway is fresh water, but there is the possibility that shiphull foulers could withstand the 8 hours of immersion during passage. Theoretically, brackish or freshwater-tolerant taxa such as *Grandidierella, Corophium, Melita,* and *Parhyale* might be able to cross the Panamic isthmus, but M. L. Jones (personal communication) has studied the fouling faunas of the lock system and found no amphipods in the upper locks, whereas the genera mentioned remain confined to lower brackish water locks. None of the very tolerant freshwater taxa of temperate climes is known to have survived the torrid tropical water.

At least one amphipod is known to have transgressed the bitter lake system of the Suez Canal and invaded the Mediterranean Sea from the Red Sea (Gilet, 1960). The percentage of species carried here and there by shipping and the percentage able to migrate passively are unclear. Undoubtedly, tests of natural selection between imported species and native Mediterranean species have occurred, but our impression is that the Mediterranean amphipod fauna are highly impoverished for several groups of amphipods and that many open niches may be present to absorb invaders. On the other hand, the Red Sea is envisioned as a dominantly corallinaceous biome and coral-inhabiting amphipods may not find good accommodations in the Mediterranean. The greatest danger to Mediterranean amphipod faunas would, in our opinion, come from the importation of mud-bottom species from warm temperate regions of Australia, where a very diverse and possibly strongly adaptive fauna occurs (Barnard and Drummond, 1978).

E. POLLUTION ECOLOGY

1. Organic Enrichment

The effect on gammaridean amphipods of the introduction of domestic sewage into the marine environment depends upon many factors, such as the amount and degree of treatment, the location (e.g., an estuary, bay, harbor, or offshore waters), the depth, as well as the particular oceanographic conditions at the discharge point. Because of the variables, it is almost as if each discharge is unique and must be considered separately. This is true, in part, but some generalities can be made. If the water circulation pattern is adequate and if the discharge is small, then the discharge of domestic wastes will have little or no effect on the environment; in fact, in some instances there will be an enrichment of the benthic fauna, particularly in terms of numbers of species and biomass. Such instances have led some people to believe that the discharge of domestic sewage is beneficial to the environment and therefore the quantity discharged can be increased. Unfortunately, a particular locality can assimilate only so much before the benthic fauna begin to deteriorate.

As discussed above, amphipods apparently are not as tolerant as some groups of animals. In a transect extending from near the Terminal Island sewage treatment plant and the fish cannery outfalls outward into Los Angeles Harbor, Reish and Ware (1976) and Ware (1978) found a total of 15 species of benthic gammaridean amphipods. The number of species and specimens was directly related to the distance from the outfalls. The benthic community near the outfalls was polluted and characterized by the polychaete *Capitella capitata*. In one survey out of eight conducted over a 13-month period, only once were amphipods taken near the discharge and rarely were specimens ever taken at the next closest station. The amphipod population became greater at the three outer stations in which the benthos contained a diversity of macroinvertebrates. McNulty (1970) found *Corophium acherusicum* and *Ericthonius brasiliensis,* both cosmopolitan species, to be abundant in polluted Biscayne Bay. Apparently, some species of amphipods are more tolerant of organic enrichment than others, but the environmental factor(s) causing this differential is unknown. Very little data are available on the effects of various parameters on amphipods (see below).

2. Oil Pollution

Studies dealing with the effects of hydrocarbons on gammaridean amphipods are limited and have been largely concerned with experimental studies. Baker (1971) reported that amphipods were observed jumping about and dying in a salt marsh community in Milford Haven following an experimental study in which fresh crude oil was sprayed over the area. A laboratory experiment which also dealt with amphipod behavior was described by Sandberg *et al.* (1972). The intertidal beachhopper, *Neohaustarius biarticulatus,* was exposed to various concentrations of No. 2 fuel oil. Amphipods were able to burrow into sand mixed with oil but within 12 hours they stopped digging, and all were dead in 24 hours in sand treated with 50 mg/liter. *Onisimus affinis* and *Gammarus oceanicus* were repelled by three crude oils; the degree of avoidance was related to the length of exposure of the oils prior to testing (Percy, 1976).

Water-soluble fractions of two petroleum oils were tested on two amphipods, *Gammarus mucronatus* and *Ampithoe valida.* The results were similar for both species, 0.8 mg/liter for No. 2 fuel oil and 2.4 mg/liter for Southern Louisiana crude oil (Lee *et al.*, 1977). Linden (1976a,b) conducted bioassays with oils on the survival of various stages in the life history and reproduction in *Gammarus oceanicus*. Juveniles were considerably more sensitive than adults. The incubation period was identical at 1.0 mg/liter oil and control, but the number of larvae produced was reduced. In higher concentrations males and females failed to enter the precopulation period.

Fragmentary at best, these data indicate that amphipods are sensitive to fresh oil and that exposure to sublethal amounts leads to various behavioral changes, i.e., burrowing and precopulation which in turn could cause a reduction or elimination of the population. Since beachhoppers are normal inhabitants of undisturbed sandy beaches (however, see section II,E,7), and since sandy beaches are often affected by oil spills, a knowledge of the sublethal effects of oils upon amphipods is a fruitful area for future investigation.

3. Heavy Metals

The discharge of heavy metals into the sea is considered a serious problem in the marine environment. Heavy metal studies have been concerned with either the body burden levels of the element or the acute toxicity of the metal to the amphipod. Since the data on both of these types of studies are meager, relating one type of study to the other is meaningless at this time. Body burden analyses data for amphipods are summarized in Table IV and toxicity results are given in Table V. Of particular interest is the findings of Ahsanullah (1976), who found *Allorchestes compressa* to be the most sensitive of seven species tested, including species of decapods, polychaetes, mollusc, and asteroids. In view of this observation and others reported in other sections herein, amphipods may be a sensitive group to toxicants.

TABLE IV

TRACE METAL CONCENTRATION IN BODY TISSUES OF GAMMARIDEAN AMPHIPODS

Species	Geographical region	Metal concentration (μg/gm dry weight)										Reference
		Ag	As	Cd	Cu	Fe	Pb	Mn	Hg	Ni	Zn	
Orchestoidea corniculata	California	1–2		10–20	15–70	50–120	0–2.5	4–5		0–5	100–120	Bender (1975)
Orchestoidea californiana	California	3		10–25	10–25	30–120	0–6	3–6		0	70–90	Bender (1975)
Lysianassidae	Canada, Arctic		7.9	7.0	26	87					43	Bohn and McElroy (1976)
Amphipods	Norway		2.3–8.9[a]	5.1–8.8	35–94		10–23				110–275	Stenner and Nickless (1974)
Amphipods	Canada, Atlantic											Kennedy (1976)
Amphipods	Finland								0.01			Nuorteva and Häsäneu (1975)

[a]Wet weight.

TABLE V

THE TOXICITY OF CADMIUM AND ZINC TO GAMMARIDEAN AMPHIPODS[a]

Species	Cd	Zn	Reference
Podocerus fulanus	0.32		Bellan-Santini and Reish (1976)
Corophium acherusicum	1.4		Bellan-Santini and Reish (1976)
Allorchestes compressa	0.2–0.4	0.58	Ahsanullah (1976)

[a]96-hour LC_{50}, milligrams per liter.

4. Detergents and Related Compounds

Studies are limited in measuring the effects of detergents or surfactants on gammarideans. Smith (1968) mentioned that gammarids which lived in cracks in the rocks or under them were able to survive the application of the detergent BP 1002 used in cleaning up the oil after the *Torrey Canyon* disaster. Whether or not the detergent came in actual contact with the amphipods is unknown.

Wildish (1970b) found that polyoxyethylene ethers were more toxic to *Gammarus oceanicus* than polyoxyethylene esters; however, it was more resistant than the salmon parr *Salmo salar*. Wildish and Zitko (1971) demonstrated uptake of PCB by the same species of amphipod. The rate of uptake was found to be related to branchial surface area, but this was not necessarily the site of entry of PCB. Uptake increased with increasing concentration, but after 4–6 hours the rate of uptake decreased one-half.

5. Radionuclides

Under experimental laboratory conditions Hoppenheit (1972) found a reduction in egg production in *Gammarus duebeni* exposed to 220 R. No effect on fecundity was noted after irradiation with 147 R or lower. No field studies have been undertaken thus far in dealing with the possible effects of radionuclides discharged in canisters at sea on amphipods.

6. Dredging

Dredging activities typically take place within a harbor in order to maintain channels of sufficient depth for shipping. Dredging affects benthic fauna by removal of the sediments and by burial of these organisms at the disposal site. The possible impact of dredging activity upon marine benthic fauna is of recent concern to environmentalists and knowledge of the effects of this change is fragmentary at best. If the quantity of sediment removed is of sufficient depth, then the entire benthic fauna will be eliminated. Since benthic amphipods reside either on the top of the sediments or in the upper few centimeters, then all these species would be eliminated by dredging. Repopulation can occur by movement of adults from nearby unaffected areas or by larval settlement (McCaulley *et al.,* 1977). Some organisms, such as the amphipod *Ampelisca agassizi,* survive transport and are able to inhabit the sediments at the spoil site (Saila *et al.,* 1972).

7. Mechanical Pollution

a. Shoreline Cleanup. Beach hoppers of the families Talitridae, Hyalellidae, and Hyalidae are widespread on sandy beaches between tidal extremes. Typically, these amphipods are concentrated in the moist zone which is usually covered by dislodged algae (wrack). Wrack assists in maintaining the moisture content of the sand and provides a possible food supply. The beach hoppers dig tunnels or depressions in the sand beneath the wrack. Although they presumably feed upon the decaying seaweed, we have been unable to identify any macroalgal species in the gut of these species. Unilocular reproductive bodies from epiphytic species growing at the stipes and blades of the macroalgal species have been the only material positively identified in the guts of these species.

Populations of beach hoppers must be transitory, and perhaps migratory, owing to changes in tidal levels, storms, general climate, as well as the quantity of wrack present. However, these species do not necessarily require wrack to exist, since populations have been found in its absence. Occasional semipermanent wrack abodes, such as those observed in Western Australia (Barnard, 1974), harbor countless numbers of individuals, in this instance, *Allorchestes* sp., a species which is normally present in subtidal waters. This Australian wrack bed was 2–3 m in depth and was constantly being bathed by spray and replenished by storms.

Along warmer coastlines near major population centers, such as southern California, the beaches are cleaned periodically of human and seacast debris by large machines that rake up the debris from the upper 5–10 cm of the beach. The net result of this activity is the absence of beach hoppers along these stretches. With the extensive sandy beaches in southern California, one ponders the possible barrier to gene flow between the northern and southern subspecies such cleaning activity may impose.

b. Recreation and Education. Rocky intertidal areas, which are limited in southern California, became the focus in the late 1940s of larger and more frequent excursion parties of children and curious adults interested in marine life. These people affected the intertidal marine life by trampling, overturning rocks, and collecting. It became increasingly difficult to see a common starfish at popular areas such as Little Corona in Orange County in the 1960s. Widdowson (1971) concluded that the decline of algal species at this locality was the result of this activity. Barnard (1961) made surveys along this and other rocky shores of southern California in the early 1960s and found the worst disturbance to be in the rocky areas of La Jolla. Evidence cited was the large number of overturned rocks, dying algae, the preponderance of domicolous amphipods over nestling species, as well as a reduction in species diversity, which may have been the result of human activity. Protection was afforded to Little Corona as well as to many other rocky shores in southern California in 1972, and the result has been the reappearance of the starfish and the general improvement of the environment. Human activities usually destroy the habitats for the nestlers and favor domicolous amphipods through a resultant increase in turbidity (Dahl, 1948).

8. Fishing

Benthic trawling, especially large rigs for shrimp on level bottoms, must be extremely deleterious to muddy habitats. In areas of heavy shrimp fisheries, such as in the Gulf of Mexico or California, trawlers systematically cover every square meter of bottom. Such activity leads to an increase in turbidity, which in turn would favor domicolous species. No studies have been undertaken to determine what effect this activity has on the amphipod population or any other animal group. Such a study, with an appropriate unaffected area as a "control" site, seems to be a fruitful area of applied research. We may be appreciably reducing the productivity of the area by this type of fishing.

9. Salinity

Laboratory survival studies have been undertaken to measure the effects of hypo- and hypersalinities on estuarine species of *Corophium*. McLusky (1967) found that *C. volutor* could tolerate salinities of 2–59‰ if supplied with sediment, but only 7.5–47.5‰ without sediment. In later studies McLusky (1970) noted that these species selected salinities of 10–30‰ in simple choice experiments. Similar results were described by Shyamasundari (1973) with *C. triaenonyx*, which would live in the salinity range of 4–55‰. Further sensitivity of this species was described by the same author (1976), who measured the effects of various salinities on development and incubation of zygotes. Some embryos were able to develop between the range of 7.5 and 37.5‰, but a satisfactory percentage of successful developments was further limited to 20–32.5‰. The embryos of both *Marinogammarus marinus* and *Orchestia gammarella* could develop in chlorinities of 4 and 7‰ but the number of juveniles able to emerge from the brood chamber was reduced at these levels (Vlasblom and Bolier, 1971).

These studies, which dealt with estuarine gammarideans, indicate that the adults have a wide range of salinity tolerances, but that the embryos and larvae are more sensitive. If these data can apply to other environmental parameters as well as to toxicants, which seems logical to assume, it is therefore important to consider developmental stages in any environmental assessment. These data further indicate that within an estuary there must be a reproductive population with a more restrictive range of distribution with a vegetative population at sublethal levels.

10. Dissolved Oxygen

The amount of dissolved oxygen present within the water mass is considered a quick and easy method of evaluating water quality. Readings over 5.0 mg/liter dissolved oxygen are generally considered adequate for fish and aquatic life. Unfortunately, very little data have been accumulated on the dissolved oxygen requirements of amphipods. Barnard (1958) found a reduction in the number of amphipods settling on test blocks whenever the dissolved oxygen content was low in Los Angeles–Long Beach Harbors. A definite reduction was noted in the number of species and specimens settling on test blocks whenever the yearly dissolved oxygen level averaged less than 2.0 mg/liter.

Vobis (1973) measured the behavior of five species of *Gammarus* to a water current under reduced dissolved oxygen conditions. Maximum upstream movement occurred at 2.7–5.3 mg/liter for four species, but could not be measured for *G. salinus;* it moved downstream whenever the concentration dropped below 2.5 mg/liter.

These limited observations indicate that amphipods are sensitive to reduced dissolved oxygen conditions. It seems to us that a fruitful area of research in light of the rapid population buildup in the warmer months would be to investigate the dissolved oxygen requirements of harbor amphipods, especially as related to reproduction.

III. Conclusions

A discussion was presented herein to examine the importance of gammaridean amphipods in the marine environment and their use in assessing the effects of pollution, and to describe the effects of natural and man-made changes on amphipods. Throughout this discussion it is apparent that there are many obvious gaps in our knowledge of this topic. It has been established that gammaridean amphipods are second only to polychaetes in their importance in and on the benthos. However, we are hampered by the small number of professional scientists who are able to identify these organisms as well as to train new ones. Advancement in our knowledge will continue to be slow until a change in philosophy occurs which recognizes the importance of systematics as a legitimate professional and scientific pursuit.

Amphipods have been shown by field and laboratory experiments to be more sensitive than some species in other groups (i.e., polychaetes and mollusks). Does this mean that we have some convenient measure to indicate, by the absence of amphipods, that a particular area is undergoing degradation? We do not know. However, on the basis of the available data, there are indications that such may be the case. One potentially fruitful area of research is to determine whether or not this is true. One must be cautious in drawing such a conclusion based on one survey; in light of the seasonal differences in amphipod populations (Reish, 1964), a completely erroneous conclusion could be drawn. McLusky (1967) noted that *Corophium voluator* was sparse in the polluted Forth estuary in Scotland, and therefore the redshank fish *Tringa totantes,* which normally feeds upon this amphipod, must seek out other organisms in this area in order to survive. While some fish can feed upon a variety of organisms (i.e., *Genyonimus lineatus* in Los Angeles Harbor; Reish and Ware, 1976), one wonders what might be the effect upon a highly selective feeder if its food organism were more sensitive to pollution and was killed. The role of amphipods in the marine food chain needs to be more clearly understood in both the natural and polluted environment.

Data on the effects of various environmental variables and potential toxicants on amphipods are limited, especially in comparison to that known for pelecypods, decapods, or polychaetes. Perhaps the reasons for the paucity of data are the lack of scientific personnel and the limited experience with culturing these organisms. Perhaps more experimental studies will be forthcoming with amphipods in light of the cultural procedures that were employed for *Ampithoe valida* by Lee (1977).

In the future we look for an expanded interest in the use of amphipods in marine environmental studies. A knowledge of their occurrence and distribution in both natural and polluted areas will give us further insight into their use as sensitive indicators of environmental changes. Laboratory studies, if cultural procedures can be successfully solved for many species, hold promise to assist us in answering more specific environmental questions.

References

Ahsanullah, M. (1976). Acute toxicity of cadmium and zinc to seven invertebrate species from Wester Port, Victoria. *Aust. J. Mar. Freshwater Res.* **27**, 187–196.

Baker, J. M. (1971). Growth stimulation following oil pollution. *In* "The Ecological Effects of Oil Pollution on Littoral Communities" (E. B. Cowell, ed.), pp. 72–77. Institute of Petroleum, London.

Barnard, J. L. (1952). Some Amphipoda from Central California. *Wasmann J. Biol.* **10**, 9–36.

Barnard, J. L. (1954). Marine Amphipods of Oregon. *Oreg. State Monogr., Stud. Zool.* No. 6, pp. 1–103.

Barnard, J. L. (1955). "The Wood Boring Habits of *Chelura terebrans* Philippi in Los Angeles Harbor," Essays Nat. Sci. Honor Capt. Allan Hancock, pp. 87–98. Hancock Found., University of Southern California, Los Angeles.

Barnard, J. L. (1958). Amphipod crustaceans as fouling organisms in Los Angeles–Long Beach Harbors, with reference to the influence of seawater turbidity. *Calif. Fish Game* **44**, 161–170.

Barnard, J. L. (1959). "Epipelagic and Under-ice Amphipods of the Central Arctic Basin," pp. 115–152. Geophys. Res. Dir., U.S. Air Force Cambridge Res. Cent., Bedford, Massachusetts.

Barnard, J. L. (1961). Relationship of California amphipod faunas in Newport Bay and in the open sea. *Pac. Nat.* **2,** 166–186.

Barnard, J. L. (1964). Marine Amphipoda of Bahia de San Quintin, Baja California. *Pac. Nat.* **4,** 55–138.

Barnard, J. L. (1965). Marine Amphipoda of atolls in Micronesia. *Proc. U.S. Natl. Mus.* **117,** 459–552.

Barnard, J. L. (1969a). Gammaridean Amphipoda of the Rocky Inter-Tidal of California: Monterey Bay to La Jolla. *U.S., Natl. Mus., Bull.* **258,** 1–230.

Barnard, J. L. (1969b). The families and genera of Marine Gammaridean Amphipoda. *U.S. Natl. Mus., Bull.* **271,** 1–535.

Barnard, J. L. (1970). Sublittoral Gammaridea (Amphipoda) of the Hawaiian Islands. *Smithson. Contrib. Zool.* **34,** 1–286.

Barnard, J. L. (1971). Keys to the Hawaiian Marine Gamaridea, 0-30 meters. *Smithson. Contrib. Zool.* **58,** 1–135.

Barnard, J. L. (1972a). The marine fauna of New Zealand: Algae-living littoral Gammaridea (Crustacea Amphipoda). *N. Z. Oceanogr. Inst.,* No. 62, pp. 7–216.

Barnard, J. L. (1972b). Gammaridean Amphipoda of Australia. Part I. *Smithson. Contrib. Zool.* **103,** 1–333.

Barnard, J. L. (1974). Gammaridean Amphipoda of Australia. Part II. *Smithson. Contrib. Zool.* **139,** 1–148.

Barnard, J. L. (1975). Identification of gammaridean amphipods. *In* "Light's Manual: Intertidal Invertebrates of the Central California Coast" (R. I. Smith and J. T. Carlton, eds.), pp. 314–366. Univ. of California Press, Berkeley.

Barnard, J. L. (1977). The cavernicolous fauna of Hawaiian lava tubes. 9. Amphipoda (Crustacea) from brackish lava ponds on Hawaii and Maui. *Pac. Insects* **17,** 267–299.

Barnard, J. L., and Drummond, M. M. (1978). Gammaridean Amphipoda of Australia. Part III. *Smithson. Contrib. Zool.* **245,** 1–555.

Barnard, K. H. (1940). Contributions to the crustacean fauna of South Africa. XII. Further additions to the Tanaidacea, Isopoda and Amphipoda, together with keys for identification of hitherto recorded marine and fresh-water species. *Ann. S. Afr. Mus.* **32,** 381–543.

Bellan-Santini, D., and Reish, D. J. (1976). Utilisation de trois espèces de crustacés comme animaux tests de lat toxicité de deux sels de métaux lourds. *C.R. Hebd. Seances Acad. Sci.* **282,** 1325–1327.

Bender, J. A. (1975). Trace metal levels in beach dipterans and amphipods. *Bull. Environ. Sci. Toxicol.* **14,** 187–192.

Bohn, A., and McElroy, R. O. (1976). Trace metals (As, Cd, Cu, Fe and Zn) in Arctic cod, *Boreogadus saida,* and selected zooplankton from Strathcona Sound, Northern Baffin Island. *J. Fish. Res. Board Can.* **33,** 2836.

Bousfield, E. L. (1973). "Shallow-water Gammaridean Amphipoda of New England." Cornell Univ. Press (Comstock), Ithaca, New York.

Chapman, J. W., and Dorman, J. A. (1975). Diagnosis, Systematics and notes on *Grandidierella japonica* (Amphipoda: Gammaridea) and its introduction to the Pacific Coast of the United States. *Bull. South. Calif. Acad. Sci.* **74,** 104–108.

Chevreux, E., and Fage, L. (1925). Amphipodes. *Faune Fr.* **9,** 1–488.

Croker, R. A. (1967a). Niche diversity in five sympatric species of intertidal amphipods (Crustacea: Haustoriidae). *Ecol. Monogr.* **37,** 173–200.

Croker, R. A. (1967b). Niche specificity of *Neohaustorius schmitzi* and *Haustorius* sp. (Crustacea: Amphipoda) in North Carolina. *Ecology* **48**, 971–975.

Dahl, E. (1948). On the smaller Arthropoda of marine algae, especially in the polyhaline waters off the Swedish west coast. *Undersokningar Oresund* **25**, 5–193.

Dexter, D. M. (1971). Life history of the sandy beach amphipod *Neohaustorius schmitzi* (Crustacea: Haustoriidae). *Mar. Biol.* **8**, 232–237.

Dunbar, M. J. (1954). The amphipod Crustacea of Ungava Bay, Canadian Eastern Arctic. *J. Fish. Res. Board Can.* **11**, 709–798.

Fenchel, T., Kofoed, L. H., and Lappalainen (1975). Particle size-selection of two deposit feeders: The amphipod *Corophium volutator* and the prosobranch *Hydobia ulvae*. *Mar. Biol.* **30**, 119–128.

Gilet, E. (1960). The benthonic Amphipoda of the Mediterranean coast of Israel. I. Notes on the geographical distribution. *Bull. Res. Counc. Isr., Sect. B* **9**, 157–166.

Greze, I. I. (1968). Feeding habits and food requirements of some amphipods in the Black Sea. *Mar. Biol.* **1**, 316–321.

Griffiths, C. L. (1975). The Amphipoda of Southern Africa. Part 5. The Gammaridea and Caprellidea of the Cape Province west of Cape Agulhas. *Ann. S. Afr. Mus.* **67**, 91–181.

Gunter, G., and Geyer, R. A. (1955). Studies on fouling organisms of the northeast Gulf of Mexico. *Publ. Inst. Mar. Sci., Univ. Tex.* **4**, 37–87.

Gurjanova, E. (1951). Bokoplavy morej SSSR i sopredel'nykh vod (Amphipoda-Gammaridea). *Akad. Nauk SSSR, Opredel. Faune* **41**, 1–1029.

Harbor Environmental Projects (1976). "Environmental Investigations and Analyses, Los Angeles-Long Beach Harbors 1973–1976," Final Rep. U.S. Army Corps Eng., Los Angeles District, Inst. Mar. Coastal Stud., Univ. of Southern California, Los Angeles.

Hoppenheit, M. (1972). Wirkungen einer einmaligen Roentgenbestrahlung auf die Fortpflanzung der Weibchen von *Gammarus duebeni* (Crustacea, Amphipoda). *Helgol. Wiss. Meeresunters.* **23**, 467–484.

Hurley, D. E. (1958). A key to the families of New Zealand Amphipods. *Tuatara* **7**, 71–83.

Kennedy, V. S. (1976). Arsenic concentrations in some coexisting marine organisms from Newfoundland and Labrador. *J. Fish. Res. Board Can.* **33**, 1388.

Lee, W. Y. (1977). Some laboratory cultured crustaceans for marine pollution studies. *Mar. Pollut. Bull.* **8**, 258–259.

Lee, W. Y., Welch, M. F., and Nicol, J. A. C. (1977). Survival of two species of amphipods in aqueous extracts of petroleum oils. *Mar. Pollut. Bull.* **8**, 92–94.

Linden, O. (1976a). Effects of oil on the reproduction of the amphipod *Gammarus oceanicus*. *Ambio* **5**, 36–37.

Linden, O. (1976b). Effects of oil on the amphipod *Gammarus oceanicus*. *Environ. Pollut.* **10**, 239–250.

Lowry, J. K. (1974). Key and checklist to the gammaridean amphipods of Kaikoura. *Mauri Ora* **2**, 95–130.

McCaulley, J. E., Parr, R. A., and Hancock, D. R. (1977). Benthic infauna and maintenance dredging. *Water Res.* **11**, 233–242.

McLusky, D. S. (1967). Some effects of salinity on the survival, moulting, and growth of *Corophium volutor* (Amphipoda). *J. Mar. Biol. Assoc. U.K.* **47**, 607–617.

McLusky, D. S. (1970). Salinity preference in *Corophium volutor*. *J. Mar. Biol. Assoc. U.K.* **50**, 747–752.

McNulty, J. K. (1970). "Effects of Abatement of Domestic Sewage Pollution in Biscayne Bay." Univ. of Miami Press, Coral Gables.

Nuorteva, P., and Häsäneu, E. (1975). Bioaccumulation of mercury in *Myoxcephalus quadricornis* (L.), (Teleostei, Cottidae) in an unpolluted area of the Baltic. *Ann. Zool. Fenn.* **12**, 247–254.

Pequegnat, W. E., and Pequegnat, L. H. (1968). "Ecological Aspects of Marine Fouling in the Northeast Gulf of Mexico," Rep. No. 68-22T. Dept. Oceanogr., Texas A&M University, College Station.

Percy, J. A. (1976). Responses of arctic marine crustaceans to crude oil and oil-tainted food. *Environ. Pollut.* **10**, 155-162.

Pinkster, S. (1975). The introduction of the alien amphipod *Gammarus tigrinus* Sexton 1939 (Crustacean: Amphipoda) in the Netherlands and its competition with indigenous species. *Hydrobiol. Bull.* **9**, 131-138.

Reish, D. J. (1959). An ecological study of pollution in Los Angeles-Long Beach Harbors, California. *Occas. Pap. Allan Hancock Found.* No. 22, pp. 1-119.

Reish, D. J. (1964). Studies on the *Mytilus edulis* community in Alamitos Bay, California. II. Population variations and discussion of the associated organisms. *Veliger* **6**, 202-207.

Reish, D. J. (1971). Effect of pollution abatement in Los Angeles Harbours. *Mar. Pollut. Bull.* **2**, 71-74.

Reish, D. J., and Ware, R. (1976). The impact of waste effluents on the benthos and food habits of fish in outer Los Angeles Harbor. *In* "Marine Studies of San Pedro Bay, California" (D. F. Soule and M. Oguri, eds.), Part 12, pp. 113-128. Harbor Environ. Proj., Univ. of Southern California, Los Angeles.

Saila, S. B., Pratt, S. D., and Polgar, T. T. (1972). Dredge spoil disposal in Rhode Island Sound. *Univ. R. I., Mar. Tech. Rep.* No. 2, pp. 1-48.

Sandberg, D. M., Michael, A. D., Brown, B., and Beebe-Center, R. (1972). Toxic effects of fuel oil on haustoriid amphipods and pagurid crabs. *Biol. Bull. (Woods Hole, Mass.)* **143**, 475-476.

Sars, G. (1895). Amphipoda. An account of the Crustacea of Norway with short descriptions and figures of all the species. *Christiana & Copenhagen* **1**, 1-711.

Sexton, E. W., and Reid, D. M. (1951). The life-history of the multiform species *Jassa falcata* (Montagu) (Crustacea Amphipoda) with a review of the bibliography of the species. *J. Linn. Soc. London, Zool.* **42**, 29-91.

Sheader, M., and Chia, F.-S. (1970). Development, fecundity and brooding behaviour of the amphipod, *Marinogammarus obtusatus*. *J. Mar. Biol. Assoc. U.K.* **50**, 1079-1099.

Shulenberger, E., and Hessler, R. R. (1974). Scavenging abyssal benthic amphipods trapped under oligotrophic central north Pacific gyre waters. *Mar. Biol.* **28**, 185-187.

Shulenberger, E., and Barnard, J. L. (1976). Amphipods from an Abyssal Trap Set in the North Pacific Gyre. *Crustaceana (Leiden)* **31**, 241-258.

Shyamasundari, K. (1973). Studies on the tube-building amphipod *Corophium triaenonyx* Stebbing from Visakhapatnam Harbor: Effect of salinity and temperature. *Biol. Bull. (Woods Hole, Mass.)* **144**, 503-510.

Shyamasundari, K. (1976). Effects of salinity and temperature on the development of eggs in the tube-building amphipod *Corophium triaenonyx* Stebbing. *Biol. Bull. (Woods Hole, Mass.)* **150**, 286-293.

Smith, J. E. (1968). "'Torrey Canyon' Pollution and Marine Life." Cambridge Univ. Press, Cambridge, Massachusetts.

Stenner, R. D., and Nickless, G. (1974). Distribution of some heavy metals in organisms in Hardangerfjord and Skjerstadfjord, Norway. *Water, Air, Soil Pollut.* **3**, 279-291.

Vlasblom, A. G., and Bolier, G. (1971). Tolerance of embryos of *Marinogammarus marinus* and *Orchestia gammarella* (Amphipoda) to lowered salinities. *Neth. J. Sea Res.* **5**, 334-341.

Vobis, H. (1973). Rheotaktisches verhalten einiger *Gammarus*-arten bei verschiedenen sauerstoffgehalt des Wassers. *Helgol. Wiss. Meeresunters.* **25**, 495-508.

Ware, R. (1978). Food habits of *Genyonemus lineatus* (Pisces: Sciaenidae) in the vicinity of the Fish Harbor and Terminal Island, California sewage outfalls. Masters' Thesis, California State University, Long Beach.

Widdowson, T. B. (1971). Changes in the intertidal algal flora of the Los Angeles area since the survey by E. Yale Dawson in 1956-1959. *Bull. South. Calif. Acad. Sci.* **70,** 2-16.

Wildish, D. J. (1970). The toxicity of polychlorinated biphenyls (PCB) in sea water to *Gammarus Oceanicus. Bull. Environ. Contam. Toxicol.* **5,** 202-204.

Wildish, D. J., and Zitko, V. (1971). Uptake of polychlorinated biphenyls from sea water by *Gammarus oceanicus. Mar. Biol.* **9,** 213-218.

Wilson, D. P. (1955). The role of microorganisms in the settlement of *Ophelia bicornis* Savigny. *J. Mar. Biol. Assoc. U.K.* **34,** 531-543.

CHAPTER 12

Clams and Snails [Mollusca: Pelecypoda (except Oysters) and Gastropoda]

WINSTON MENZEL

I. Introduction

The estuary has been called the "septic tank of the megalopsis" (De Falco, 1967). We have always had wastes that had to be disposed of in some manner, and with the increased population and permanent community living, the many settlements on estuaries allowed these to be convenient places to dump our wastes, letting the tides and currents disperse them. Industries were located on estuaries because they met the water needs of the manufacturing processes and because they provided a means of transportation. In addition, our increasing

371

urbanization and agricultural practices resulted in increased runoff of rainfall, carrying the wastes from land.

Life forms in estuaries are adapted to withstand many stresses such as variable salinities and temperatures and often considerable amounts of silt in the water. An estuary is not a stable environment, but because of the continual addition of nutrients from land as well as the bringing in of nutrients from the near shore sea, it is very productive. Those organisms adapted to live under these conditions are great in biomass. An estuary has been termed a nutrient trap, and the same factors that make it a nutrient trap also operate to make it a pollution trap. Mollusks, especially the filter feeding bivalves, are dominant organisms of most estuaries and aid in recycling the nutrients. The mollusks serve as indicators of the nutrient and detritus cycle (Hedgpeth, 1967) and also are indicators of the many pollutants that occur.

Of the generally recognized six classes of mollusks, only two will be treated here. Of these two, the pelecypods (including clams and mussels) will be dealt with more fully than the gastropods, partly because more investigations have been made of bivalves than snails. Oysters (*Crassostrea* and *Ostrea*) will not be discussed, although they may be referred to from time to time. Such extensive investigations have been made of the effects of pollution on oysters, especially the estuarine genus *Crassostrea,* that a discussion would extend this chapter beyond the allowable limits. No attempt will be made to cover all the published data, but it is hoped the coverage will be representive. The systematics of the species will follow R. Tucker Abbott (1974).

Bivalve mollusks are filter feeders, mostly on phytoplankton, and hence can be termed primary consumers, although some feed on detritus. A current is created by the cilia on the gills and large amounts of water are "pumped" through the animal, with the removal of the food particles. Most of the bivalves in the estuaries have the ability to remove selectively those particles that are acceptable and discard the others as psuedofeces. These discarded particles are bound in mucus and usually become part of the bottom sediment, that may be reworked by other organisms. Many bivalves have the mantle modified into restricted inhalent and exhalent openings, the siphon, which allows them to live many centimeters buried in the bottom, with the siphons extended to the bottom water surface. Although the adults are mostly sessile, there is a planktonic larval stage. Many species have external fertilization and development in the water column after the spawning of the eggs and sperm, whereas in some the eggs are retained in the mantle chamber where early development occurs before the shortened planktonic life.

Snails are generally benthic and slow moving as adults. There is a free-living, or planktonic, stage for most species after the eggs are hatched, although some emerge as crawling juvenile snails. Many snails are vegetarians, scraping algae from structures or organisms with their radulas, or eating macroalgae and other plants. Some snails are carnivorous and use the radulas to drill holes in other shelled animals, mainly bivalves, and consume the tissues. Others use their foot and shell to chip the edges of the shells of bivalves to gain access to the inner tissues. Some snails consume other gastropods whole, digest out the tissue, and regurgitate the empty shells.

Pollution here is defined as the introduction of substances into or altering of the environment so that it is made less suitable for existing life forms. It is implied that these changes come about by man's activities, but forces of nature, such as storms that open or close barrier islands or shift bottom sediment, affect the ecology and could be termed pollution. Previous activities by man may intensify or lessen the forces of nature.

II. Types and Sources of Pollution and Its Ecological Effects

There are many types and sources of pollution, and these will be categorized broadly. Often the same type of pollution will be from several sources, e.g., excessive silting from dredging, agricultural practices, storm runoff, etc. The ecological effects of pollution may be synergistic, in that the combined effects of two or more pollutants is more than the sum of each. As stated previously, organisms living in estuaries are able to withstand many stresses but additional stresses, even of the types encountered normally, will be detrimental. The stresses that result from man's activities are likely to be permanent.

The young planktonic stages are usually more sensitive to stress than the adult. Although the adult may be able to survive a stress there may be sublethal effects on growth rate, feeding, and sexual maturation. The pollutant may destroy or curtail a link in the food web or some less resistant organism that may be a competitor or predator, which would lead to an ecological imbalance.

A. DOMESTIC SEWAGE

The type of pollution usually associated with mollusks is domestic sewage, and with it there is the possibility of typhoid, hepatitis, cholera, or other illnesses associated with eating, especially raw, contaminated shellfish. This is a real threat and considerable effort is expended to ensure that the clams and oysters we eat are fit for human consumption. The continued surveillance by health authorities is strict, although it has not always been so. In former years, through neglect or ignorance, situations existed that would be unthinkable now. At one time clams and oysters were held for future marketing in floats that were in close proximity to open privies (Cummings, 1916). Shellfish areas are still being condemned for health reasons, but the rate is decreasing despite the increasing population, developing of resort areas, and increased water activities such as boating. The rate of closure for the period 1971–1974 in the United States was one-half the rate in the previous 5 years (Report to Congress on Molluscan Shellfisheries, 1977).

I have observed on numerous occasions the closure of shellfish areas that were in close proximity to concentrations of domestic animals, especially after heavy rains with the resulting runoff. Large numbers of migrating Canada geese have caused closure because of high bacterial count (Ritchie, 1977). The method used to determine if human wastes are present is a relatively simple test for *Eschericha coli,* a common nonpathogenic bacterium found in the intestine, but positive tests are also obtained from other warm-blooded animals. It is recognized that the tests leave something to be desired, but better ones, which are still practical, have not yet been developed (Report to Congress on Molluscan Shellfisheries, 1977).

Human wastes are fertilizers and the nutrients promote the growth of plant life, with the resulting growth of animal life. The mussel *Mytilus galloprovincalis* (Lamarck) increases in polluted waters (Bellan and Bellan-Santani, 1971), and I have observed luxuriant growths of mussels and oysters on sewer outfalls; however these seemingly healthy and fat mollusks are not recommended for human consumption, especially when raw. Excess sewage may be detrimental in that the environment may not be able to assimilate all of the organic matter, especially in areas of poor circulation, with a resulting increase in biochemical oxygen demand (BOD) and reduced dissolved oxygen (DO). At an outfall near Copenhagen, Denmark, the bottom fauna were denuded within a radius of 100 to 200 m, but beyond this the fauna were very abundant, dominated by the bivalves *Cardium edule* Linné, *Macoma balthica* Linné, and *Mya arenaria* Linné. The pH may be changed also and it has been found that the larvae of the hard clam, *Mercenaria mercenaria* (Linné), only develop normally in a 7–8 pH range (Calabrese, 1972).

Nutrients (nitrates and phosphates) from secondary treated sewage have been used in a recycling experiment to culture bivalve mollusks. Phytoplankton are grown in the wastes and then these are fed to scallops [*Argopectin irradians* (Lamarck)], hard clams (*M. mercenaria*) and oysters (*Crassostrea virginica* (Gmelin), and *C. gigas* (Thunberg)) (Ryther and Mann, 1977). There is a precedent for using human feces as fertilizers; the use of nightsoil has long been a practice in the Orient.

Domestic sewage not only has the possibility of transmitting disease organisms and causing a high organic load that changes the DO and pH but also contains other substances that disturb the ecology and may be toxic. The widespread use of detergents has been shown to be toxic to bivalve larvae. The biodegradable detergents (alkylate sulfonates) are just as toxic as the so-called hard detergents (alkyl benzene sulfonates) but the action is not as long lasting. In addition, many of the detergents contain phosphates, which promote plant growth and accelerate eutrophication.

Studies have shown that with the abatement of domestic pollution the water column improves rapidly, but the bottom, with its associated fauna, may not. The improvement in the bottom ecology is dependent on the amount of pollution locked up in the bottom sediments as well as the circulation (McNulty, 1966; Simon, 1974). The Raritan River and Bay in New Jersey was grossly polluted and after the installation of a trunk line, which improved conditions, the larvae of the soft-shell clam *Mya arenaria* and the hard shell clam *Mercenaria mercenaria* increased many times (Jefferies, 1964). Later, the soft-shell clam extended its range considerably (Dean and Haskins, 1964). The mud snail *Ilynassa obsoleta* (Say) and the ubiquitous bivalve *Mulinia lateralis* (Say) were found in considerable numbers in formerly impoverished areas.

In many areas the same system used for domestic sewage is used for storm water runoff as well as small industries. The systems become overloaded, especially after heavy rainfall, resulting in the release of untreated sewage into the estuaries. In addition, the storm water contains other pollutants that affect the ecology of the estuary, including pesticides and wastes from automobiles and trucks. Other wastes included in this category would be plastic wrappings and other discarded containers and unused food products, all of which add additional stresses to the estuarine system.

B. INDUSTRIAL POLLUTION

Industrial pollution covers a myriad of aspects, from the release of toxic chemicals to the release of heated water used in the manufacturing processes.

Paper mills have long been recognized as sources of pollution in estuaries and have been investigated extensively because of the detrimental effects on the biota, especially mollusks. Toxic chemicals are released as well as wood and bark which change the pH and DO. There are two processes in paper making. In one the spent sulfite liquor is acid, whereas the Kraft method uses an alkaline cooking liquor, both of which are toxic.

A sulfite paper mill caused the disappearance of all intertidal mollusks in the vicinity of the effluents (Waldichuk, 1959). Laboratory experiments with sulfite liquor caused high mortality to the eggs, embryos, and larvae of the hard clam *Mercenaria mercenaria,* with the eggs being the most sensitive (Woelke, 1960). Paper mill wastes cause a high BOD as well as are toxic; both factors are reduced by dilution according to the distance from the source. There is, however, apparently no relationship between BOD and toxicity (O'Neal, 1966), as determined by experiments with the blue mussel *Mytilus edulis* (Linné). Pulp mill wastes sometimes discolor water so that photosynthesis is reduced, and I have seen oysters (*C. virginica*) colored an unpalatable brown by such effluents.

In addition, there are usually heavy metals in pulp mill and other industrial effluents. These metals are usually present under natural conditions as traces, and such minute amounts may even be necessary for the physiological functions of the mollusks. Overabundance, however, definitely causes malfunctions or out-right mortality, as has been amply demonstrated. In addition, mollusks accumulate metals, either from the food they eat and/or absorption. This further adds to the problems. The soft shell clam *Mya arenaria* has been found to contain silver in concentrations as high as 17 ppm wet weight near a source of pollution (Parsons *et al.,* 1973). The accumulation of copper in the visceral mass of mollusks almost always indicates industrial pollution, and I have noticed that such mollusks are in poorer condition.

Studies on the mud snail *Ilynassa obselata,* a common inhabitant of the estuarine areas of the East Coast, were made with the salts of arsenic, cadmium, copper, silver, and zinc and a combination of cadmium and copper (MacInnes and Thurberg, 1973). Most of the metals caused death at 5 ppm, and sublethal effects occurred at lesser concentrations. Some of the metals caused a depression of oxygen consumption and some caused an elevation. Either could be stressful under certain conditions. The combination of cadmium and copper caused a greater depression of oxygen consumption than the same amount of copper alone, although cadmium by itself had caused an elevation.

Laboratory studies on adult blue mussels (*Mytilus edulis*), hard clams (*Mercenaria mercenaria*), and soft-shell clams (*Mya arenaria*) showed that copper depressed oxygen consumption and silver increased it. Silver caused a proportionately higher increase with decreasing salinities in the soft-shell clam (Thurberg *et al.,* 1974). There is a "normal" increase in oxygen consumption at decreasing salinities because of the additional "work" to maintain the proper osmotic balance, and silver adds additional stress to an already stressful situation.

Metals affect the young planktonic stages also. The effects of the salts of mercury, silver, zinc, and lead were tested on the eggs, embryos, and larvae of the hard clam. Mercury was most toxic followed by silver, zinc, and lead. The advanced larvae were generally more tolerant to higher concentrations than the eggs and embryos (Calabrese and Nelson, 1974).

Zinc and other metals are taken up rapidly by mollusks, with hard clams accumulating zinc more rapidly than either the scallop or oyster (Chipman *et al.,* 1958; Duke, 1967). Zinc and other metals can be depurated. The blue mussel, which had accumulated zinc-65 at the mouth of the Columbia River, lost the metal in several months when moved to California (Young and Folsom, 1967).

Thermal pollution has received much attention in recent years, especially after the proliferation of nuclear power plants, which require tremendous amounts of water as a coolant. Thermal pollution has been termed thermal enrichment when the warmed water is used to accelerate growth, especially in aquaculture. Thermal enrichment has been implicated in the establishment of the exotic hard clam *Mercenaria mercenaria* in certain areas of the British Isles where the water would normally have been too cold before the construction of nuclear power plants (Ansell *et al.*, 1964; Korringa, 1971).

In the more tropical regions many animals are living near their upper thermal limit, and a few degrees' increase in temperature would be fatal. Probably of equal importance, even when death does not occur, is the interruption of the normal maturation and spawning cycles. In the more temperate and colder regions mollusks are geared to a cycle of growth and buildup of reserves to be converted to gametes and subsequent spawning. Temperature is probably the primary factor in this cycle. A change in the temperature regimes could result in an imbalance and prevent spawning or even gonadal maturation. A continuous production of gametes might occur, which would result in little or no growth because all the energy is used for producing the gametes. Shell growth of the hard clam (native to New York), when transplanted to Florida waters, ceases when the temperature reaches about 27°C (Menzel, 1962). Warm water might cause spawning when the proper kind of phytoplankton food is not present in sufficient abundance, or the larvae might be carried by the currents to colder water, which would be detrimental to them.

The soft-shell clam *Mya arenaria* is abundant in Chesapeake Bay and supports a large fishery in Maryland, although it is a "cold water" species and is near its upper temperature tolerance. A warming of the water by only several degrees could eliminate the soft-shell clam from the Bay. This is not likely at the present time because of the good circulation—but the increase of industry, discharging warm water, could be serious in the future.

The fouling of the cooling pipes by many organisms, including mollusks, is a serious economic factor in the industries needing the water to help dissipate excess heat. One method that is often used to control this fouling is treatment with chlorine. The chlorine kills many of the animals and is especially destructive of the phytoplankton. Chlorine is the common treatment for treating swimming pools to keep down algal growth, as well as disinfecting the pools. In a power plant on Long Island there was an 83% decrease in phytoplankton production when chlorine was used in the coolant lines (Carpenter *et al.*, 1972), with no apparent decrease when chlorine was not used. Another study showed a 91% decrease in primary productivity when chlorine was used as an antifoulant measure (Hamilton *et al.*, 1970). Because the phytoplankton are the food of the filter feeding mollusks one can see the ecological consequences. In addition to the use of chlorine as an antifouling measure, it is used extensively in the treatment of domestic sewage. It has been found that chlorine caused the increased oxidization of organic compounds, increasing the BOD and also releasing absorbed heavy metals. Consideration is now being given to alternate methods of fouling prevention and sewage treatment. Because of the known ecological damage this is an imperative issue.

C. Radiation Pollution

The advent of the atomic age and the occurrence of radioactive components in the aquatic realm from fallout, leakages, and accidental spills, as well as ionization of elements in the coolant water of nuclear plants, have caused considerable concern. As mollusks are known to accumulate metals and certain chemicals and are an important component of our seafood diet, it is understandable that there is concern that we may become contaminated.

One federal laboratory devoted most of their resources and time to the monitoring of field conditions and to the performance of laboratory experiments to determine possible dangers. These studies have been published in several reports (e.g., Chipman, 1958, 1960; Chipman *et al.,* 1958; Duke, 1967; Rice *et al.,* 1965). Some of the mollusks that have been investigated included hard clams (*Mercenaria mercenaria*), soft-shell clams (*Mya arenaria*), bay scallops (*Argopectin irradians*), and blue mussels (*Mytilus edulis*)—as well as oysters.

The studies showed that the accumulation of fission products by the shellfish was very rapid and, coincidently perhaps, those isotopes with a shorter half life were taken up more rapidly than those with longer periods of decay. Mollusks, especially the hard clam, accumulated zinc-65 more rapidly than either crustaceans or fish. The bay scallop accumulated high concentrations within 1 day of exposure. Strontium-90, which acts physiologically as calcium, was deposited in the shells, with no residue in the soft tissues. Cesium-137, on the other hand, was concentrated in the adductor muscle of scallops. Cobalt-60 also was concentrated in the tissues.

Radioactivity induces more rapid genetic mutation and has long been a tool of geneticists. Irradiation was used to induce mutation in oysters, but without conclusive results that could be used in aquaculture (Longwell and Stiles, 1972). There is always the possibility that mutations would be induced in mollusks, but as a general rule mutations are deleterious and would be eliminated in a natural population. No studies have been done on the effects of radionuclides on the ecology of mollusks.

There have been no widespread closures of shellfish areas because of radioactivity, although there have been "scares" in the past which resulted in a depressed market. From the standpoint of the mollusks this was beneficial in that harvesting was decreased. The continued surveillance and the controls to prevent contamination of the environment (much stricter than for other kinds of pollution) have been successful; however, the ever-present threat remains.

D. PETROLEUM POLLUTION

Pollution from petroleum has been prominent in the public eye in recent years because of the increase in use, especially from foreign sources and the proliferation of offshore wells. Every accident of tankers or wild wells along our coast or elsewhere results in alarm and dire predictions about ecological disaster. There have been many investigations on the consequences of the loss of petroleum products in the environment, and no attempt will be made here to give a complete coverage. Many of the investigations have resulted in conflicting results and conclusions. This is partly understandable when one considers that crude oil, as it comes from the well, is made up of thousands of compounds, that crudes from some areas are less toxic than others, and that the refined products introduce other variable factors. Crude oil mixtures are mainly hydrocarbons, but may have impurities in the form of nitrogen, sulfur, and other chemicals that the petrochemical industry adds. Investigations have shown that the lighter fractions, those more soluble in water, as well as the refined products, are more toxic than the crude oils.

Oil pollution enters the estuaries from many sources; the most spectacular are from tanker accidents or from wild wells. These two sources, however, account for only a small fraction of the worldwide oil pollution (Rhode Island Newsletter, 1977). In addition, if at all possible, efforts are made to remove such spills as soon as possible. It is the continuous pollution from many sources that is more serious. Waste motor oil may contribute as much as 30% of the total pollution in the oceans (Porricelli *et al.*, 1971). Much of this is from land runoff and through the sewer systems. Other sources include motor-driven boats, from the outboard to the battleship. The increasing numbers of pleasure craft and the associated marinas add greatly to the overall pollution.

During World War II many tankers were sunk because of enemy action; newspapers report that some of these are still leaking oil. There was no opportunity to try to clean up this oil and the environment had to absorb the full impact of thousands of barrels. As far as is known there was no long-term ecological damage, although the immediate and long-term effects were not documented.

In many rather recent spills detergents or emulisifiers were used to help disperse the oil. Careful studies have shown that the dispersing agents caused more damage than the oil itself, i.e., they killed more of the biota (George, 1961; Nelson-Smith, 1968; Odum *et al.*, 1974). The use of detergents has been curtailed.

Mollusks can become so contaminated with oil that they taste of it. They take in the oil adhering to the particles of silt or food or as small droplets. A spill in

Massachusetts caused the closure of large shellfish areas for the soft-shell clam and scallops that had become "tainted" (Blumer *et al.*, 1971). Clams were placed in flowing water without oil and they rid themselves of the oil, but oil could be detected chemically for as long as 6 months. There have been many studies that show mollusks can depurate oil and some do so rather quickly (Lee, 1977; Lee *et al.*, 1972; Michell *et al.*, 1969; Stainken, 1977).

Oil can cause mortality in adult mollusks. The spill from the oil well in Santa Barbara, California did not cause immediate damage, but caused (*Mytilus californianus* Conrad) continued mortalities for several weeks (Michell *et al.*, 1969). Later there was seemingly complete ecological recovery. A spill of refined oil in Maine resulted in immediate mortality of soft-shell clams, with some continued mortality for as long as 3½ years (Dow and Hurst, 1976). A tanker wreck in the Baja California area partially blocked a cove and spilled oil for 8 months. Despite this incident the intertidal gastropod *Littorina planaxis* Philippi survived, but abalones (*Haliotis* spp.) and subtidal gastropods were killed and the reduction in numbers lasted for 7 years (North *et al.*, 1965).

Intertidal mollusks are usually more resistant to oil pollution than those that are subtidal (Taylor and Karinen, 1977). The clam *Macoma balthica* was very resistant to short-term exposures to Prudhoe Bay, Alaska oil, but long exposures caused the clams to come out of the substrate. The oil may have acted as a irritant to the nervous or muscular systems or may have changed the pH or reduced the DO. Surfacing would be deleterious because the clams would be more vulnerable to predators. As the oil weathered there was less effect on the clams. In California there was evidence that populations of mussels which had been subjected to chronic oil pollution, either from natural seeps or man-made, were more resistant to oil in laboratory experiments than those from populations which had never been subjected to oil pollution (Kanter, 1974).

The incidence of gonad tumors was much higher in soft-shell clams from those areas with oil contamination than from uncontaminated areas (LaRoche, 1973). After a spill of No. 2 fuel oil the blue mussel did not have gonadal maturation and there was a decrease in respiration and feeding as well as poor byssal attachment (Blumer *et al.*, 1971; Cooley, 1977). The hard clam has been observed to accumulate black tarlike discolorations, probably a polymer of petroleum, in areas of chronic oil pollution. These clams had a high incidence of mud worm infestation in the shells, which is an indication that the clams were not normally buried in the substrate (Jefferies, 1972). Contamination by No. 6 fuel oil caused a reduction in the carbon flux in the soft-shell clam (Gilfillan *et al.*, 1976).

A carcinogen, benzo[*a*]pyrene, an abundant product of petroleum combustion, has been found in the butter clam *Saxidomus giganteus* (Deshayes), the gaper clam *Tresus nutalli* (Conrad), the soft-shell clam *Mya arenaria*, the blue mussel *Mytilus edulis*, and the Pacific oyster *Crassostrea gigas* in Oregon estuaries where industries occur, but not in those areas where there are no industries.

Creosote is a source of the carcinogen but mussels attached to creosoted pilings did not contain it (Mix *et al.*, 1977).

An intensive study was done of the effects of crude oil pollution in the Delta Region of Louisiana (Mackin and Hopkins, 1962). The main emphasis was on oysters, but many other ecological observations were made. Several accidental spills occurred during the investigations, and oil pollution was evident, as shown by the detection of an oil film on the water surface when the bottom was stirred, especially in the vicinity of the oil fields. The oil pollution seemingly caused no diminution in the numbers and growth of oysters or in the other organisms associated with oysters. *Rangia cuneata* (Sowerby) and *Mytilopsis leucophaeta* (Conrad) were abundant mollusks in the less saline areas. In the higher salinities the gastropods *Thais haemastoma* Linné, *Nassarius vibex* (Say), *Anachis obesa* (C. B. Adams), and *Neritana reclivata* (Say) were abundant, as well as the bivalve *Ischadium recurvum* (Rafinesque). A subsequent investigation was made in the vicinity of a spill that persisted for some time (Mackin and Sparks, 1962). In addition to the above higher-salinity species, the oyster shells were heavily infested with the boring clam *Diplothyra smithii* (Tryon).

Oil may serve as a carrier or solvent for other substances that may be harmful to mollusks. Such metals as zinc, nickel, and cadmium were found in many thousand times greater concentrations in the oil phase than in the water column (Walker and Colwell, 1977).

Oil pollution has long been surmised to be toxic to the planktonic stages of mollusks (Lane *et al.*, 1925). Since the development of reliable techniques to rear the larvae of mollusks (Loosanoff and Davis, 1963), controlled laboratory experiments have been conducted on the effects of various types and concentrations of oils.

The spermatozoa of the bivalve *Mulinia lateralis* were subjected to the water-soluble fraction of an oil and then used to fertilize untreated eggs, with a resulting reduction in the percentage of fertilization. Treatment with the water-soluble fraction caused a 62% mortality in the early ciliated embryos and a 59% mortality in older swimming larvae (Renzoni, 1975). The planktonic stages of the hybrids between the northern hard clam *Mercenaria mercenaria* and the southern *M. campechiensis* (Gmelin) were used to determine the toxicity and other effects of water soluble fractions of several types of oil. The most toxic was the fraction from used motor oil, followed by "Jay Field" crude oil from Florida. The water-soluble fraction from No. 2 fuel oil was next in toxicity, and then Bunker C. The other two, both crudes, were Kuwait and Southern Louisiana, with the latter being the least toxic. It was concluded that the used motor oil had picked up added impurities from the engines of the automobiles. As stated previously, refined oils are generally more toxic than the crude oil as it comes from the well. The crude oil from Florida is known as a "sour oil" and must be scrubbed to get rid of some of the impurities, especially sulfur. Number 2 fuel oil is more refined than Bunker C. The amounts of the crudes from Kuwait, and especially Louisiana, necessary to cause toxicity probably never would occur except possibly in the middle of a spill. However, even when significant mortalities did not occur, there was a reduction in growth rates in all the oil soluble fractions, even at low concentrations (Byrne and Calder, 1977).

The ecological consequences of oil pollution in estuaries are very complex—and seemingly some mollusks, with certain types of oil, suffer no damage. The refined products are usually more toxic than the crude oil. Weathering results in oils, especially crudes, loosing most of their toxicity, probably because the more toxic fractions evaporate or are oxidized. The young planktonic stages are more sensitive than the adult mollusks. The recent sinking of the *Argo Merchant* off Nantuckett Island loaded with No. 6 fuel oil was spectacular, but through a fortuitious concourse of climatic circumstances the predicted ecological damage seemed nonexistant or slight. The greatest danger appears to be from the continuous additions of oil, especially the waste oils, through runoff and in sewage.

E. PHYSICAL ALTERATIONS

There are few estuaries in the United States that have not been altered in some manner to "improve conditions." These alterations include dredging for navigation and for the use of the dredged material or spoil to build land for home and factory sites, various activities for flood control, and the damming and channelization of the rivers that flow into the estuaries. In addition, marshlands have been drained or filled or made into freshwater lakes. Our agricultural and forestry practices, as well as urbanization, have affected or changed the "virgin" conditions of the estuaries. These activities have caused drastic or subtle changes in the current and salinity patterns, increased the siltation, etc. (Copeland, 1966; Cronin 1967; Cronin *et al.*, 1971). On the West Coast hydraulic mining deposited layers of sediment, and was partly responsible for the decline of the mussel industry (Lutz, 1977).

Flood control is one of the activities that has altered estuaries, often to the detriment of the biota. Flooding is a normal sequence and the ecological cycle in most estuaries is geared to periodic freshets (Gunter *et al.*, 1974). Hurricane Agnes in 1972 killed many of the soft-shell clams *Mya arenaria* in the upper Chesapeake Bay (Haven *et al.*, 1975), but there was a recovery (Shaw and Hamons, 1974). This was so despite the fact that man's activities in the watershed undoubtly increased the severity of this very intense storm. Most mollusks are able to live through freshets. It was found in South Carolina that the hard-shell clam *Mercenaria mercenaria* was able to live through very severe lessening of the salinity by respiring anerobically (Burrell, 1977).

Flood control has been going on for a long period in the construction of levees along the Mississippi River. Formerly the periodic floods, usually in the spring, would spread out at the mouth of the river and the silt load would be deposited, building the very productive delta. Since levee construction, the water and its silt load is channeled into the open Gulf of Mexico, 100 miles south of New Orleans. The delta has been subsiding and there is no continual nourishment of sediment. This has allowed the intrusion of salt water much further inland, to the detriment of those organisms adapted to live in the formerly brackish water. The ecology of the area is being changed with the increase of more stenohaline animals such as the predacious snail, *Thais haemastoma* (Gunter, 1953, 1957; Gunter *et al.,* 1974).

Construction and maintenance of navigation channels represent the largest alteration of the estuarine system in the Gulf of Mexico area (Indall and Saloman, 1977), mainly by the United States Corps of Engineers. The actual removal of the bottom destroys the habitats and the biota living in them. Shell collectors have recovered prize specimens of mollusks from the spoil of dredging operations. The spoil must be placed somewhere. If placed in the adjacent water, especially if it extends above the tidal range, the area is destroyed forever as an aquatic habitat. If placed on the marshland this productive area is then destroyed. Other disposal in the water may smother the bottom fauna or cause changes in currents. The silt that is suspended from dredging operations will settle and may do so over the benthic organisms and smother them. When the silt is disposed in the water, the usually high organic content will increase the BOD and lower the DO as well as cause changes in the pH. In addition, metals, pesticides, and nutrients that are bound up in the bottom sediments will be relased into the water column (Brown and Clark, 1968). In some instances dredged areas quickly become recolonized with the same ensemble of organisms or with different ones, and in other instances the dredged portions may become anerobic with a complete absence of macroscopic fauna.

One of the most ecologically damaging types of dredging is the construction of "finger canals." Canals are dug into the land or marsh or in shallow bays with the spoil used to make land so that every home or factory site is waterfront property. Such practices are especially detrimental in the Gulf of Mexico area, which has little tidal amplitude and hence has little water movement in the canals, causing stagnation and anerobic conditions. A series of investigations were made in the Tampa Bay area of Florida (Sykes, 1971; Sykes and Hall, 1970; Taylor *et al.,* 1970; Taylor and Saloman 1968). Mollusks were the dominant large animals found in both the dredged canals and adjacent bay area. An average of 0.6 species of mollusks and 1.1 individuals were collected per unit sample in the canals in contrast to 3.8 species and 60.5 individuals in undredged areas. A total of 116 specimens of snails *Nassarius vibex* Say and *Haminoea antillarum* (Orbigny) and bivalves *Brachidontes exustus* Linné, *Anomalocardia* suberiana (Conrad), and *Mercenaria campechiensis* were found in 14 stations in the canals, in contrast to the above species and 151 others—a total of 5621 mollusks in 93 stations in the undredged area. Some of the control stations were in the vicinity of sewer outfalls with obvious pollution, but these areas were healthier ecologically than the finger canals. Probably the primary factor that prevented the occurrence of mollusks in the canals was the soft sediments. The lack of current allowed the accumulation of semifluid ooze, high in hydrogen sulfide, up to a depth of 4 m in canals that had been dredged 15 years previously. The phosphorous content in the water column was higher in proportion to nitrates than is known to be conducive to good phytoplankton growth.

Some interesting results were obtained from a study where a settling pond broke and allowed the accumulated silt to flow over a bay area. The silt was dredged from part of the area for the purpose of restoring the original ecological conditions. Three areas were sampled immediately after the "clean-up" dredging: dredged silted area; undredged silted area; and nearby control area (Mahadevan *et al.*, 1976). The control area yielded the greatest numbers and most species of mollusks but the undisturbed silted area was next in productivity. One gastropod, *Haminoea succinea* (Conrad), was found only in the dredged area. This condition may be similar to some plants colonizing disturbed land and then decreasing or disappearing as conditions become more stable. Unfortunately there was no follow-up study to determine the extent of the recolonization of the dredged silted portion.

Considerable dredging, especially of canals in connection with the petroleum industry, has been done in the Delta in Louisiana. The dredging has interrupted the circulation pattern, destroyed portions of productive marshland and caused more ecological disturbances than that caused by petroleum losses (St. Amant, 1972). Dredging in West Bay, Texas, reduced the acreage of the natural shoreline and marshland vegetation, changed the drainage pattern and water depth, and reduced the water exchange. All of the changes caused the nutrient level to become higher with the natural ecosystem getting out of balance (Corlis and Trent, 1971). No specific studies were made of mollusks in the above two investigations, but changes occurred.

An extensive study was made in San Antonio Bay, Texas, on the effects of shell dredging on the biota (Harper and Hopkins, 1976). Shell dredging involves the removal of dead shells, mainly oysters, for industrial purposes and the redeposit of the overburden into the cut trenches. During this process the water column becomes very turbid and some of the suspended silt settles into adjacent areas. The trenches remain deeper than the adjacent bottom for considerable periods of time, partly dependent on how deep the cuts are made. One such trench could be detected that had been made 22 years previously. The less compact sediments in the trenches prevent recolonization immediately, but it was found that recolonization occurred within about 1 year. Natural conditions, flooding after a prolonged drought of several years, and seasonal variation of temperature had more effect on the abundance of the mollusks than did the dredging. Strong winds in the shallow bay cause as much turbidity as that from the dredging operations. No stratification was found in salinity, except on occasions, and the DO never went below 4.4 ppm.

Studies of dredging in Mobile Bay, Alabama, showed that there were no long-term effects on the biota (May, 1973, 1974; Vitter, 1974). The sediments put into suspension quickly settled and there was no significant loss in habitat. There was a slight rise in nitrates and phosphates during the dredging operations, but no increase in heavy metals in the water column. The pesticide content actually decreased, probably because of a burial in the deeper sediments. Overall there was little ecological disturbance, and benthic mollusks survived in abundance. A study in Tampa Bay, Florida, showed that the benthic fauna returned to normal within 12 months after dredging had ceased (Simon *et al.*, 1976).

A large dredging operation for navigation purpose was done in the upper Chesapeake Bay and was monitored and studied from many aspects (Chesapeake Biological Laboratory, 1967, 1968; Cronin, 1970; Flemer *et al.*, 1968; Pfitzen-meyer, 1970). The spoil was deposited in open water and spread over an area to a depth of 1 foot, five times as large as the designated area. This resulted in smothering of the benthic biota in an area larger than had been anticipated. No long-term effects were noted. The bivalves *Macoma phenax* Dall and *M. balthica* recovered and were as numerous after the dredging as before. *Mytilopsis leucophaetata* (Conrad) were found throughout the dredged area. Recolonization occurred within about 18 months in the dredged area and within 1 year in the disposal area. The dominant species of mollusks, *Nucula proxima* Say and *Tellina agilis* Stimpson, declined significantly in dredged and spoil disposal areas at the mouth of the Delaware Bay (Maurer *et al.*, 1974). It is not known whether this area was recolonized later with the same or different species.

Thus many of the studies show that dredging causes only a temporary effect, and the areas become recolonized with the endemic species, especially if there is good circulation. If there is little or no circulation, such as in finger canals, the areas become and remain depopulated. In the cases where the water circulation would be altered and with changes in salinity, the ecological effects would be considerable. Unless the conditions are rendered too unsatisfactory, the areas would be recolonized but with species that are more tolerant to the changed conditions.

Another aspect that usually is not considered as alteration of the environment is shellfish harvesting. The mollusks are removed by dredges, tongs, hand digging, and picking—and these operations certainly affect the ecology of the mollusks. Some mollusks are planted, e.g., oysters, and others are harvested from natural recruitment. The practices of harvest not only remove the mollusks sought after but also associated organisms, including many mollusks. In fact in some areas shellfish growers spend considerable efforts to control predators and competitors of the mollusk they are culturing. Needless to say the efforts are not always too successful and the "weeds" grow in abundance. The ecological implications of harvesting with dredges are recognized and such practices are restricted in many areas, especially those under government control, although in some instances the restriction of dredging is in the interest of conservation in that it is too efficient and would cause rapid depletion of the resource.

I have observed the practice in which old shell reefs of oysters and other species are dredged with open bag dredges. Such a method turns over the shells and exposes new surfaces, killing the associated organisms, many of which are mollusks, including mussels, jingles, and others. Organisms quickly colonize the clean shell clutch, but oysters at least have a chance to compete with the many predators and competitors that have accumulated on the reef. These are reefs that have few oysters, partly because of removal by man in a discrimatory fashion.

The traditional method of harvesting the soft-shell clam *Mya arenaria* along the New England coast is by hand digging forks at low tides. The soft-shell clam is abundant in the Chesapeake Bay but is subtidal; another method, the Hanks' escalator dredge, has been used successfully since the mid-1950s (Manning, 1959). The escalator dredge is used to harvest hard clams *Mercenaria mercenaria* also, especially in Great South Bay, New York. This type of dredge has been tried in other areas, more or less experimentally, and studies made of the ecological impact, mainly on the clams themselves (Baptist, 1955; Godcharles, 1971; Godcharles and Japp, 1973; Haven, 1970; Kyte and Chew, 1975; Medcof, 1961; Pfitzenmeyer and Drobeck, 1967). The dredge does cause alterations in that a trench is cut in the bottom.

The escalator dredge is very efficient and actually causes less shell breakage and mortality of the soft-shell clam than hand digging. Studies in the Chesapeake Bay showed that the numbers of juvenile clams in the cuts, after a period, were as numerous or more so than the adjacent bottom. The survival rate was better, which may have been because of less competition. It was found that the dredged trenches were less compact and did not become as compact as the undredged bottom for a year or more. In Florida the trenches could be readily observed for as long as 2 years, partly because the marine vegetation was eradicated, and the use of the dredge has been restricted.

There are many other examples of physical alterations that affect the ecology of mollusks. One such instance was reported in Louisiana, where an inner coastal area was impounded as a source of fresh water. In the impoundment was a population of the brackish water clam, *Rangia cuneata* (Sowerby), which can survive fresh water and live for 10 years or more in these conditions (Gunter and Shell, 1958). This clam will live in fresh water but needs brackish water for gonadal maturation and spawning (Cain, 1972). The clam population in the fresh-water impoundment is doomed.

In the construction of the levees on the lower Mississippi River, spillways were built that could be opened during times of abnormal flooding and thus eliminate the possibility of the levees being breached. The Bonnet Carré Spillway above New Orleans has been opened several times to protect the city from possible floods. One such opening was investigated (Gunter, 1953). The oysters were killed by the fresh water for some distance from the spillway, as well as the associated fauna. Oysters are more euryhaline than some of the predators and competitors, such as the boring clam *Diplothyra smithii,* which lives in the oyster shells, and the snail *Thais haemastoma,* which preys on the oysters. The fresh water affected these mollusks more than the oysters and also at greater distances from the source of the fresh water from the spillway. The fecundity of the oysters enabled them to repopulate the old reefs rapidly as soon as conditions became stabilized to the original salinity conditions. A management method beneficial to oysters would be to open the spillway every several years. Such periodic flooding probably occurred before the building of the levees.

High turbidity is characteristic of estuaries, but man's activities have added to the turbidity. Mollusks that live in estuaries are adapted to relatively high turbidities normally, and with excess have developed adaptations such as closure of the shells and anaerobic respiration. Excessive turbidities are stresses to the mollusks, and continued high turbidities render the habitat less suitable. Turbidity and siltation have increased in our estuaries, as determined by several investigations. In the York River, Virginia, the net accumulation of sediments average 104 foot-acres/year during the period from 1857 to 1911 but increased to 576 foot-acres/year during the period from 1911 to 1938 (Brown *et al.,* 1938). This was a fivefold increase in the last 27 years compared to the previous 57 years. Similar increasing sedimentation has been noted in other areas (Gottschalk, 1945; Fuller and Hart, 1972; Hart and Fuller, 1972). It has been said that the acceleration of sedimentation started with the clearing of land to grow tobacco in the early colonization days (Schubel and Meade, 1977).

Heavy sedimentation can smother benthic mollusks, as has been noted before. There are many other effects both on the adults and the planktonic stages. Laboratory studies showed that increased tubidities increases the formation of psuedofeces and decreases the amount of water that is pumped. A small amount of silt, however, stimulates the amount of water pumped and filtered by mollusks. Hard clam eggs to not develop normally in heavy turbidity. Particles of small size seem more harmful than larger particles because the larvae ingest them. Depending on the degree of turbidity, the effects range from lessened growth to mortality (Davis and Hidu, 1969a; Loosanoff, 1962; Rice and Smith, 1958). Field observations revealed that bottoms with heavy deposits of sediments, mainly clays, have smaller populations of *Mercenaria mercenaria* and *Pitar morrhuanus* Lindsley. The smaller populations may be due to smothering or decreased DO due to high BOD (Pratt, 1953; Sanders, 1958).

I have experienced a situation where man's activities resulted in an ecological disturbance. Adjacent to the marine laboratory where rearing of the larvae of the hard clam was being conducted are extensive managed forest lands. The practice is to clear-cut the timber (pine trees for the paper industry), disk the area to control the undergrowth, and replant. The trees were planted in the winter and the adjacent property was almost devoid of vegetation the following early spring, which is the period of the most success for inducing spawning and rearing of larval clams. The rains during this period and the denuded land allowed heavy runoff, which discolored the water in the adjacent bay from which the saltwater system of the laboratory is supplied. The salinity was lowered several parts per thousand, but was still higher than the lower critical limit for clam larvae. Clams were induced to spawn but the planktonic forms could not be reared. The precise factor or factors that caused the adverse condition was not determined, but it was assumed to be from the runoff. Several weeks later the rains ceased to some extent, and together with the spring growth of vegetation on the cleared land the water returned to normal, both in color and salinity, and clams were reared successfully. Wild oysters abound in the area and there was no noticeable mortality. The normal spawning time of the oysters was several months later and the adverse conditions experienced earlier had no effect on the recruitment of these mollusks.

In summary, physical alterations rank with domestic and industrial sewage in the disturbance of the ecology of our coastal areas, and perhaps is even more serious. Many of these alterations are permanent, and although mollusks are adapted to withstand stresses caused by these alterations the permanence causes changes that sometimes may be more subtle but just as lethal.

F. PESTICIDES

Various pesticides are used in the control of noxious organisms of all kinds and the use is widespread. This chemical control has accelerated since World War II and various chemical companies have concocted new pesticides or killers, many of which are long lasting. The former massive uncontrolled use of pesticides has been regulated somewhat after the recognition of some of the ecological damage being done; however, the damage is still extensive. Everyone is aware of the consequences of the massive use of DDT, which together with its breakdown products, which are just as toxic, causes widespread damage to many organisms, especially birds.

Mollusks concentrate some of the pesticides in a way analogous to the concentration of metals. A field survey showed that the mud snail, *Ilyanassa obsolata,* accumulated 0.26 ppm and the hard clam 0.42 ppm of DDT in an area where DDT had been used (Woodwell *et al.,* 1967). In California the products of DDT, dieldrin, and endrin have been found in mollusks in concentrations of from 10 to 3,600 ppb (Modin, 1969). Mollusks can depurate pesticides after accumulation, again analogous to what happens with metals. In the laboratory the mussel

Brachidontes exustus (Linné) eliminated pesticides within 15 days that had been at a level of 24 ppm, when placed in clean flowing water. The hard clam reduced from a level of 6 to 0.5 ppm within 7 days. Two mussels, *Modiolus modiolus* (Linné) and *Mytilus edulis,* were also able to clear themselves of pesticides when placed in uncontaminated water. The soft-shell clam, which had accumulated DDT to 25 ppb, lost half of the amount within 1 week (Butler, 1971, 1973).

One investigator found that adult hard clams were not affected when subjected to water containing 1 ppb of DDT for 3 months, but did have considerable mortality in the fourth month (Butler, 1965). The same investigator found that shell growth in oysters was retarded in the presence of pesticides. It has been stated that the sublethal effects are more significant than the acute toxicity in adult mollusks (Butler, 1971).

Investigations have shown that certain pesticides can be used to control certain oyster enemies. The oyster drill *Urosalpinx cinerea* Say is a very serious oyster enemy wherever it occurs, and Sevin and orthodichlorobenzene has been used as a chemical control method (Castagna *et al.,* 1969; Haven *et al.,* 1966; Loosanoff, 1961; Shaw and Griffith, 1967). It was shown, however, that residues of the pesticides could be detected for 1 year after application, and because of the complicated ecological balance and the possibility of biological magnification the use of these pesticides could not be advocated.

Some of the pesticides are very toxic to the planktonic stages of mollusks. Sevin and its first hydrolytic product 1-napthol were toxic to the young stages of the blue mussel *Mytilus edulis.* The young embryo, just after fertilization, showed more sensitivity than the older swimming larvae. It was deduced that the pesticide damaged the spindle formation, which resulted in malfunction of cell cleavage (Armstrong and Milleman, 1974). A laboratory study of the effect of nearly 50 pesticides of different types on the larvae of several bivalves showed that some were more toxic than others, as would be expected. Generally the eggs and the newly fertilized eggs were less tolerant than the older larvae. It was noted that the solvent for the pesticide was often toxic or could act in a synergistic fashion with the pesticide. All of the pesticides had some effect, from mortality in large quantities to the lessening of the growth rate in smaller concentrations (Davis and Hidu, 1969b).

Another control is the use of antifouling and antiboring chemicals. Creosote is the common preservative for wooden pilings to prevent the ravages of shipworms, *Bankia* and *Teredo.* Copper paints are applied to the hulls of ships, both to prevent the borings of shipworms in wood and the attachment of the many sessile organisms. Many new antifouling paints have been devised which work on the principle of killing the planktonic forms of organisms before they can attach and cause problems. It is not known what overall effect these potent chemicals have on the normal occurrence of organisms, for instance in a busy harbor or marina. The planktonic stages of mollusks are known to be vulnerable.

Conclusions

The estuary has been called a fragile ecosystem. In some respects this is true but in others the estuary has proven to be very resistant and resilient. It has to be so to withstand all the abuses it has been subjected to, truly the septic tank of the magalopsis. Mollusks are mostly sessile, at least in the adult stages, and hence cannot escape situations that are antagonistic, but must remain and cope. Many have developed anatomical and physiological attributes that allow them to cope, often very successfully. Those mollusks that are adapted to tolerate the very dynamic situations in an estuary of changing salinities, temperatures, currents, and turbidities (and, with the intertidal forms, dessication) often occur in huge numbers, reflecting the productivity of the estuary.

Occasionally large molluscan populations are destroyed in the sequence of natural occurrences. When and if the situation becomes stable again, the natural fecundity of most bivalves and those gastropods with pelagic larval stages results in recolonization, sometimes very rapidly. Man-made changes, especially from alterations that cause chemical changes or from pollution from toxic chemicals, may prevent the recolonization. Such changes may be only sublethal to the adults, but the planktonic stages are generally less tolerant. In addition the man-made pollution may prevent gonadal maturation and spawning. Many of the pollution factors do not cause drastic kills but slow and steady deterioration. These sublethal factors are probably the cause for the most alarm because large mortalities cause reaction followed by action.

Fortunately in the United States, at least, we are becoming educated concerned citizens about the dangers of pollution, and there are ever-increasing efforts to eliminate, or at least control, all types. The technology exists to eliminate or control many types of pollution, but it would be very expensive and would take time to construct, the construction itself causing pollution. Because of the adaptability of mollusks and our efforts to curb pollution and maintain the estuaries, the mollusks will probably survive even if not in their pristine state.

References

Abbott, R. T. (1974). "American Seashells," 2nd ed. Van Nostrand-Reinhold, Princeton, New Jersey.

Ansell, A. D., Lander, K. F., Coughlan, J., and Loosemore, F. A. (1964). Studies on hard shell clam, *Venus mercenaria* in British waters. I. Growth and reproduction in natural and experimental colonies. *J. Appl. Ecol.* **1,** 63–83.

Armstrong, D. A., and Milleman, R. E. (1974). Effects of insecticide Sevin and its first hydrolytic product, 1-napthol, on some early development stages of the bay mussel. *Mytilus edulis. Mar. Biol.* **28,** 11–15.

Baptist, J. P. (1955). Burrowing ability of juvenile clams. *U. S., Fish Wildl. Serv., Spec. Sci. Rep.—Fish.* **140,** 1–13.

Bellan, G., and Bellan-Santini, G. (1971). "Influence of Pollution in the Marseilles Region," Report of the FAO Technical Conference on Marine Pollution and its Effects on Living Resources and Fishing, FAO Fish. Rep. No. 99, pp. 124–125. Ford Agric. Organ. V. N., Rome.

Blegvard, H. (1932). Investigations of the bottom fauna at outfalls and drains in the Sound. *Rep. Dan. Biol. Stn.* **37**, 1–20.

Blumer, M., Sanders, H. L., Grassle, J. F., and Hampson, G. R. (1971). A small oil spill. *Environment* **13**, 2–12.

Brown, C. B., Seavy, L. M., and Rittenhouse, G. (1938). Advance report on an investigation of silting in the York River, Virginia. *U. S. Soil Conserv. Serv., Div. Res.* **SCS-SS-32**, 1–12.

Brown, C. L., and Clark, R. (1968). Observations on dredging and dissolved oxygen in a tidal waterway. *Eng. Geol. (Amsterdam)* **4**, 1381–1384.

Butler, P. A. (1965). Reaction of some estuarine mollusks to environmental factors. *U.S., Public Health Serv. Publ.* **999-WP-25.**

Butler, P. A. (1966). Fixation of DDT in estuaries. *Trans. North Am. Wildl. Conf.* **21**, 184–189.

Butler, P. A. (1971). Influence of pesticides on marine ecosystems. *Proc. R. Soc. London, Ser. B* **177**, 321–329.

Butler, P. A. (1973). Residues in fish, wildlife and estuaries. *Pestic. Monit. J.* **6**, 238–262.

Burrell, V. G., Jr. (1977). Mortalities of oysters and clams associated with heavy runoff in Santee River System, South Carolina in spring of 1975. *Proc. Natl. Shellfish. Assoc.* **67**, 35–43.

Byrne, C. J., and Calder, J. A. (1977). Effect of water-soluble fractions of crude, refined and waste oils on the embryonic and larval stages of the quahog clam *Mercenaria* sp. *Mar. Biol.* **40**, 225–231.

Cain, T. D. (1972). The reproductive cycle and larval tolerances of *Rangia cuneata* in the James River, Virginia. Ph.D. Thesis, University of Virginia, Charlottesville.

Calabrese, A. (1972). How some pollutants affect embryos & larvae of American oyster & hard-shell clam. *Mar. Fish. Rev.* **34**, 66–67.

Calabrese, A., and Nelson, D. A. (1974). Inhibition of embryonic development of hard clam, *Mercenaria mercenaria,* by heavy metals. *Bull. Environ. Contam. Toxicol.* **11**, 92–97.

Carpenter, E. J., Peck, B. B., and Anderson, S. J. (1972). Cooling water chlorination and productivity of entrained phytoplankton. *Mar. Biol.* **16**, 37–40.

Castagna, M., Haven, D. S., and Whitcomb, J. B. (1969). Treatment of shell clutch with Polystream to increase yield of seed oysters, *Crassostrea virginica. Proc. Natl. Shellfish. Assoc.* **59**, 84–90.

Chesapeake Biological Laboratory (1967). "Interim Report on Gross Physical and Biological Effects of Overboard Spill Disposals," No. 67-34. Dept. Nat. Resour. University of Maryland, College Park.

Chesapeake Biological Laboratory. (1968). "Biological and Geological Research on the Effects of Dredging and Spoil Disposal in the Upper Chesapeake Bay," 7th Prog. Rep., Ref. No. 68-2-A. University of Maryland, College Park.

Chipman, W. A. (1958). Accumulation of radioactive materials by fishery organisms. *Gulf Caribb. Fish. Inst., Univ. Miami, Proc.* **11**, 97–110.

Chipman, W. A. (1960). "Biological Aspects of Disposal of Radioactive Wastes in Marine Environments." IAEA, Vienna.

Chipman, W. A., Rice, T. R., and Price, T. J. (1958). Uptake and accumulation of radioactive zinc by marine plankton, fish and shellfish. *Fish. Bull.* **58**, 278–292.

Cooley, J. F. (1977). Oil inhibits reproduction in test mussels. *Maritimes, Univ. R. I.* pp. 12–14.

Copeland, B. J. (1966). Effects of decreased river flow on estuarine ecology. *J. Water Pollut. Control Fed.* **38**, 1831–1839.

Corlis, J., and Trent, L. (1971). Comparison of phytoplankton production between natural and altered areas in West Bay, Texas. *Fish. Bull.* **69**, 829–832.

Cronin, L. E. (1967). The role of man in estuarine processes. *Publ. Am. Assoc. Adv. Sci.* **83**, 667–689.

Cronin, L. E. (1970). Summary, conclusions and recommendations. *In* "Gross Physical and Biological Effects of Overboard Disposal in Upper Chesapeake Bay," Final Report to U. S. Dept. of Sport Fisheries and Wildlife, Contract No. 14-16-005-2096. Nat. Resour. Inst., University of Maryland, College Park.

Cronin, L. E., Gunter, G., and Hopkins, S. H. (1971). "Effects of Engineering Activities on Coastal Ecology." Report to Chief of U. S. Corps of Engineers.

Cummings, H. S. (1916). Investigations of the pollution of tidal waters of Maryland Virginia. *U. S. Public Health Bull.* **74**, 199p.

Davis, H. C., and Hidu, H. (1969a). Effects of turbidity-producing substances in sea water on eggs and larvae of three genera of bivalve mollusks. *Veliger* **11**, 316–323.

Davis, H. C., and Hidu, H. (1969b). Effects of pesticides on embryonic development of clams and oysters and on the survival and growth of the larvae. *Fish. Bull.* **67**, 393–398.

Dean, D., and Haskins, H. H. (1964). Benthic repopulation of the Raritan River estuary following pollution abatement. *Limnol. Oceanogr.* **9**, 551–563.

De Falco, P., Jr. (1967). The estuary—septic tank of the megalopolis. *Publ. Am. Assoc. Adv. Sci.* **83**, 701–703.

Dow, R. L., and Hurst, J. W., Jr. (1976). The ecological, chemical and histopathologic evaluation of an oil spill site. Part 1. Ecological studies. *Mar. Pollut. Bull.* **6**, 164–166.

Duke, T. W. (1967). Possible routes of zinc-65 from an experimental estuarine environment to man. *J. Water Pollut. Control Fed.* **39**, 536–542.

Flemer, D. A., Dovel, W. L., Pfitzenmeyer, H. T., and Richie, D. E., Jr. (1968). Biological effects of spoil disposal in Chesapeake Bay. *J. Sanit. Eng.* **81**, 683–706.

Fuller, S., and Hart, C. W., Jr. (1972). Changes along the Patuxent. *Frontiers* **36**, 2–7.

George, M. (1961). Oil pollution of marine organisms. *Nature (London)* **192**, 1209.

Gilfillan, E. S., Mayo, D., Hanson, S., Donovan, D., and Jiang, L. C. (1976). Reduction in carbon flux in *Mya arenaria* caused by a spill of No. 6 fuel oil. *Mar. Biol.* **37**, 115–123.

Godcharles, M. F. (1971). A study of the effects of a commercial hydraulic clam dredge on benthic communities in estuarine areas. *Fla., Dep. Nat. Resour., Tech. Ser.* **64**, 1–51.

Godcharles, M. F., and Japp, W. C. (1973). Exploratory clam survey of Florida nearshore and estuarine waters with commercial hydraulic dredging gear. *Fla., Dep. Nat. Resour., Prof. Pap. Ser.* **21**, 1–77.

Gottschalk, L. C. (1945). Effects of soil erosion on navigation in upper Chesapeake Bay. *Geogr. Rev.* **35**, 219–238.

Greffard, J., and Meury, L. J. (1967). Carcinogenic hydrocarbon pollution in Toulen Harbour. *Cah. Oceanogr.* **19**, 457–468.

Gunter, G. (1953). The relationship of the Bonnet Carré Spillway to oyster beds in Mississippi Sound, and the "Louisiana Marsh", with a report on the 1950 opening. *Publ. Inst. Mar. Sci., Univ. Tex.* **3**, 21–71.

Gunter, G. (1957). How does siltation effect fish populations? *Nat. Fisherman* **38**, 18–19.

Gunter, G., and Shell, W. E. (1958). A study of an estuarine area with water-level control in a Louisiana marsh. *Proc. La. Acad. Sci.* **21**, 5–34.

Gunter, G., Ballard, B. S., and Venkataramiah, A. (1974). A review of salinity problems of organisms in continental coastal areas. subject to engineering works. *Gulf Res. Rep.* **4**, 38–475.

Hamilton, D. H., Jr., Flemer, D. A., Keefe, C. W., and Milhursky, J. A. (1970). Power plants: Effect of chlorination on estuarine primary production. *Science* **169**, 197–198.

Harper, D. E., Jr., and Hopkins, S. H. (1976). The effects of oyster shell dredging on macrobenthic and nektonic organisms in San Antonio Bay. *In* "Shell Dredging and its Influence on Gulf

Coast Environments'' (A. H. Bouma, ed.), Chapter 12, pp. 232-279. Gulf Pub. Co., Houston, Texas.

Hart, C. W., Jr., and Fuller, S. L. H. (1972). Environmental degradation on the Patuxent River estuary, Maryland. *Contrib. Dep. Limnol., Acad. Nat. Sci. Philadelphia* **1**, 1-15.

Haven, D. (1970). ''A Study of Hard and Soft-shell Clam Resources of Virginia,'' Annu. Rep. (Manuscript). U. S. Fish Wildl. Serv., Washington, D.C.

Haven, D., Castagna, M., Chanley, P., Wass, M., and Whitcomb, J. (1966). Effects of treatment of an oyster bed with Polystream and Sevin. *Chesapeake Sci.* **7**, 179-188.

Haven, D. S., Hargis, W. J., Jr., Loesch, J. G., and Whitcomb, J. P. (1975). The effect of tropical storm Agnes on oysters, hard clams and oyster drills. *Chesapeake Res. Consortium, Publ.* **34**, 017-0200.

Hedgpeth, J. W. (1967). The sense of the meeting. *Publ. Am. Assoc. Adv. Sci.* **83**, 707-710.

Hidu, H. (1965). Effects of synthetic surfactants on the larvae of clams (*M. mercenaria*) and oysters (*C. virginica*). *J. Water Pollut. Control Fed.* **37**, 262-270.

Indall, W. N., Jr., and Saloman, C. H. (1977). Alteration and destruction of estuaries affecting fishery resources in the Gulf of Mexico. *Mar. Fish. Rev.* **39**, 1-7.

Jefferies, H. P. (1964). Comparative studies on estuarine zooplankton. *Limnol. Oceanogr.* **9**, 354-364.

Jefferies, H. P. (1972). A stress syndrome in the hard clam, *Mercenaria mercenaria*. *J. Invertebr. Pathol.* **20**, 242-251.

Kanter, K. (1974). Susceptibility to crude oil with respect to size, season and geographic location in *Mytilus califorianus* (Bivalvia). *Univ. Calif. Sea Grant* **USC-SG-4-74**, 1-43.

Kennish, M. J. (1976). Monitoring thermal discharges: A natural method. *Underwater Nat.* **9**, 8-11.

Kerr, R. A. (1977). Oil in the ocean: Circumstances control its impact. *Science* **198**, 1134-1136.

Korringa, P. (1971). Marine pollution and its biological consequences. *In* ''Fertility of the Sea'' (J. F. Costlow, ed.), pp. 215-223.

Kyte, M. A., and Chew, K. K. (1975). A review of hydraulic escalator shellfish harvester and its known effects in relation to the soft shell clam, *Mya arenaria*. *Univ. Wash. Sea Grant Publ.* **WSG-75-2**, 1-32.

Lane, F. W., Bauer, A. D., Fisher, H. F., and Harding, P. N. (1925). Effects of oil pollution on marine and wildlife. *U. S. Bur. Fish., Doc.* **995**, 171-181.

La Roche, G. (1973). Analytical approach in the evaluation of biological damage resulting from spilled oil. *Natl. Acad. Sci. Proc. (Oil Pollut.)* pp. 347-374.

Lee, R. F. (1977). Accumulation and turnover of petroleum hydrocarbons in marine organisms. *In* ''Fate and Effects of Petroleum Hydrocarbons in Marine Ecosystems and Organisms'' (D. A. Wolfe, ed.), pp. 60-70. Pergamon, Oxford.

Lee, R. F., Sauerheber, R., and Benson, A. A. (1972). Petroleum hydrocarbons: Uptake and discharge by the marine mussel, *Mytilus edulis*. *Science* **177**, 344-346.

Lindall, W. N., Jr., and Salomon, C. H. (1972). Alteration of estuaries affecting fisheries resources of the Gulf of Mexico. *Mar. Fish. Rev.* **39**, 1-7.

Longwell, A. C., and Stiles, S. S. (1972). Breeding response of the commercial oyster in ionizing radiation. *Radiat. Res.* **51**, No. 2 (abstr.).

Loosanoff, V. L. (1961). Recent advances in the control of shellfish predators and competitors. *Gulf Caribb. Fish. Inst., Univ. Miami, Proc.* **13**, 113-127.

Loosanoff, V. L. (1962). Effect of turbidity on some larval and adult bivalves. *Gulf Caribb. Fish. Inst., Univ. Miami, Proc.* **14**, 80-95.

Loosanoff, V. L., and Davis, H. S. (1963). Rearing of bivalve mollusks. *Adv. Mar. Biol.* **1**, 1-136.

Lutz, R. A. (1977). ''A Comprehensive Review of the Commercial Mussel Industries in the United States,'' Stock No. 003-020-00133-5. Ira C. Farling Cent., University of Maine, Orono [to National Marine Fisheries Service (NOAA)].

MacInnes, J. R., and Thurberg, F. P. (1973). Effects of metals on the behaviour and oxygen consumption of the mud snail. *Mar. Pollut. Bull.* **4**, 185–186.

Mackin, J. C., and Hopkins, S. H. (1962). Studies on oyster mortality in relation to natural environments and oil fields in Louisiana. *Publ. Ins. Mar. Sci., Univ. Tex.* **7**, 1–131.

Mackin, J. C., and Sparks, A. K. (1962). A study of the effects of crude oil loss from a wild well. *Pub. Inst. Mar. Sci., Univ. Tex.* **7**, 230–261.

McKinney, L. D., Bedinger, C. A., and Hopkins, S. H. (1976). The effects of shell dredging and siltation from dredging on organisms associated with oyster reefs. *In* "Shell Dredging and its Influence on Gulf Coast Environments" (A. H. Bouma, ed.), Chapter 14, pp. 280–303. Gulf Publ. Co., Houston, Texas.

McNulty, J. K. (1966). Recovery of Biscayne Bay from pollution. Ph.D. Thesis, University of Miami, Miami, Florida.

Mahadevan, S., Culter, J., Hoover, S., Murdoch, G., Reeves, F., and Schulze, R. (1976). " A Study of the Effects of Silt Spill and Subsequent Dredging on Benthic Infauna at Apolla Beach Embayment." Conservation Consultants, Inc., Palmetto, Florida.

Manning, J. H. (1959). Commercial and biological uses of the Maryland soft-shell clam dredge. *Gulf Caribb. Fish. Inst., Univ. Miami, Proc.* **12**, 61–67.

Maurer, D., Biggs, R., Leathen, W., Kinner, P., Treasure, W., Otley, M., Watling, L., and Klemas, V. (1974). "Effects of Spoil Disposal on Benthic Communities Near the Mouth of the Delaware Bay." Coll. Mar. Stud., University of Delaware, Newark.

May, E. B. (1973). Environmental effects of hydraulic dredging in estuaries. *Ala. Mar. Res. Bull.* **9**, 1–85.

May, E. B. (1974). Effects on water quality when dredging polluted harbor using confined spoil disposal. *Ala. Mar. Resour. Bull.* **10**, 1–8.

Medcof, J. C. (1961). Effects of hydraulic escalator harvestors on undersized soft-shell clams. *Proc. Natl. Shellfish. Assoc.* **50**, 151–161.

Menzel, R. W. (1962). Seasonal growth of the Northern and Southern quahogs *Mercenaria mercenaria* and *M. campechiensis* and their hybrids in Florida. *Proc. Natl. Shellfish. Assoc.* **53**, 111–119.

Michell, C. T., Anderson, E. K., Jones, L. G., and North, W. J. (1969). Ecological effects of oil spillage in the sea. *Water Pollut. Conf., 42nd, 1969* pp. 1–10.

Milhursky, J. A. (1976). Thermal discharges and estuarine systems. *In* "Estuarine Pollution: National Assessment." Environ. Prot. Agency, Washington, D.C.

Mix, M. C., Riley, R. T., King, K. I., Trenholm, S. R., and Schaffer, R. L. (1977). Chemical carcinogens in the marine environment. Benzo (a) Pyrene in economically important mollusks from Oregon Estuaries. *In* "Fate and Effects of Petroleum Hydrocarbons in Marine Organisms and Ecosystems" (D. A. Wolfe, ed.), pp. 421–431. Pergamon, Oxford.

Modin, J. C. (1969). Chlorinated hydrocarbons pesticides in California bays and estuaries. *Pestic. Monit. J.* **3**, 1–7.

Moore, S. F., Dyer, R. L., and Katz, A. M. (1973). "A Preliminary Assessment of the Environmental Vulnerability of Machias Bay, Maine to Oil Super Tankers," Rep. No. MITSG 73-6. Massachusetts Institute of Technology, Cambridge.

Nelson-Smith, A. (1968). The effects of oil pollution and emulsifier cleansing on shore life in southwest Britain. *J. Appl. Ecol.* **5**, 97–107.

North, W. J., Neushal, M., and Clendenning, K. A. (1965). "Successive Biological Changes Observed in a Marine Cove Exposed to a Large Spillage of Mineral Oil," pp. 333–354. Comm. Explor. Sci. Medit., Monaco.

Odum, H. T., Copeland, B. J., and Mahan, E. A. (1974). "Coastal Ecological Systems of the United States," Vol. III. Conserv. Found., Washington, D.C.

O'Neal, C. L. (1966). Degradation of Kraft pulping wastes in Estuarine waters. M.S. Thesis, Oregon State University, Corvallis.

Parsons, T. R., Bawden, C. A., and Heath, W. A. (1973). Preliminary survey of mercury and other metals contained in animals from Fraser River mudflats. *J. Fish. Res. Board Can.* **30**, 1014–1016.

Pfitzenmeyer, H. T. (1970). Benthos. *In* "Gross Physical and Biological Effects of Overboard Spoil Disposal in Upper Chesapeake Bay," Spec. Rep., Vol. 3, pp. 26–38. Nat. Resour. Inst., University of Maryland, College Park.

Pfitzenmeyer, H. T., and Drobeck, K. C. (1967). Some factors influencing reburrowing activity of soft-shell clam *Mya areanaia*. *Chesapeake Sci.* **8**, 193–199.

Pilpel, N. (1968). Natural fate of oil in the sea. *Endeavour* **27**, 11–13.

Porricelli, J. D., Keith, V. F., and Storch, R. L. (1971). Tankers and ecology. *Soc. Nav. Archit. Mar. Eng., Trans.* **79**, 169–221.

Pratt, D. M. (1953). Abundance and growth of *Venus mercenaria* and *Callocardia morrhuana* in relation to the character of the bottom sediments. *J. Mar. Res.* **12**, 60–74.

Rawls, C. K. (1965). Field tests of herbicide toxicity to certains estuarine animals. *Chesapeake Sci.* **6**, 150–161.

Renzoni, A. (1975). Toxicity of three oils to bivalve gametes and larvae. *Mar. Pollut. Bull.* **6**, 125–128.

Report to Congress on Molluscan Shellfisheries (1977). "The Molluscan Shellfish Industries and Water Quality: Problems and Opportunities," Stock No. 003-020-00142-4. Nat. Mar. Fish. Serv., NOAA, Washington, D.C.

Rhode Island Newsletter (1977). "Marine Advisory Service Newsletter," September/October, No. 63, pp. 2–3. University of Rhode Island, Kingston.

Rice, T. R., and Smith, R. J. (1958). Filtering rate of the hard clam (*Venus mercenaria*) determined with radioactive phytoplankton. *Fish. Bull.* **58**, 73–82.

Rice, T. R., Baptist, J. P., and Price, T. J. (1965). Accumulation of mixed fission products by marine organisms. *Int. Conf. Water Pollut. Res., 1964* Vol. 3, pp. 263–286.

Ritchie, T. P. (1977). A comprehensive review of the commercial clam industries in the United States. *Del. Sea Grant Program* **DEL-SC-26-76**, 1–106.

Ryther, J. H., and Mann, R. (1977). "Bivalve Mollusk Culture in a Waste Recycling Aquaculture System," Report 77-59. Dept. of Commerce, NOAA, Woods Hole Oceanogr. Inst. Woods Hole, Massachusetts.

Sanders, H. L. (1958). Benthic studies in Buzzards Bay. I. Animal sediments relationships. *Limnol. Oceanogr.* **3**, 245–258.

Schubel, J. R., and Meade, R. H. (1977). Man's Impact on estuarine sedimentation. *Estuarine Pollut. Control Assess., Proc. Con., 1975* Vol. I, pp. 193–209.

Shaw, W. N., and Griffith, C. T. (1967). Effects of Polystream and Drillex on oyster setting in Chesapeake Bay and Chincoteague Bay. *Proc. Natl. Shellfish. Assoc.* **57**, 17–23.

Shaw, W. N., and Hamons, F. (1974). The present status of the soft-shell clam in Maryland. *Proc. Natl. Shellfish. Assoc.* **64**, 38–44.

Simon, J. L. (1974). Tampa Bay estuarine system—a synopsis. *Fla. Sci., Fla. Acad. Sci. J.* **37**, 217–244.

Simon, J. L., Doyle, L. J., and Conner, W. G. (1976). "Environmental Impact of Oyster Shell Dredging in Tampa Bay, Florida," Final Report on the Long Term Effects of Oyster Shell Dredging in Tampa Bay. Report to Florida Department of Environmental Regulations.

Stainken, D. (1977). The accumulation and depuration of No. 2 fuel oil by the soft-shell clam. *Mya areanaria* L. *In* "Fate and Effects of Petroleum Hydrocarbons in Marine Organisms and Ecosystems" (D. A. Wolfe, ed.), pp. 313–322. Pergamon, Oxford.

St.Amant, L. S. (1972). The petroleum industry as it affects marine estuarine ecology. *J. Pet. Technol.* pp. 288–289.

Sykes, J. E. (1971). Implications of dredging and filling in Boca Ciega Bay, Florida. *Environ. Lett.* **1**, 151–156.

Sykes, J. E., and Hall, J. R. (1970). Comparative distribution of mollusks in dredged and undredged portions of an estuary, with a systematic list of species. *Fish. Bull.* **68**, 299–306.

Taylor, J. L., and Saloman, C. H. (1968). Some effects of hydraulic dredging and coastal development in Boca Ciega Bay, Florida. *Fish. Bull.* **67**, 213–241.

Taylor, J. L., Hall, J. R., and Saloman, C. H. (1970). Mollusks and benthic environments in Hillsborough Bay, Florida. *Fish. Bull.* **68**, 191–202.

Taylor, T. L., and Karinen, J. F. (1977). Response of the clam *Macoma balthica* (Linnaeus) exposed to Prudhoe Bay crude oil, as unmixed oil, water-soluble fraction and oil-contained sediment in the laboratory. *In* "Fate and Effects of Petroleum Hydrocarbons in Marine Organisms and Ecosystems" (D. A. Wolfe, ed.), pp. 229–237. Pergamon, Oxford.

Thurberg, F. P., Calabrese, A., and Dawson, M. A. (1974). Effects of silver on oxygen consumption of bivalves at various salinities. *In* "Pollution and Physiology of Marine Organisms" (F. J. and W. B. Vernberg, eds.), pp. 67–78. Academic Press, New York.

Vitter, B. A. (1974). Effects of channel dredging on biota of a shallow Alabama estuary. *J. Mar. Sci.* **2**, 111–133.

Waldichuk, M. (1959). Effects of pulp and paper mill wastes on the marine environment. *In* "Second Seminar on Biological Problems in Water Pollution." R. A. Taft Sanit. Eng. Cent. (Processed).

Walker, J. D., and Colwell, R. R. (1977). Mercury resistant bacteria and petroleum degradation. *Appl. Microbiol.* **27**, 285–287.

Woelke, C. E. (1960). The effects of sulfite waste liquor on the development of eggs and larvae of two marine molluscs and three of their food organisms. *Wash., Dep. Fish. Resour., Bull.* No. 6, pp. 86–106.

Woodwell, G. M., Wurster, C. F., Jr., and Isaacson, P. A. (1967). DDT residues in an east coast estuary: A case of biological concentration of a persistent pesticide. *Science* **156**, 821–824.

Young, D. R., and Folsom, T. R. (1967). Loss of zinc-65 from the California sea mussel, *Mytilus califorianus. Biol. Bull. (Woods Hole, Mass.)* **133**, 443–447.

Index